Base SAS® 9.2 Procedures Guide
Statistical Procedures
Second Edition

SAS® Documentation

The correct bibliographic citation for this manual is as follows: SAS Institute Inc. 2009. *Base SAS® 9.2 Procedures Guide: Statistical Procedures, Second Edition*. Cary, NC: SAS Institute Inc.

Base SAS® 9.2 Procedures Guide: Statistical Procedures, Second Edition

Copyright © 2009, SAS Institute Inc., Cary, NC, USA

ISBN 978-1-60764-417-0

All rights reserved. Produced in the United States of America.

For a hard-copy book: No part of this publication may be reproduced, stored in a retrieval system, or transmitted, in any form or by any means, electronic, mechanical, photocopying, or otherwise, without the prior written permission of the publisher, SAS Institute Inc.

For a Web download or e-book: Your use of this publication shall be governed by the terms established by the vendor at the time you acquire this publication.

U.S. Government Restricted Rights Notice: Use, duplication, or disclosure of this software and related documentation by the U.S. government is subject to the Agreement with SAS Institute and the restrictions set forth in FAR 52.227-19, Commercial Computer Software-Restricted Rights (June 1987).

SAS Institute Inc., SAS Campus Drive, Cary, North Carolina 27513.

1st electronic book, September 2009

1st printing, September 2009

SAS® Publishing provides a complete selection of books and electronic products to help customers use SAS software to its fullest potential. For more information about our e-books, e-learning products, CDs, and hard-copy books, visit the SAS Publishing Web site at **support.sas.com/publishing** or call 1-800-727-3228.

SAS® and all other SAS Institute Inc. product or service names are registered trademarks or trademarks of SAS Institute Inc. in the USA and other countries. ® indicates USA registration.

Other brand and product names are registered trademarks or trademarks of their respective companies.

Contents

Chapter 1. What's New in the Base SAS Statistical Procedures 1

Chapter 2. The CORR Procedure 3

Chapter 3. The FREQ Procedure 63

Chapter 4. The UNIVARIATE Procedure 221

Subject Index   **465**

Syntax Index   **473**

Chapter 1

What's New in the Base SAS Statistical Procedures

CORR Procedure

The new ID statement specifies one or more additional tip variables to identify observations in scatter plots and scatter plot matrices.

FREQ Procedure

The FREQ procedure can now produce frequency plots, cumulative frequency plots, deviation plots, odds ratio plots, and kappa plots by using ODS Graphics. The crosstabulation table now has an ODS template that you can customize with the TEMPLATE procedure. Equivalence and noninferiority tests are now available for the binomial proportion and the proportion difference. New confidence limits for the binomial proportion include Agresti-Coull, Jeffreys, and Wilson (score) confidence limits. The RISKDIFF option in the EXACT statement provides unconditional exact confidence limits for the proportion (risk) difference. The EQOR option in the EXACT statement provides Zelen's exact test for equal odds ratios.

UNIVARIATE Procedure

The UNIVARIATE procedure now produces graphs that conform to ODS styles, so that creating consistent output is easier. Also, you now have two alternative methods for producing graphs. With traditional graphics you can control every detail of a graph through familiar procedure syntax and GOPTION and SYMBOL statements. With ODS Graphics (experimental for the UNIVARIATE procedure in SAS 9.2), you can obtain the highest quality output with minimal syntax and full compatibility with graphics produced by SAS/STAT and SAS/ETS procedures.

The new CDFPLOT statement plots the observed cumulative distribution function (cdf) of a variable and enables you to superimpose a fitted theoretical distribution on the graph. The new PPPLOT statement creates a probability-probability plot (also referred to as a P-P plot or percent plot), which compares the empirical cumulative distribution function (ecdf) of a variable with a specified theoretical cumulative distribution function. The beta, exponential, gamma, lognormal, normal, and Weibull distributions are available in both statements.

Chapter 2
The CORR Procedure

Contents

Overview: CORR Procedure	**4**
Getting Started: CORR Procedure	**5**
Syntax: CORR Procedure	**8**
PROC CORR Statement	8
BY Statement	14
FREQ Statement	15
ID Statement	15
PARTIAL Statement	15
VAR Statement	16
WEIGHT Statement	16
WITH Statement	16
Details: CORR Procedure	**17**
Pearson Product-Moment Correlation	17
Spearman Rank-Order Correlation	19
Kendall's Tau-b Correlation Coefficient	20
Hoeffding Dependence Coefficient	21
Partial Correlation	22
Fisher's z Transformation	24
Cronbach's Coefficient Alpha	27
Confidence and Prediction Ellipses	28
Missing Values	29
Output Tables	30
Output Data Sets	31
ODS Table Names	32
ODS Graphics	33
Examples: CORR Procedure	**33**
Example 2.1: Computing Four Measures of Association	33
Example 2.2: Computing Correlations between Two Sets of Variables	38
Example 2.3: Analysis Using Fisher's z Transformation	42
Example 2.4: Applications of Fisher's z Transformation	45
Example 2.5: Computing Cronbach's Coefficient Alpha	48
Example 2.6: Saving Correlations in an Output Data Set	51
Example 2.7: Creating Scatter Plots	53
Example 2.8: Computing Partial Correlations	58
References	**60**

Overview: CORR Procedure

The CORR procedure computes Pearson correlation coefficients, three nonparametric measures of association, and the probabilities associated with these statistics. The correlation statistics include the following:

- Pearson product-moment correlation
- Spearman rank-order correlation
- Kendall's tau-b coefficient
- Hoeffding's measure of dependence, D
- Pearson, Spearman, and Kendall partial correlation

Pearson product-moment correlation is a parametric measure of a linear relationship between two variables. For nonparametric measures of association, Spearman rank-order correlation uses the ranks of the data values and Kendall's tau-b uses the number of concordances and discordances in paired observations. Hoeffding's measure of dependence is another nonparametric measure of association that detects more general departures from independence. A partial correlation provides a measure of the correlation between two variables after controlling the effects of other variables.

With only one set of analysis variables specified, the default correlation analysis includes descriptive statistics for each analysis variable and Pearson correlation statistics for these variables. You can also compute Cronbach's coefficient alpha for estimating reliability.

With two sets of analysis variables specified, the default correlation analysis includes descriptive statistics for each analysis variable and Pearson correlation statistics between these two sets of variables.

For a Pearson or Spearman correlation, the Fisher's z transformation can be used to derive its confidence limits and a p-value under a specified null hypothesis $H_0: \rho = \rho_0$. Either a one-sided or a two-sided alternative is used for these statistics.

You can save the correlation statistics in a SAS data set for use with other statistical and reporting procedures.

When the relationship between two variables is nonlinear or when outliers are present, the correlation coefficient might incorrectly estimate the strength of the relationship. Plotting the data enables you to verify the linear relationship and to identify the potential outliers. If the `ods graphics on` statement is specified, scatter plots and a scatter plot matrix can be created via the Output Delivery System (ODS). Confidence and prediction ellipses can also be added to the scatter plot. See the section "Confidence and Prediction Ellipses" on page 28 for a detailed description of the ellipses.

Getting Started: CORR Procedure

The following statements create the data set Fitness, which has been altered to contain some missing values:

```
*----------------- Data on Physical Fitness -----------------*
| These measurements were made on men involved in a physical |
| fitness course at N.C. State University.                   |
| The variables are Age (years), Weight (kg),                |
| Runtime (time to run 1.5 miles in minutes), and            |
| Oxygen (oxygen intake, ml per kg body weight per minute)   |
| Certain values were changed to missing for the analysis.   |
*------------------------------------------------------------*;
data Fitness;
   input Age Weight Oxygen RunTime @@;
   datalines;
44 89.47 44.609 11.37    40 75.07 45.313 10.07
44 85.84 54.297  8.65    42 68.15 59.571  8.17
38 89.02 49.874    .     47 77.45 44.811 11.63
40 75.98 45.681 11.95    43 81.19 49.091 10.85
44 81.42 39.442 13.08    38 81.87 60.055  8.63
44 73.03 50.541 10.13    45 87.66 37.388 14.03
45 66.45 44.754 11.12    47 79.15 47.273 10.60
54 83.12 51.855 10.33    49 81.42 49.156  8.95
51 69.63 40.836 10.95    51 77.91 46.672 10.00
48 91.63 46.774 10.25    49 73.37   .    10.08
57 73.37 39.407 12.63    54 79.38 46.080 11.17
52 76.32 45.441  9.63    50 70.87 54.625  8.92
51 67.25 45.118 11.08    54 91.63 39.203 12.88
51 73.71 45.790 10.47    57 59.08 50.545  9.93
49 76.32   .       .     48 61.24 47.920 11.50
52 82.78 47.467 10.50
;
```

The following statements invoke the CORR procedure and request a correlation analysis:

```
ods graphics on;
proc corr data=Fitness plots=matrix(histogram);
run;
ods graphics off;
```

The "Simple Statistics" table in Figure 2.1 displays univariate statistics for the analysis variables.

Figure 2.1 Univariate Statistics

```
                         The CORR Procedure

            4   Variables:    Age      Weight     Oxygen    RunTime

                             Simple Statistics

  Variable      N       Mean      Std Dev         Sum      Minimum     Maximum

  Age          31    47.67742     5.21144        1478     38.00000    57.00000
  Weight       31    77.44452     8.32857        2401     59.08000    91.63000
  Oxygen       29    47.22721     5.47718        1370     37.38800    60.05500
  RunTime      29    10.67414     1.39194    309.55000      8.17000    14.03000
```

By default, all numeric variables not listed in other statements are used in the analysis. Observations with nonmissing values for each variable are used to derive the univariate statistics for that variable.

The "Pearson Correlation Coefficients" table in Figure 2.2 displays the Pearson correlation, the p-value under the null hypothesis of zero correlation, and the number of nonmissing observations for each pair of variables.

Figure 2.2 Pearson Correlation Coefficients

```
                  Pearson Correlation Coefficients
                     Prob > |r| under H0: Rho=0
                        Number of Observations

                    Age         Weight       Oxygen       RunTime

     Age         1.00000       -0.23354     -0.31474       0.14478
                                 0.2061       0.0963        0.4536
                      31            31           29            29

     Weight     -0.23354        1.00000     -0.15358       0.20072
                  0.2061                      0.4264        0.2965
                      31            31           29            29

     Oxygen     -0.31474       -0.15358      1.00000      -0.86843
                  0.0963         0.4264                    <.0001
                      29            29           29            28

     RunTime     0.14478        0.20072     -0.86843       1.00000
                  0.4536         0.2965      <.0001
                      29            29           28            29
```

By default, Pearson correlation statistics are computed from observations with nonmissing values for each pair of analysis variables. Figure 2.2 displays a correlation of −0.86843 between Runtime and Oxygen, which is significant with a p-value less than 0.0001. That is, there exists an inverse linear relationship between these two variables. As Runtime (time to run 1.5 miles in minutes) increases, Oxygen (oxygen intake, ml per kg body weight per minute) decreases.

This graphical display is requested by specifying the `ods graphics on` statement and the PLOTS option. For more information about the `ods graphics` statement, see Chapter 21, "Statistical Graphics Using ODS" (*SAS/STAT User's Guide*).

When you use the PLOTS=MATRIX(HISTOGRAM) option, the CORR procedure displays a symmetric matrix plot for the analysis variables in Figure 2.3. The histograms for these analysis variables are also displayed on the diagonal of the matrix plot. This inverse linear relationship between the two variables, Oxygen and Runtime, is also shown in the plot.

Figure 2.3 Symmetric Matrix Plot

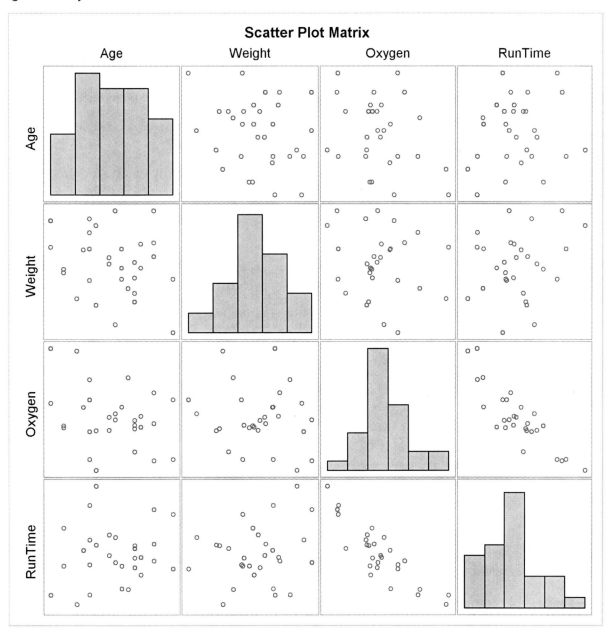

Syntax: CORR Procedure

The following statements are available in PROC CORR:

PROC CORR < *options* > ;
    BY *variables* ;
    FREQ *variable* ;
    ID *variables* ;
    PARTIAL *variables* ;
    VAR *variables* ;
    WEIGHT *variable* ;
    WITH *variables* ;

The BY statement specifies groups in which separate correlation analyses are performed.

The FREQ statement specifies the variable that represents the frequency of occurrence for other values in the observation.

The ID statement specifies one or more additional tip variables to identify observations in scatter plots and scatter plot matrices.

The PARTIAL statement identifies controlling variables to compute Pearson, Spearman, or Kendall partial-correlation coefficients.

The VAR statement lists the numeric variables to be analyzed and their order in the correlation matrix. If you omit the VAR statement, all numeric variables not listed in other statements are used.

The WEIGHT statement identifies the variable whose values weight each observation to compute Pearson product-moment correlation.

The WITH statement lists the numeric variables with which correlations are to be computed.

The PROC CORR statement is the only required statement for the CORR procedure. The rest of this section provides detailed syntax information for each of these statements, beginning with the PROC CORR statement. The remaining statements are presented in alphabetical order.

PROC CORR Statement

PROC CORR < *options* > ;

Table 2.1 summarizes the options available in the PROC CORR statement.

Table 2.1 Summary of PROC CORR Options

Option	Description
Data Sets	
DATA=	specifies input data set
OUTH=	specifies output data set with Hoeffding's D statistics

Table 2.1 *continued*

Option	Description
OUTK=	specifies output data set with Kendall correlation statistics
OUTP=	specifies output data set with Pearson correlation statistics
OUTS=	specifies output data set with Spearman correlation statistics
Statistical Analysis	
EXCLNPWGT	excludes observations with nonpositive weight values from the analysis
FISHER	requests correlation statistics using Fisher's z transformation
HOEFFDING	requests Hoeffding's measure of dependence, D
KENDALL	requests Kendall's tau-b
NOMISS	excludes observations with missing analysis values from the analysis
PEARSON	requests Pearson product-moment correlation
SPEARMAN	requests Spearman rank-order correlation
Pearson Correlation Statistics	
ALPHA	computes Cronbach's coefficient alpha
COV	computes covariances
CSSCP	computes corrected sums of squares and crossproducts
FISHER	computes correlation statistics based on Fisher's z transformation
NOMISS	excludes missing values
SINGULAR=	specifies singularity criterion
SSCP	computes sums of squares and crossproducts
VARDEF=	specifies the divisor for variance calculations
ODS Output Graphics	
PLOTS=MATRIX	displays scatter plot matrix
PLOTS=SCATTER	displays scatter plots for pairs of variables
Printed Output	
BEST=	displays a specified number of ordered correlation coefficients
NOCORR	suppresses Pearson correlations
NOPRINT	suppresses all printed output
NOPROB	suppresses p-values
NOSIMPLE	suppresses descriptive statistics
RANK	displays ordered correlation coefficients

The following options can be used in the PROC CORR statement. They are listed in alphabetical order.

ALPHA

calculates and prints Cronbach's coefficient alpha. PROC CORR computes separate coefficients using raw and standardized values (scaling the variables to a unit variance of 1). For each VAR statement variable, PROC CORR computes the correlation between the variable and the total of the remaining variables. It also computes Cronbach's coefficient alpha by using only the remaining variables.

If a WITH statement is specified, the ALPHA option is invalid. When you specify the ALPHA option, the Pearson correlations will also be displayed. If you specify the OUTP= option, the output data set also contains observations with Cronbach's coefficient alpha. If you use the PARTIAL statement, PROC CORR calculates Cronbach's coefficient alpha for partialled variables. See the section "Partial Correlation" on page 22 for details.

BEST=n

prints the n highest correlation coefficients for each variable, $n \geq 1$. Correlations are ordered from highest to lowest in absolute value. Otherwise, PROC CORR prints correlations in a rectangular table, using the variable names as row and column labels.

If you specify the HOEFFDING option, PROC CORR displays the D statistics in order from highest to lowest.

COV

displays the variance and covariance matrix. When you specify the COV option, the Pearson correlations will also be displayed. If you specify the OUTP= option, the output data set also contains the covariance matrix with the corresponding _TYPE_ variable value 'COV.' If you use the PARTIAL statement, PROC CORR computes a partial covariance matrix.

CSSCP

displays a table of the corrected sums of squares and crossproducts. When you specify the CSSCP option, the Pearson correlations will also be displayed. If you specify the OUTP= option, the output data set also contains a CSSCP matrix with the corresponding _TYPE_ variable value 'CSSCP.' If you use a PARTIAL statement, PROC CORR prints both an unpartial and a partial CSSCP matrix, and the output data set contains a partial CSSCP matrix.

DATA=*SAS-data-set*

names the SAS data set to be analyzed by PROC CORR. By default, the procedure uses the most recently created SAS data set.

EXCLNPWGT

EXCLNPWGTS

excludes observations with nonpositive weight values from the analysis. By default, PROC CORR treats observations with negative weights like those with zero weights and counts them in the total number of observations.

FISHER < (*fisher-options*) >

requests confidence limits and *p*-values under a specified null hypothesis, $H_0: \rho = \rho_0$, for correlation coefficients by using Fisher's z transformation. These correlations include the Pearson correlations and Spearman correlations.

The following *fisher-options* are available:

ALPHA=α

specifies the level of the confidence limits for the correlation, $100(1-\alpha)\%$. The value of the ALPHA= option must be between 0 and 1, and the default is ALPHA=0.05.

BIASADJ=YES | NO

specifies whether or not the bias adjustment is used in constructing confidence limits. The BIASADJ=YES option also produces a new correlation estimate that uses the bias adjustment. By default, BIASADJ=YES.

RHO0=ρ_0

specifies the value ρ_0 in the null hypothesis $H_0: \rho = \rho_0$, where $-1 < \rho_0 < 1$. By default, RHO0=0.

TYPE=LOWER | UPPER | TWOSIDED

specifies the type of confidence limits. The TYPE=LOWER option requests a lower confidence limit from the lower alternative $H_1: \rho < \rho_0$, the TYPE=UPPER option requests an upper confidence limit from the upper alternative $H_1: \rho > \rho_0$, and the default TYPE=TWOSIDED option requests two-sided confidence limits from the two-sided alternative $H_1: \rho \neq \rho_0$.

HOEFFDING

requests a table of Hoeffding's D statistics. This D statistic is 30 times larger than the usual definition and scales the range between -0.5 and 1 so that large positive values indicate dependence. The HOEFFDING option is invalid if a WEIGHT or PARTIAL statement is used.

KENDALL

requests a table of Kendall's tau-b coefficients based on the number of concordant and discordant pairs of observations. Kendall's tau-b ranges from -1 to 1.

The KENDALL option is invalid if a WEIGHT statement is used. If you use a PARTIAL statement, probability values for Kendall's partial tau-b are not available.

NOCORR

suppresses displaying of Pearson correlations. If you specify the OUTP= option, the data set type remains CORR. To change the data set type to COV, CSSCP, or SSCP, use the TYPE= data set option.

NOMISS

excludes observations with missing values from the analysis. Otherwise, PROC CORR computes correlation statistics by using all of the nonmissing pairs of variables. Using the NOMISS option is computationally more efficient.

NOPRINT

suppresses all displayed output, which also includes output produced with ODS Graphics. Use the NOPRINT option if you want to create an output data set only.

NOPROB

suppresses displaying the probabilities associated with each correlation coefficient.

NOSIMPLE

suppresses printing simple descriptive statistics for each variable. However, if you request an output data set, the output data set still contains simple descriptive statistics for the variables.

OUTH=*output-data-set*

creates an output data set containing Hoeffding's D statistics. The contents of the output data set are similar to those of the OUTP= data set. When you specify the OUTH= option, the Hoeffding's D statistics will be displayed.

OUTK=output-data-set

creates an output data set containing Kendall correlation statistics. The contents of the output data set are similar to those of the OUTP= data set. When you specify the OUTK= option, the Kendall correlation statistics will be displayed.

OUTP=output-data-set

OUT=output-data-set

creates an output data set containing Pearson correlation statistics. This data set also includes means, standard deviations, and the number of observations. The value of the _TYPE_ variable is 'CORR.' When you specify the OUTP= option, the Pearson correlations will also be displayed. If you specify the ALPHA option, the output data set also contains six observations with Cronbach's coefficient alpha.

OUTS=SAS-data-set

creates an output data set containing Spearman correlation coefficients. The contents of the output data set are similar to those of the OUTP= data set. When you specify the OUTS= option, the Spearman correlation coefficients will be displayed.

PEARSON

requests a table of Pearson product-moment correlations. The correlations range from −1 to 1. If you do not specify the HOEFFDING, KENDALL, SPEARMAN, OUTH=, OUTK=, or OUTS= option, the CORR procedure produces Pearson product-moment correlations by default. Otherwise, you must specify the PEARSON, ALPHA, COV, CSSCP, SSCP, or OUT= option for Pearson correlations. Also, if a scatter plot or a scatter plot matrix is requested, the Pearson correlations will be displayed.

PLOTS < (ONLY) > < = plot-request >

PLOTS < (ONLY) > < = (plot-request < ... plot-request >) >

requests statistical graphics via the Output Delivery System (ODS). To request these graphs, you must specify the `ods graphics on` statement in addition to the following options in the PROC CORR statement. For more information about the `ods graphics` statement, see Chapter 21, "Statistical Graphics Using ODS" (*SAS/STAT User's Guide*).

The global plot option ONLY suppresses the default plots, and only plots specifically requested are displayed. The plot request options include the following:

ALL

produces all appropriate plots.

MATRIX < (matrix-options) >

requests a scatter plot matrix for variables. That is, the procedure displays a symmetric matrix plot with variables in the VAR list if a WITH statement is not specified. Otherwise, the procedure displays a rectangular matrix plot with the WITH variables appearing down the side and the VAR variables appearing across the top.

NONE

suppresses all plots.

SCATTER < (*scatter-options*) >

requests scatter plots for pairs of variables. That is, the procedure displays a scatter plot for each applicable pair of distinct variables from the VAR list if a WITH statement is not specified. Otherwise, the procedure displays a scatter plot for each applicable pair of variables, one from the WITH list and the other from the VAR list.

By default, PLOTS=MATRIX, a scatter plot matrix for all variables is displayed. When a scatter plot or a scatter plot matrix is requested, the Pearson correlations will also be displayed.

The available *matrix-options* are as follows:

HIST | HISTOGRAM

displays histograms of variables in the VAR list in the symmetric matrix plot.

NVAR=ALL | *n*

specifies the maximum number of variables in the VAR list to be displayed in the matrix plot, where $n > 0$. The NVAR=ALL option uses all variables in the VAR list. By default, NVAR=5.

NWITH=ALL | *n*

specifies the maximum number of variables in the WITH list to be displayed in the matrix plot, where $n > 0$. The NWITH=ALL option uses all variables in the WITH list. By default, NWITH=5.

The available *scatter-options* are as follows:

ALPHA=α

specifies the α values for the confidence or prediction ellipses to be displayed in the scatter plots, where $0 < \alpha < 1$. For each α value specified, a $(1 - \alpha)$ confidence or prediction ellipse is created. By default, $\alpha = 0.05$.

ELLIPSE=PREDICTION | CONFIDENCE | NONE

requests prediction ellipses for new observations (ELLIPSE=PREDICTION), confidence ellipses for the mean (ELLIPSE=CONFIDENCE), or no ellipses (ELLIPSE=NONE) to be created in the scatter plots. By default, ELLIPSE=PREDICTION.

NOINSET

suppresses the default inset of summary information for the scatter plot. The inset table contains the number of observations (Observations) and correlation.

NVAR=ALL | *n*

specifies the maximum number of variables in the VAR list to be displayed in the plots, where $n > 0$. The NVAR=ALL option uses all variables in the VAR list. By default, NVAR=5.

NWITH=ALL | *n*

specifies the maximum number of variables in the WITH list to be displayed in the plots, where $n > 0$. The NWITH=ALL option uses all variables in the WITH list. By default, NWITH=5.

RANK
    displays the ordered correlation coefficients for each variable. Correlations are ordered from highest to lowest in absolute value. If you specify the HOEFFDING option, the D statistics are displayed in order from highest to lowest.

SINGULAR=p
    specifies the criterion for determining the singularity of a variable if you use a PARTIAL statement. A variable is considered singular if its corresponding diagonal element after Cholesky decomposition has a value less than p times the original unpartialled value of that variable. The default value is 1E−8. The range of p is between 0 and 1.

SPEARMAN
    requests a table of Spearman correlation coefficients based on the ranks of the variables. The correlations range from −1 to 1. If you specify a WEIGHT statement, the SPEARMAN option is invalid.

SSCP
    displays a table of the sums of squares and crossproducts. When you specify the SSCP option, the Pearson correlations will also be displayed. If you specify the OUTP= option, the output data set contains a SSCP matrix and the corresponding _TYPE_ variable value is 'SSCP.' If you use a PARTIAL statement, the unpartial SSCP matrix is displayed, and the output data set does not contain an SSCP matrix.

VARDEF=DF | N | WDF | WEIGHT | WGT
    specifies the variance divisor in the calculation of variances and covariances. The default is VARDEF=DF.

    Table 2.2 displays available values and associated divisors for the VARDEF= option, where n is the number of nonmissing observations, k is the number of variables specified in the PARTIAL statement, and w_j is the weight associated with the jth nonmissing observation.

Table 2.2 Possible Values for the VARDEF= Option

Value	Description	Divisor
DF	degrees of freedom	$n - k - 1$
N	number of observations	n
WDF	sum of weights minus one	$\sum_j^n w_j - k - 1$
WEIGHT | WGT	sum of weights	$\sum_j^n w_j$

BY Statement

    BY *variables* ;

You can specify a BY statement with PROC CORR to obtain separate analyses on observations in groups defined by the BY variables. If a BY statement appears, the procedure expects the input data set to be sorted in order of the BY variables.

If your input data set is not sorted in ascending order, use one of the following alternatives:

- Sort the data by using the SORT procedure with a similar BY statement.
- Specify the BY statement option NOTSORTED or DESCENDING in the BY statement for the CORR procedure. The NOTSORTED option does not mean that the data are unsorted but rather that the data are arranged in groups (according to values of the BY variables) and that these groups are not necessarily in alphabetical or increasing numeric order.
- Create an index on the BY variables by using the DATASETS procedure.

For more information about the BY statement, see *SAS Language Reference: Concepts*. For more information about the DATASETS procedure, see the *Base SAS Procedures Guide*.

FREQ Statement

FREQ *variable* ;

The FREQ statement lists a numeric variable whose value represents the frequency of the observation. If you use the FREQ statement, the procedure assumes that each observation represents n observations, where n is the value of the FREQ variable. If n is not an integer, SAS truncates it. If n is less than 1 or is missing, the observation is excluded from the analysis. The sum of the frequency variable represents the total number of observations.

The effects of the FREQ and WEIGHT statements are similar except when calculating degrees of freedom.

ID Statement

ID *variables* ;

The ID statement specifies one or more additional tip variables to identify observations in scatter plots and scatter plot matrix. For each plot, the tip variables include the X-axis variable, the Y-axis variable, and the variable for observation numbers. The ID statement names additional variables to identify observations in scatter plots and scatter plot matrices.

PARTIAL Statement

PARTIAL *variables* ;

The PARTIAL statement lists variables to use in the calculation of partial correlation statistics. Only the Pearson partial correlation, Spearman partial rank-order correlation, and Kendall's partial tau-b can be computed. When you use the PARTIAL statement, observations with missing values are excluded.

With a PARTIAL statement, PROC CORR also displays the partial variance and standard deviation for each analysis variable if the PEARSON option is specified.

VAR Statement

VAR *variables* ;

The VAR statement lists variables for which to compute correlation coefficients. If the VAR statement is not specified, PROC CORR computes correlations for all numeric variables not listed in other statements.

WEIGHT Statement

WEIGHT *variable* ;

The WEIGHT statement lists weights to use in the calculation of Pearson weighted product-moment correlation. The HOEFFDING, KENDALL, and SPEARMAN options are not valid with the WEIGHT statement.

The observations with missing weights are excluded from the analysis. By default, for observations with nonpositive weights, weights are set to zero and the observations are included in the analysis. You can use the EXCLNPWGT option to exclude observations with negative or zero weights from the analysis.

WITH Statement

WITH *variables* ;

The WITH statement lists variables with which correlations of the VAR statement variables are to be computed. The WITH statement requests correlations of the form $r(X_i, Y_j)$, where $X_1, \ldots, X_m$ are analysis variables specified in the VAR statement, and $Y_1, \ldots, Y_n$ are variables specified in the WITH statement. The correlation matrix has a rectangular structure of the form

$$\begin{bmatrix} r(Y_1, X_1) & \cdots & r(Y_1, X_m) \\ \vdots & \ddots & \vdots \\ r(Y_n, X_1) & \cdots & r(Y_n, X_m) \end{bmatrix}$$

For example, the statements

```
proc corr;
    var x1 x2;
    with y1 y2 y3;
run;
```

produce correlations for the following combinations:

$$\begin{bmatrix} r(Y1, X1) & r(Y1, X2) \\ r(Y2, X1) & r(Y2, X2) \\ r(Y3, X1) & r(Y3, X2) \end{bmatrix}$$

Details: CORR Procedure

Pearson Product-Moment Correlation

The Pearson product-moment correlation is a parametric measure of association for two variables. It measures both the strength and the direction of a linear relationship. If one variable X is an exact linear function of another variable Y, a positive relationship exists if the correlation is 1 and a negative relationship exists if the correlation is -1. If there is no linear predictability between the two variables, the correlation is 0. If the two variables are normal with a correlation 0, the two variables are independent. However, correlation does not imply causality because, in some cases, an underlying causal relationship might not exist.

The scatter plot matrix in Figure 2.4 displays the relationship between two numeric random variables in various situations.

Figure 2.4 Correlations between Two Variables

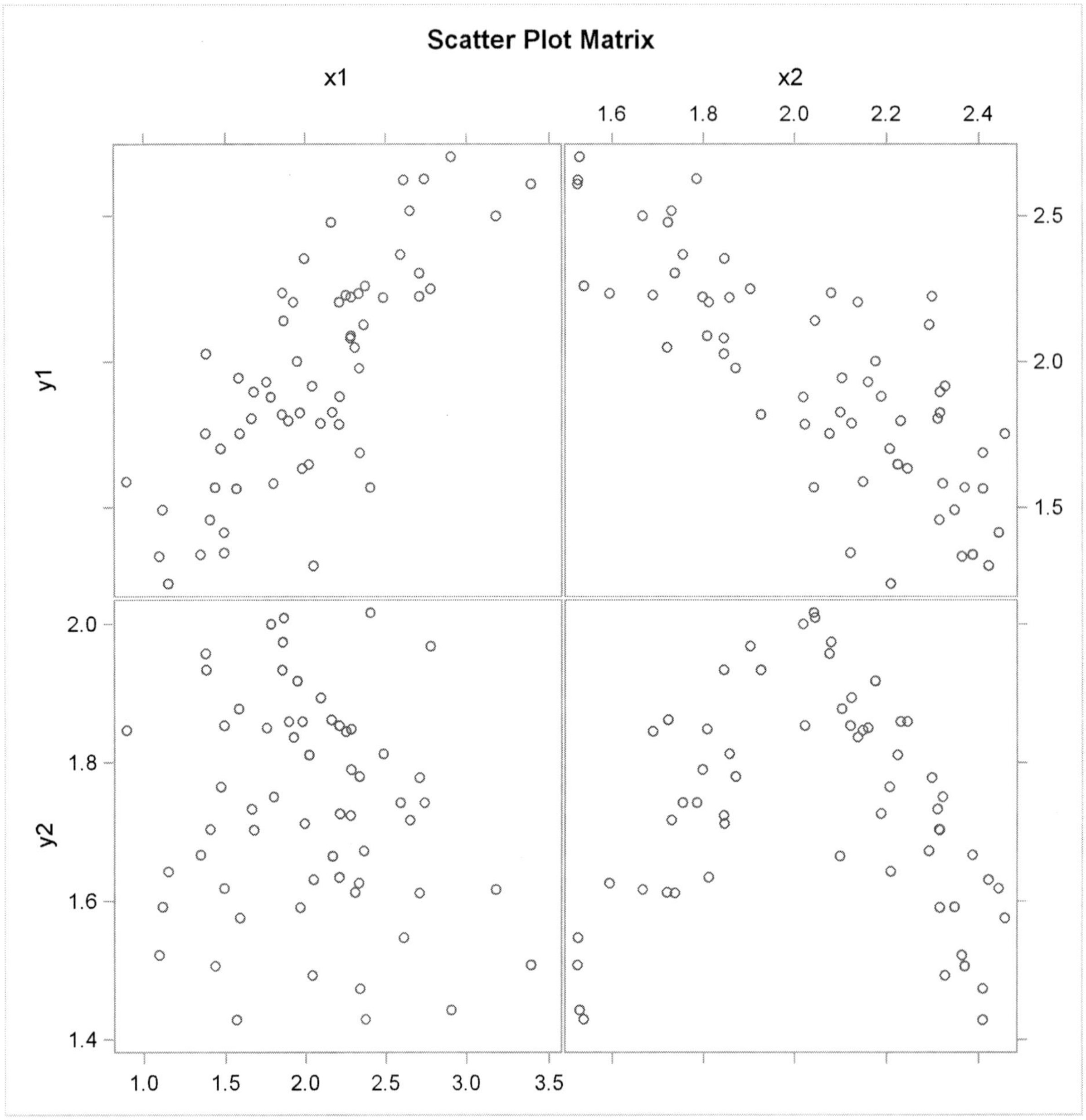

[Scatter Plot Matrix showing pairwise scatter plots of variables y1, y2 against x1, x2]

The scatter plot matrix shows a positive correlation between variables Y1 and X1, a negative correlation between Y1 and X2, and no clear correlation between Y2 and X1. The plot also shows no clear linear correlation between Y2 and X2, even though Y2 is dependent on X2.

The formula for the population Pearson product-moment correlation, denoted ρ_{xy}, is

$$\rho_{xy} = \frac{\text{Cov}(x, y)}{\sqrt{\text{V}(x)\text{V}(y)}} = \frac{\text{E}((x - \text{E}(x))(y - \text{E}(y)))}{\sqrt{\text{E}(x - \text{E}(x))^2 \, \text{E}(y - \text{E}(y))^2}}$$

The sample correlation, such as a Pearson product-moment correlation or weighted product-moment correlation, estimates the population correlation. The formula for the sample Pearson product-moment correlation is

$$r_{xy} = \frac{\sum_i ((x_i - \bar{x})(y_i - \bar{y}))}{\sqrt{\sum_i (x_i - \bar{x})^2 \sum_i (y_i - \bar{y})^2}}$$

where $\bar{x}$ is the sample mean of x and $\bar{y}$ is the sample mean of y. The formula for a weighted Pearson product-moment correlation is

$$r_{xy} = \frac{\sum_i w_i (x_i - \bar{x}_w)(y_i - \bar{y}_w)}{\sqrt{\sum_i w_i (x_i - \bar{x}_w)^2 \sum_i w_i (y_i - \bar{y}_w)^2}}$$

where w_i is the weight, $\bar{x}_w$ is the weighted mean of x, and $\bar{y}_w$ is the weighted mean of y.

Probability Values

Probability values for the Pearson correlation are computed by treating

$$t = (n - 2)^{1/2} \left(\frac{r^2}{1 - r^2} \right)^{1/2}$$

as coming from a t distribution with $(n - 2)$ degrees of freedom, where r is the sample correlation.

Spearman Rank-Order Correlation

Spearman rank-order correlation is a nonparametric measure of association based on the ranks of the data values. The formula is

$$\theta = \frac{\sum_i ((R_i - \bar{R})(S_i - \bar{S}))}{\sqrt{\sum_i (R_i - \bar{R})^2 \sum (S_i - \bar{S})^2}}$$

where R_i is the rank of x_i, S_i is the rank of y_i, $\bar{R}$ is the mean of the R_i values, and $\bar{S}$ is the mean of the S_i values.

PROC CORR computes the Spearman correlation by ranking the data and using the ranks in the Pearson product-moment correlation formula. In case of ties, the averaged ranks are used.

Probability Values

Probability values for the Spearman correlation are computed by treating

$$t = (n - 2)^{1/2} \left(\frac{r^2}{1 - r^2} \right)^{1/2}$$

as coming from a t distribution with $(n - 2)$ degrees of freedom, where r is the sample Spearman correlation.

Kendall's Tau-b Correlation Coefficient

Kendall's tau-b is a nonparametric measure of association based on the number of concordances and discordances in paired observations. Concordance occurs when paired observations vary together, and discordance occurs when paired observations vary differently. The formula for Kendall's tau-b is

$$\tau = \frac{\sum_{i<j} (\text{sgn}(x_i - x_j)\text{sgn}(y_i - y_j))}{\sqrt{(T_0 - T_1)(T_0 - T_2)}}$$

where $T_0 = n(n-1)/2$, $T_1 = \sum_k t_k(t_k-1)/2$, and $T_2 = \sum_l u_l(u_l-1)/2$. The t_k is the number of tied x values in the kth group of tied x values, u_l is the number of tied y values in the lth group of tied y values, n is the number of observations, and $\text{sgn}(z)$ is defined as

$$\text{sgn}(z) = \begin{cases} 1 & \text{if } z > 0 \\ 0 & \text{if } z = 0 \\ -1 & \text{if } z < 0 \end{cases}$$

PROC CORR computes Kendall's tau-b by ranking the data and using a method similar to Knight (1966). The data are double sorted by ranking observations according to values of the first variable and reranking the observations according to values of the second variable. PROC CORR computes Kendall's tau-b from the number of interchanges of the first variable and corrects for tied pairs (pairs of observations with equal values of X or equal values of Y).

Probability Values

Probability values for Kendall's tau-b are computed by treating

$$\frac{s}{\sqrt{V(s)}}$$

as coming from a standard normal distribution where

$$s = \sum_{i<j} (\text{sgn}(x_i - x_j)\text{sgn}(y_i - y_j))$$

and $V(s)$, the variance of s, is computed as

$$V(s) = \frac{v_0 - v_t - v_u}{18} + \frac{v_1}{2n(n-1)} + \frac{v_2}{9n(n-1)(n-2)}$$

where

$v_0 = n(n-1)(2n+5)$

$v_t = \sum_k t_k(t_k-1)(2t_k+5)$

$v_u = \sum_l u_l(u_l-1)(2u_l+5)$

$$v_1 = \left(\sum_k t_k(t_k - 1)\right)\left(\sum u_i(u_l - 1)\right)$$

$$v_2 = \left(\sum_l t_i(t_k - 1)(t_k - 2)\right)\left(\sum u_l(u_l - 1)(u_l - 2)\right)$$

The sums are over tied groups of values where t_i is the number of tied x values and u_i is the number of tied y values (Noether 1967). The sampling distribution of Kendall's partial tau-b is unknown; therefore, the probability values are not available.

Hoeffding Dependence Coefficient

Hoeffding's measure of dependence, D, is a nonparametric measure of association that detects more general departures from independence. The statistic approximates a weighted sum over observations of chi-square statistics for two-by-two classification tables (Hoeffding 1948). Each set of (x, y) values are cut points for the classification. The formula for Hoeffding's D is

$$D = 30 \frac{(n-2)(n-3)D_1 + D_2 - 2(n-2)D_3}{n(n-1)(n-2)(n-3)(n-4)}$$

where $D_1 = \sum_i (Q_i - 1)(Q_i - 2)$, $D_2 = \sum_i (R_i - 1)(R_i - 2)(S_i - 1)(S_i - 2)$, and $D_3 = \sum_i (R_i - 2)(S_i - 2)(Q_i - 1)$. R_i is the rank of x_i, S_i is the rank of y_i, and Q_i (also called the bivariate rank) is 1 plus the number of points with both x and y values less than the ith point.

A point that is tied on only the x value or y value contributes 1/2 to Q_i if the other value is less than the corresponding value for the ith point.

A point that is tied on both x and y contributes 1/4 to Q_i. PROC CORR obtains the Q_i values by first ranking the data. The data are then double sorted by ranking observations according to values of the first variable and reranking the observations according to values of the second variable. Hoeffding's D statistic is computed using the number of interchanges of the first variable. When no ties occur among data set observations, the D statistic values are between -0.5 and 1, with 1 indicating complete dependence. However, when ties occur, the D statistic might result in a smaller value. That is, for a pair of variables with identical values, the Hoeffding's D statistic might be less than 1. With a large number of ties in a small data set, the D statistic might be less than -0.5. For more information about Hoeffding's D, see Hollander and Wolfe (1999).

Probability Values

The probability values for Hoeffding's D statistic are computed using the asymptotic distribution computed by Blum, Kiefer, and Rosenblatt (1961). The formula is

$$\frac{(n-1)\pi^4}{60}D + \frac{\pi^4}{72}$$

which comes from the asymptotic distribution. If the sample size is less than 10, refer to the tables for the distribution of D in Hollander and Wolfe (1999).

Partial Correlation

A partial correlation measures the strength of a relationship between two variables, while controlling the effect of other variables. The Pearson partial correlation between two variables, after controlling for variables in the PARTIAL statement, is equivalent to the Pearson correlation between the residuals of the two variables after regression on the controlling variables.

Let $\mathbf{y} = (y_1, y_2, \ldots, y_v)$ be the set of variables to correlate and $\mathbf{z} = (z_1, z_2, \ldots, z_p)$ be the set of controlling variables. The population Pearson partial correlation between the ith and the jth variables of $\mathbf{y}$ given $\mathbf{z}$ is the correlation between errors $(y_i - E(y_i))$ and $(y_j - E(y_j))$, where

$$E(y_i) = \alpha_i + \mathbf{z}\boldsymbol{\beta}_i \quad \text{and} \quad E(y_j) = \alpha_j + \mathbf{z}\boldsymbol{\beta}_j$$

are the regression models for variables y_i and y_j given the set of controlling variables $\mathbf{z}$, respectively.

For a given sample of observations, a sample Pearson partial correlation between y_i and y_j given $\mathbf{z}$ is derived from the residuals $y_i - \hat{y}_i$ and $y_j - \hat{y}_j$, where

$$\hat{y}_i = \hat{\alpha}_i + \mathbf{z}\hat{\boldsymbol{\beta}}_i \quad \text{and} \quad \hat{y}_j = \hat{\alpha}_j + \mathbf{z}\hat{\boldsymbol{\beta}}_j$$

are fitted values from regression models for variables y_i and y_j given $\mathbf{z}$.

The partial corrected sums of squares and crossproducts (CSSCP) of $\mathbf{y}$ given $\mathbf{z}$ are the corrected sums of squares and crossproducts of the residuals $\mathbf{y} - \hat{\mathbf{y}}$. Using these partial corrected sums of squares and crossproducts, you can calculate the partial covariances and partial correlations.

PROC CORR derives the partial corrected sums of squares and crossproducts matrix by applying the Cholesky decomposition algorithm to the CSSCP matrix. For Pearson partial correlations, let S be the partitioned CSSCP matrix between two sets of variables, $\mathbf{z}$ and $\mathbf{y}$:

$$S = \begin{bmatrix} S_{zz} & S_{zy} \\ S'_{zy} & S_{yy} \end{bmatrix}$$

PROC CORR calculates $S_{yy \cdot z}$, the partial CSSCP matrix of $\mathbf{y}$ after controlling for $\mathbf{z}$, by applying the Cholesky decomposition algorithm sequentially on the rows associated with $\mathbf{z}$, the variables being partialled out.

After applying the Cholesky decomposition algorithm to each row associated with variables $\mathbf{z}$, PROC CORR checks all higher-numbered diagonal elements associated with $\mathbf{z}$ for singularity. A variable is considered singular if the value of the corresponding diagonal element is less than ε times the original unpartialled corrected sum of squares of that variable. You can specify the singularity criterion ε by using the SINGULAR= option. For Pearson partial correlations, a controlling variable $\mathbf{z}$ is considered singular if the R^2 for predicting this variable from the variables that are already partialled out exceeds $1 - \varepsilon$. When this happens, PROC CORR excludes the variable from the analysis. Similarly, a variable is considered singular if the R^2 for predicting this variable from the controlling variables exceeds $1 - \varepsilon$. When this happens, its associated diagonal element and all higher-numbered elements in this row or column are set to zero.

After the Cholesky decomposition algorithm is applied to all rows associated with **z**, the resulting matrix has the form

$$T = \begin{bmatrix} \mathbf{T}_{zz} & \mathbf{T}_{zy} \\ 0 & \mathbf{S}_{yy.z} \end{bmatrix}$$

where T_{zz} is an upper triangular matrix with $T'_{zz}T_{zz} = S'_{zz}$, $T'_{zz}T_{zy} = S'_{zy}$, and $S_{yy.z} = S_{yy} - T'_{zy}T_{zy}$.

If S_{zz} is positive definite, then $T_{zy} = {T'_{zz}}^{-1} S'_{zy}$ and the partial CSSCP matrix $S_{yy.z}$ is identical to the matrix derived from the formula

$$S_{yy.z} = S_{yy} - S'_{zy} S'^{-1}_{zz} S_{zy}$$

The partial variance-covariance matrix is calculated with the variance divisor (VARDEF= option). PROC CORR then uses the standard Pearson correlation formula on the partial variance-covariance matrix to calculate the Pearson partial correlation matrix.

When a correlation matrix is positive definite, the resulting partial correlation between variables x and y after adjusting for a single variable z is identical to that obtained from the first-order partial correlation formula

$$r_{xy.z} = \frac{r_{xy} - r_{xz}r_{yz}}{\sqrt{(1-r_{xz}^2)(1-r_{yz}^2)}}$$

where r_{xy}, r_{xz}, and r_{yz} are the appropriate correlations.

The formula for higher-order partial correlations is a straightforward extension of the preceding first-order formula. For example, when the correlation matrix is positive definite, the partial correlation between x and y controlling for both z_1 and z_2 is identical to the second-order partial correlation formula

$$r_{xy.z_1 z_2} = \frac{r_{xy.z_1} - r_{xz_2.z_1}r_{yz_2.z_1}}{\sqrt{(1-r_{xz_2.z_1}^2)(1-r_{yz_2.z_1}^2)}}$$

where $r_{xy.z_1}$, $r_{xz_2.z_1}$, and $r_{yz_2.z_1}$ are first-order partial correlations among variables x, y, and z_2 given z_1.

To derive the corresponding Spearman partial rank-order correlations and Kendall partial tau-b correlations, PROC CORR applies the Cholesky decomposition algorithm to the Spearman rank-order correlation matrix and Kendall's tau-b correlation matrix and uses the correlation formula. That is, the Spearman partial correlation is equivalent to the Pearson correlation between the residuals of the linear regression of the ranks of the two variables on the ranks of the partialled variables. Thus, if a PARTIAL statement is specified with the CORR=SPEARMAN option, the residuals of the ranks of the two variables are displayed in the plot. The partial tau-b correlations range from −1 to 1. However, the sampling distribution of this partial tau-b is unknown; therefore, the probability values are not available.

Probability Values

Probability values for the Pearson and Spearman partial correlations are computed by treating

$$\frac{(n-k-2)^{1/2} r}{(1-r^2)^{1/2}}$$

as coming from a t distribution with $(n-k-2)$ degrees of freedom, where r is the partial correlation and k is the number of variables being partialled out.

Fisher's z Transformation

For a sample correlation r that uses a sample from a bivariate normal distribution with correlation $\rho = 0$, the statistic

$$t_r = (n-2)^{1/2} \left(\frac{r^2}{1-r^2} \right)^{1/2}$$

has a Student's t distribution with $(n-2)$ degrees of freedom.

With the monotone transformation of the correlation r (Fisher 1921)

$$z_r = \tanh^{-1}(r) = \frac{1}{2} \log \left(\frac{1+r}{1-r} \right)$$

the statistic z has an approximate normal distribution with mean and variance

$$E(z_r) = \zeta + \frac{\rho}{2(n-1)}$$

$$V(z_r) = \frac{1}{n-3}$$

where $\zeta = \tanh^{-1}(\rho)$.

For the transformed z_r, the approximate variance $V(z_r) = 1/(n-3)$ is independent of the correlation ρ. Furthermore, even the distribution of z_r is not strictly normal, it tends to normality rapidly as the sample size increases for any values of ρ (Fisher 1970, pp. 200–201).

For the null hypothesis $H_0: \rho = \rho_0$, the p-values are computed by treating

$$z_r - \zeta_0 - \frac{\rho_0}{2(n-1)}$$

as a normal random variable with mean zero and variance $1/(n-3)$, where $\zeta_0 = \tanh^{-1}(\rho_0)$ (Fisher 1970, p. 207; Anderson 1984, p. 123).

Note that the bias adjustment, $\rho_0/(2(n-1))$, is always used when computing p-values under the null hypothesis $H_0: \rho = \rho_0$ in the CORR procedure.

The ALPHA= option in the FISHER option specifies the value α for the confidence level $1 - \alpha$, the RHO0= option specifies the value ρ_0 in the hypothesis $H_0: \rho = \rho_0$, and the BIASADJ= option specifies whether the bias adjustment is to be used for the confidence limits.

The TYPE= option specifies the type of confidence limits. The TYPE=TWOSIDED option requests two-sided confidence limits and a *p*-value under the hypothesis $H_0: \rho = \rho_0$. For a one-sided confidence limit, the TYPE=LOWER option requests a lower confidence limit and a *p*-value under the hypothesis $H_0: \rho <= \rho_0$, and the TYPE=UPPER option requests an upper confidence limit and a *p*-value under the hypothesis $H_0: \rho >= \rho_0$.

Confidence Limits for the Correlation

The confidence limits for the correlation ρ are derived through the confidence limits for the parameter ζ, with or without the bias adjustment.

Without a bias adjustment, confidence limits for ζ are computed by treating

$$z_r - \zeta$$

as having a normal distribution with mean zero and variance $1/(n - 3)$.

That is, the two-sided confidence limits for ζ are computed as

$$\zeta_l = z_r - z_{(1-\alpha/2)} \sqrt{\frac{1}{n-3}}$$

$$\zeta_u = z_r + z_{(1-\alpha/2)} \sqrt{\frac{1}{n-3}}$$

where $z_{(1-\alpha/2)}$ is the $100(1 - \alpha/2)$ percentage point of the standard normal distribution.

With a bias adjustment, confidence limits for ζ are computed by treating

$$z_r - \zeta - \text{bias}(r)$$

as having a normal distribution with mean zero and variance $1/(n - 3)$, where the bias adjustment function (Keeping 1962, p. 308) is

$$\text{bias}(r_r) = \frac{r}{2(n-1)}$$

That is, the two-sided confidence limits for ζ are computed as

$$\zeta_l = z_r - \text{bias}(r) - z_{(1-\alpha/2)} \sqrt{\frac{1}{n-3}}$$

$$\zeta_u = z_r - \text{bias}(r) + z_{(1-\alpha/2)} \sqrt{\frac{1}{n-3}}$$

These computed confidence limits of ζ_l and ζ_u are then transformed back to derive the confidence limits for the correlation ρ:

$$r_l = \tanh(\zeta_l) = \frac{\exp(2\zeta_l) - 1}{\exp(2\zeta_l) + 1}$$

$$r_u = \tanh(\zeta_u) = \frac{\exp(2\zeta_u) - 1}{\exp(2\zeta_u) + 1}$$

Note that with a bias adjustment, the CORR procedure also displays the following correlation estimate:

$$r_{adj} = \tanh(z_r - \text{bias}(r))$$

Applications of Fisher's z Transformation

Fisher (1970, p. 199) describes the following practical applications of the z transformation:

- testing whether a population correlation is equal to a given value
- testing for equality of two population correlations
- combining correlation estimates from different samples

To test if a population correlation ρ_1 from a sample of n_1 observations with sample correlation r_1 is equal to a given ρ_0, first apply the z transformation to r_1 and ρ_0: $z_1 = \tanh^{-1}(r_1)$ and $\zeta_0 = \tanh^{-1}(\rho_0)$.

The p-value is then computed by treating

$$z_1 - \zeta_0 - \frac{\rho_0}{2(n_1 - 1)}$$

as a normal random variable with mean zero and variance $1/(n_1 - 3)$.

Assume that sample correlations r_1 and r_2 are computed from two independent samples of n_1 and n_2 observations, respectively. To test whether the two corresponding population correlations, ρ_1 and ρ_2, are equal, first apply the z transformation to the two sample correlations: $z_1 = \tanh^{-1}(r_1)$ and $z_2 = \tanh^{-1}(r_2)$.

The p-value is derived under the null hypothesis of equal correlation. That is, the difference $z_1 - z_2$ is distributed as a normal random variable with mean zero and variance $1/(n_1 - 3) + 1/(n_2 - 3)$.

Assuming further that the two samples are from populations with identical correlation, a combined correlation estimate can be computed. The weighted average of the corresponding z values is

$$\bar{z} = \frac{(n_1 - 3)z_1 + (n_2 - 3)z_2}{n_1 + n_2 - 6}$$

where the weights are inversely proportional to their variances.

Thus, a combined correlation estimate is $\bar{r} = \tanh(\bar{z})$ and $V(\bar{z}) = 1/(n_1 + n_2 - 6)$. See Example 2.4 for further illustrations of these applications.

Note that this approach can be extended to include more than two samples.

Cronbach's Coefficient Alpha

Analyzing latent constructs such as job satisfaction, motor ability, sensory recognition, or customer satisfaction requires instruments to accurately measure the constructs. Interrelated items can be summed to obtain an overall score for each participant. Cronbach's coefficient alpha estimates the reliability of this type of scale by determining the internal consistency of the test or the average correlation of items within the test (Cronbach 1951).

When a value is recorded, the observed value contains some degree of measurement error. Two sets of measurements on the same variable for the same individual might not have identical values. However, repeated measurements for a series of individuals will show some consistency. Reliability measures internal consistency from one set of measurements to another. The observed value Y is divided into two components, a true value T and a measurement error E. The measurement error is assumed to be independent of the true value; that is,

$$Y = T + E \quad \mathrm{Cov}(T, E) = 0$$

The reliability coefficient of a measurement test is defined as the squared correlation between the observed value Y and the true value T; that is,

$$r^2(Y, T) = \frac{\mathrm{Cov}(Y, T)^2}{\mathrm{V}(Y)\mathrm{V}(T)} = \frac{\mathrm{V}(T)^2}{\mathrm{V}(Y)\mathrm{V}(T)} = \frac{\mathrm{V}(T)}{\mathrm{V}(Y)}$$

which is the proportion of the observed variance due to true differences among individuals in the sample. If Y is the sum of several observed variables measuring the same feature, you can estimate $V(T)$. Cronbach's coefficient alpha, based on a lower bound for $V(T)$, is an estimate of the reliability coefficient.

Suppose p variables are used with $Y_j = T_j + E_j$ for $j = 1, 2, \ldots, p$, where Y_j is the observed value, T_j is the true value, and E_j is the measurement error. The measurement errors (E_j) are independent of the true values (T_j) and are also independent of each other. Let $Y_0 = \sum_j Y_j$ be the total observed score and let $T_0 = \sum_j T_j$ be the total true score. Because

$$(p-1) \sum_j V(T_j) \geq \sum_{i \neq j} \mathrm{Cov}(T_i, T_j)$$

a lower bound for $V(T_0)$ is given by

$$\frac{p}{p-1} \sum_{i \neq j} \mathrm{Cov}(T_i, T_j)$$

With $\mathrm{Cov}(Y_i, Y_j) = \mathrm{Cov}(T_i, T_j)$ for $i \neq j$, a lower bound for the reliability coefficient, $V(T_0)/V(Y_0)$, is then given by the Cronbach's coefficient alpha:

$$\alpha = \left(\frac{p}{p-1}\right) \frac{\sum_{i \neq j} \mathrm{Cov}(Y_i, Y_j)}{V(Y_0)} = \left(\frac{p}{p-1}\right) \left(1 - \frac{\sum_j V(Y_j)}{V(Y_0)}\right)$$

If the variances of the items vary widely, you can standardize the items to a standard deviation of 1 before computing the coefficient alpha. If the variables are dichotomous (0,1), the coefficient alpha is equivalent to the Kuder-Richardson 20 (KR-20) reliability measure.

When the correlation between each pair of variables is 1, the coefficient alpha has a maximum value of 1. With negative correlations between some variables, the coefficient alpha can have a value less than zero. The larger the overall alpha coefficient, the more likely that items contribute to a reliable scale. Nunnally and Bernstein (1994) suggests 0.70 as an acceptable reliability coefficient; smaller reliability coefficients are seen as inadequate. However, this varies by discipline.

To determine how each item reflects the reliability of the scale, you calculate a coefficient alpha after deleting each variable independently from the scale. Cronbach's coefficient alpha from all variables except the kth variable is given by

$$\alpha_k = \left(\frac{p-1}{p-2}\right)\left(1 - \frac{\sum_{i \neq k} V(Y_i)}{V(\sum_{i \neq k} Y_i)}\right)$$

If the reliability coefficient increases after an item is deleted from the scale, you can assume that the item is not correlated highly with other items in the scale. Conversely, if the reliability coefficient decreases, you can assume that the item is highly correlated with other items in the scale. Refer to *SAS Communications* (1994) for more information about how to interpret Cronbach's coefficient alpha.

Listwise deletion of observations with missing values is necessary to correctly calculate Cronbach's coefficient alpha. PROC CORR does not automatically use listwise deletion if you specify the ALPHA option. Therefore, you should use the NOMISS option if the data set contains missing values. Otherwise, PROC CORR prints a warning message indicating the need to use the NOMISS option with the ALPHA option.

Confidence and Prediction Ellipses

When the relationship between two variables is nonlinear or when outliers are present, the correlation coefficient might incorrectly estimate the strength of the relationship. Plotting the data enables you to verify the linear relationship and to identify the potential outliers.

The partial correlation between two variables, after controlling for variables in the PARTIAL statement, is the correlation between the residuals of the linear regression of the two variables on the partialled variables. Thus, if a PARTIAL statement is also specified, the residuals of the analysis variables are displayed in the scatter plot matrix and scatter plots.

The CORR procedure optionally provides two types of ellipses for each pair of variables in a scatter plot. One is a confidence ellipse for the population mean, and the other is a prediction ellipse for a new observation. Both assume a bivariate normal distribution.

Let $\bar{\mathbf{Z}}$ and $\mathbf{S}$ be the sample mean and sample covariance matrix of a random sample of size n from a bivariate normal distribution with mean μ and covariance matrix Σ. The variable $\bar{\mathbf{Z}} - \mu$ is distributed as a bivariate normal variate with mean zero and covariance $(1/n)\Sigma$, and it is independent of $\mathbf{S}$. Using Hotelling's T^2 statistic, which is defined as

$$T^2 = n(\bar{\mathbf{Z}} - \mu)'\mathbf{S}^{-1}(\bar{\mathbf{Z}} - \mu)$$

a $100(1 - \alpha)\%$ confidence ellipse for μ is computed from the equation

$$\frac{n}{n-1}(\bar{Z} - \mu)'S^{-1}(\bar{Z} - \mu) = \frac{2}{n-2}F_{2,n-2}(1 - \alpha)$$

where $F_{2,n-2}(1 - \alpha)$ is the $(1 - \alpha)$ critical value of an F distribution with degrees of freedom 2 and $n - 2$.

A prediction ellipse is a region for predicting a new observation in the population. It also approximates a region containing a specified percentage of the population.

Denote a new observation as the bivariate random variable Z_{new}. The variable

$$Z_{new} - \bar{Z} = (Z_{new} - \mu) - (\bar{Z} - \mu)$$

is distributed as a bivariate normal variate with mean zero (the zero vector) and covariance $(1 + 1/n)\Sigma$, and it is independent of S. A $100(1 - \alpha)\%$ prediction ellipse is then given by the equation

$$\frac{n}{n-1}(\bar{Z} - \mu)'S^{-1}(\bar{Z} - \mu) = \frac{2(n+1)}{n-2}F_{2,n-2}(1 - \alpha)$$

The family of ellipses generated by different critical values of the F distribution has a common center (the sample mean) and common major and minor axis directions.

The shape of an ellipse depends on the aspect ratio of the plot. The ellipse indicates the correlation between the two variables if the variables are standardized (by dividing the variables by their respective standard deviations). In this situation, the ratio between the major and minor axis lengths is

$$\sqrt{\frac{1 + |r|}{1 - |r|}}$$

In particular, if $r = 0$, the ratio is 1, which corresponds to a circular confidence contour and indicates that the variables are uncorrelated. A larger value of the ratio indicates a larger positive or negative correlation between the variables.

Missing Values

PROC CORR excludes observations with missing values in the WEIGHT and FREQ variables. By default, PROC CORR uses *pairwise deletion* when observations contain missing values. PROC CORR includes all nonmissing pairs of values for each pair of variables in the statistical computations. Therefore, the correlation statistics might be based on different numbers of observations.

If you specify the NOMISS option, PROC CORR uses *listwise deletion* when a value of the VAR or WITH statement variable is missing. PROC CORR excludes all observations with missing values from the analysis. Therefore, the number of observations for each pair of variables is identical.

The PARTIAL statement always excludes the observations with missing values by automatically invoking the NOMISS option. With the NOMISS option, the data are processed more efficiently because fewer resources are needed. Also, the resulting correlation matrix is nonnegative definite.

In contrast, if the data set contains missing values for the analysis variables and the NOMISS option is not specified, the resulting correlation matrix might not be nonnegative definite. This leads to several statistical difficulties if you use the correlations as input to regression or other statistical procedures.

Output Tables

By default, PROC CORR prints a report that includes descriptive statistics and correlation statistics for each variable. The descriptive statistics include the number of observations with nonmissing values, the mean, the standard deviation, the minimum, and the maximum.

If a nonparametric measure of association is requested, the descriptive statistics include the median. Otherwise, the sample sum is included. If a Pearson partial correlation is requested, the descriptive statistics also include the partial variance and partial standard deviation.

If variable labels are available, PROC CORR labels the variables. If you specify the CSSCP, SSCP, or COV option, the appropriate sums of squares and crossproducts and covariance matrix appear at the top of the correlation report. If the data set contains missing values, PROC CORR prints additional statistics for each pair of variables. These statistics, calculated from the observations with nonmissing row and column variable values, might include the following:

- SSCP('W','V'), uncorrected sums of squares and crossproducts
- USS('W'), uncorrected sums of squares for the row variable
- USS('V'), uncorrected sums of squares for the column variable
- CSSCP('W','V'), corrected sums of squares and crossproducts
- CSS('W'), corrected sums of squares for the row variable
- CSS('V'), corrected sums of squares for the column variable
- COV('W','V'), covariance
- VAR('W'), variance for the row variable
- VAR('V'), variance for the column variable
- DF('W','V'), divisor for calculating covariance and variances

For each pair of variables, PROC CORR prints the correlation coefficients, the number of observations used to calculate the coefficient, and the *p*-value.

If you specify the ALPHA option, PROC CORR prints Cronbach's coefficient alpha, the correlation between the variable and the total of the remaining variables, and Cronbach's coefficient alpha by using the remaining variables for the raw variables and the standardized variables.

Output Data Sets

If you specify the OUTP=, OUTS=, OUTK=, or OUTH= option, PROC CORR creates an output data set containing statistics for Pearson correlation, Spearman correlation, Kendall's tau-b, or Hoeffding's D, respectively. By default, the output data set is a special data set type (TYPE=CORR) that many SAS/STAT procedures recognize, including PROC REG and PROC FACTOR. When you specify the NOCORR option and the COV, CSSCP, or SSCP option, use the TYPE= data set option to change the data set type to COV, CSSCP, or SSCP.

The output data set includes the following variables:

- BY variables, which identify the BY group when using a BY statement
- _TYPE_ variable, which identifies the type of observation
- _NAME_ variable, which identifies the variable that corresponds to a given row of the correlation matrix
- INTERCEPT variable, which identifies variable sums when specifying the SSCP option
- VAR variables, which identify the variables listed in the VAR statement

You can use a combination of the _TYPE_ and _NAME_ variables to identify the contents of an observation. The _NAME_ variable indicates which row of the correlation matrix the observation corresponds to. The values of the _TYPE_ variable are as follows:

- SSCP, uncorrected sums of squares and crossproducts
- CSSCP, corrected sums of squares and crossproducts
- COV, covariances
- MEAN, mean of each variable
- STD, standard deviation of each variable
- N, number of nonmissing observations for each variable
- SUMWGT, sum of the weights for each variable when using a WEIGHT statement
- CORR, correlation statistics for each variable

If you specify the SSCP option, the OUTP= data set includes an additional observation that contains intercept values. If you specify the ALPHA option, the OUTP= data set also includes observations with the following _TYPE_ values:

- RAWALPHA, Cronbach's coefficient alpha for raw variables
- STDALPHA, Cronbach's coefficient alpha for standardized variables

- RAWALDEL, Cronbach's coefficient alpha for raw variables after deleting one variable

- STDALDEL, Cronbach's coefficient alpha for standardized variables after deleting one variable

- RAWCTDEL, the correlation between a raw variable and the total of the remaining raw variables

- STDCTDEL, the correlation between a standardized variable and the total of the remaining standardized variables

If you use a PARTIAL statement, the statistics are calculated after the variables are partialled. If PROC CORR computes Pearson correlation statistics, MEAN equals zero and STD equals the partial standard deviation associated with the partial variance for the OUTP=, OUTK=, and OUTS= data sets. Otherwise, PROC CORR assigns missing values to MEAN and STD.

ODS Table Names

PROC CORR assigns a name to each table it creates. You must use these names to reference tables when using the Output Delivery System (ODS). These names are listed in Table 2.3 and Table 2.4. For more information about ODS, see Chapter 20, "Using the Output Delivery System" (*SAS/STAT User's Guide*).

Table 2.3 ODS Tables Produced by PROC CPRR

ODS Table Name	Description	Option
Cov	Covariances	COV
CronbachAlpha	Coefficient alpha	ALPHA
CronbachAlphaDel	Coefficient alpha with deleted variable	ALPHA
Csscp	Corrected sums of squares and crossproducts	CSSCP
FisherPearsonCorr	Pearson correlation statistics using Fisher's z transformation	FISHER
FisherSpearmanCorr	Spearman correlation statistics using Fisher's z transformation	FISHER SPEARMAN
HoeffdingCorr	Hoeffding's D statistics	HOEFFDING
KendallCorr	Kendall's tau-b coefficients	KENDALL
PearsonCorr	Pearson correlations	PEARSON
SimpleStats	Simple descriptive statistics	
SpearmanCorr	Spearman correlations	SPEARMAN
Sscp	Sums of squares and crossproducts	SSCP
VarInformation	Variable information	

Table 2.4 ODS Tables Produced with the PARTIAL Statement

ODS Table Name	Description	Option
FisherPearsonPartialCorr	Pearson partial correlation statistics using Fisher's z transformation	FISHER
FisherSpearmanPartialCorr	Spearman partial correlation statistics using Fisher's z transformation	FISHER SPEARMAN
PartialCsscp	Partial corrected sums of squares and crossproducts	CSSCP
PartialCov	Partial covariances	COV
PartialKendallCorr	Partial Kendall tau-b coefficients	KENDALL
PartialPearsonCorr	Partial Pearson correlations	
PartialSpearmanCorr	Partial Spearman correlations	SPEARMAN

ODS Graphics

PROC CORR assigns a name to each graph it creates using ODS. You can use these names to reference the graphs when using ODS. The names are listed in Table 2.5.

To request these graphs, you must specify the **ods graphics on** statement in addition to the options indicated in Table 2.5. For more information about the **ods graphics** statement, see Chapter 21, "Statistical Graphics Using ODS" (*SAS/STAT User's Guide*).

Table 2.5 ODS Graphics Produced by PROC CORR

ODS Graph Name	Plot Description	Option
ScatterPlot	Scatter plot	PLOTS=SCATTER
MatrixPlot	Scatter plot matrix	PLOTS=MATRIX

Examples: CORR Procedure

Example 2.1: Computing Four Measures of Association

This example produces a correlation analysis with descriptive statistics and four measures of association: the Pearson product-moment correlation, the Spearman rank-order correlation, Kendall's tau-b coefficients, and Hoeffding's measure of dependence, D.

The Fitness data set created in the section "Getting Started: CORR Procedure" on page 5 contains measurements from a study of physical fitness of 31 participants. The following statements request all four measures of association for the variables Weight, Oxygen, and Runtime:

```
ods graphics on;
title 'Measures of Association for a Physical Fitness Study';
proc corr data=Fitness pearson spearman kendall hoeffding
          plots=matrix(histogram);
   var Weight Oxygen RunTime;
run;
ods graphics off;
```

Note that Pearson correlations are computed by default only if all three nonparametric correlations (SPEARMAN, KENDALL, and HOEFFDING) are not specified. Otherwise, you need to specify the PEARSON option explicitly to compute Pearson correlations.

The "Simple Statistics" table in Output 2.1.1 displays univariate descriptive statistics for analysis variables. By default, observations with nonmissing values for each variable are used to derive the univariate statistics for that variable. When nonparametric measures of association are specified, the procedure displays the median instead of the sum as an additional descriptive measure.

Output 2.1.1 Simple Statistics

Measures of Association for a Physical Fitness Study

The CORR Procedure

3 Variables: Weight Oxygen RunTime

Simple Statistics

Variable	N	Mean	Std Dev	Median	Minimum	Maximum
Weight	31	77.44452	8.32857	77.45000	59.08000	91.63000
Oxygen	29	47.22721	5.47718	46.67200	37.38800	60.05500
RunTime	29	10.67414	1.39194	10.50000	8.17000	14.03000

The "Pearson Correlation Coefficients" table in Output 2.1.2 displays Pearson correlation statistics for pairs of analysis variables. The Pearson correlation is a parametric measure of association for two continuous random variables. When there are missing data, the number of observations used to calculate the correlation can vary.

Output 2.1.2 Pearson Correlation Coefficients

```
              Pearson Correlation Coefficients
                 Prob > |r| under H0: Rho=0
                    Number of Observations

                    Weight        Oxygen        RunTime

    Weight         1.00000       -0.15358        0.20072
                                  0.4264         0.2965
                     31             29             29

    Oxygen        -0.15358        1.00000       -0.86843
                   0.4264                        <.0001
                     29             29             28

    RunTime        0.20072       -0.86843        1.00000
                   0.2965         <.0001
                     29             28             29
```

The table shows that the Pearson correlation between Runtime and Oxygen is −0.86843, which is significant with a *p*-value less than 0.0001. This indicates a strong negative linear relationship between these two variables. As Runtime increases, Oxygen decreases linearly.

The Spearman rank-order correlation is a nonparametric measure of association based on the ranks of the data values. The "Spearman Correlation Coefficients" table in Output 2.1.3 displays results similar to those of the "Pearson Correlation Coefficients" table in Output 2.1.2.

Output 2.1.3 Spearman Correlation Coefficients

```
              Spearman Correlation Coefficients
                 Prob > |r| under H0: Rho=0
                    Number of Observations

                    Weight        Oxygen        RunTime

    Weight         1.00000       -0.06824        0.13749
                                  0.7250         0.4769
                     31             29             29

    Oxygen        -0.06824        1.00000       -0.80131
                   0.7250                        <.0001
                     29             29             28

    RunTime        0.13749       -0.80131        1.00000
                   0.4769         <.0001
                     29             28             29
```

Kendall's tau-b is a nonparametric measure of association based on the number of concordances and discordances in paired observations. The "Kendall Tau b Correlation Coefficients" table in Output 2.1.4 displays results similar to those of the "Pearson Correlation Coefficients" table in Output 2.1.2.

Output 2.1.4 Kendall's Tau-b Correlation Coefficients

```
          Kendall Tau b Correlation Coefficients
               Prob > |tau| under H0: Tau=0
                    Number of Observations

                    Weight          Oxygen         RunTime

    Weight         1.00000         -0.00988         0.06675
                                    0.9402          0.6123
                     31               29              29

    Oxygen        -0.00988          1.00000        -0.62434
                   0.9402                           <.0001
                     29               29              28

    RunTime        0.06675         -0.62434         1.00000
                   0.6123           <.0001
                     29               28              29
```

Hoeffding's measure of dependence, D, is a nonparametric measure of association that detects more general departures from independence. Without ties in the variables, the values of the D statistic can vary between -0.5 and 1, with 1 indicating complete dependence. Otherwise, the D statistic can result in a smaller value. The "Hoeffding Dependence Coefficients" table in Output 2.1.5 displays Hoeffding dependence coefficients. Since ties occur in the variable Weight, the D statistic for the Weight variable is less than 1.

Output 2.1.5 Hoeffding's Dependence Coefficients

```
            Hoeffding Dependence Coefficients
                Prob > D under H0: D=0
                  Number of Observations

                    Weight          Oxygen         RunTime

    Weight         0.97690         -0.00497        -0.02355
                   <.0001           0.5101          1.0000
                     31               29              29

    Oxygen        -0.00497          1.00000         0.23449
                   0.5101                           <.0001
                     29               29              28

    RunTime       -0.02355          0.23449         1.00000
                   1.0000           <.0001
                     29               28              29
```

When you use the PLOTS=MATRIX(HISTOGRAM) option, the CORR procedure displays a symmetric matrix plot for the analysis variables listed in the VAR statement (Output 2.1.6).

Output 2.1.6 Symmetric Scatter Plot Matrix

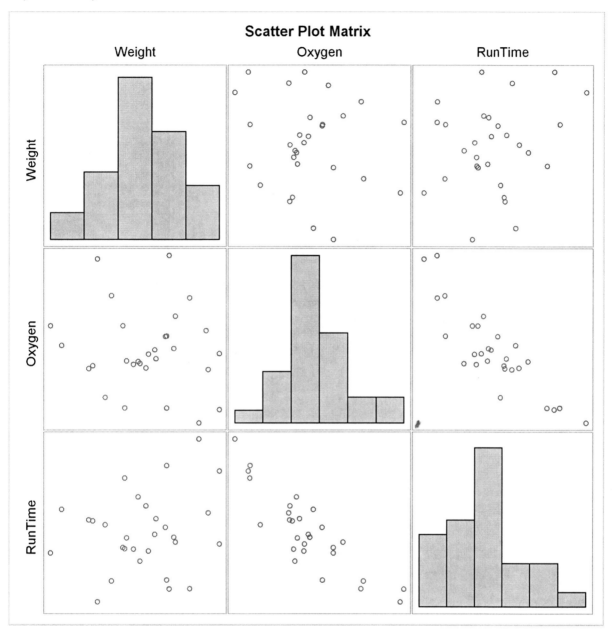

The strong negative linear relationship between Oxygen and Runtime is evident in Output 2.1.6.

Note that this graphical display is requested by specifying the **ods graphics on** statement and the PLOTS option. For more information about the **ods graphics** statement, see Chapter 21, "Statistical Graphics Using ODS" (*SAS/STAT User's Guide*).

Example 2.2: Computing Correlations between Two Sets of Variables

The following statements create a data set which contains measurements for four iris parts from Fisher's iris data (1936): sepal length, sepal width, petal length, and petal width. Each observation represents one specimen.

```
*-------------------- Data on Iris Setosa --------------------*
| The data set contains 50 iris specimens from the species    |
| Iris Setosa with the following four measurements:           |
| SepalLength (sepal length)                                  |
| SepalWidth  (sepal width)                                   |
| PetalLength (petal length)                                  |
| PetalWidth  (petal width)                                   |
| Certain values were changed to missing for the analysis.    |
*-------------------------------------------------------------*;
data Setosa;
   input SepalLength SepalWidth PetalLength PetalWidth @@;
   label sepallength='Sepal Length in mm.'
         sepalwidth='Sepal Width in mm.'
         petallength='Petal Length in mm.'
         petalwidth='Petal Width in mm.';
   datalines;
50 33 14 02   46 34 14 03   46 36  . 02
51 33 17 05   55 35 13 02   48 31 16 02
52 34 14 02   49 36 14 01   44 32 13 02
50 35 16 06   44 30 13 02   47 32 16 02
48 30 14 03   51 38 16 02   48 34 19 02
50 30 16 02   50 32 12 02   43 30 11  .
58 40 12 02   51 38 19 04   49 30 14 02
51 35 14 02   50 34 16 04   46 32 14 02
57 44 15 04   50 36 14 02   54 34 15 04
52 41 15  .   55 42 14 02   49 31 15 02
54 39 17 04   50 34 15 02   44 29 14 02
47 32 13 02   46 31 15 02   51 34 15 02
50 35 13 03   49 31 15 01   54 37 15 02
54 39 13 04   51 35 14 03   48 34 16 02
48 30 14 01   45 23 13 03   57 38 17 03
51 38 15 03   54 34 17 02   51 37 15 04
52 35 15 02   53 37 15 02
;
```

The following statements request a correlation analysis between two sets of variables, the sepal measurements (length and width) and the petal measurements (length and width):

```
ods graphics on;
title 'Fisher (1936) Iris Setosa Data';
proc corr data=Setosa sscp cov plots;
   var   sepallength sepalwidth;
   with petallength petalwidth;
run;
ods graphics off;
```

The "Simple Statistics" table in Output 2.2.1 displays univariate statistics for variables in the VAR and WITH statements.

Output 2.2.1 Simple Statistics

```
                    Fisher (1936) Iris Setosa Data

                          The CORR Procedure

           2  With Variables:    PetalLength PetalWidth
           2       Variables:    SepalLength SepalWidth


                           Simple Statistics

    Variable            N          Mean       Std Dev           Sum

    PetalLength        49      14.71429       1.62019     721.00000
    PetalWidth         48       2.52083       1.03121     121.00000
    SepalLength        50      50.06000       3.52490          2503
    SepalWidth         50      34.28000       3.79064          1714


                           Simple Statistics

    Variable         Minimum       Maximum    Label

    PetalLength     11.00000      19.00000    Petal Length in mm.
    PetalWidth       1.00000       6.00000    Petal Width in mm.
    SepalLength     43.00000      58.00000    Sepal Length in mm.
    SepalWidth      23.00000      44.00000    Sepal Width in mm.
```

When the WITH statement is specified together with the VAR statement, the CORR procedure produces rectangular matrices for statistics such as covariances and correlations. The matrix rows correspond to the WITH variables (PetalLength and PetalWidth), while the matrix columns correspond to the VAR variables (SepalLength and SepalWidth). The CORR procedure uses the WITH variable labels to label the matrix rows.

The SSCP option requests a table of the uncorrected sum-of-squares and crossproducts matrix, and the COV option requests a table of the covariance matrix. The SSCP and COV options also produce a table of the Pearson correlations.

The sum-of-squares and crossproducts statistics for each pair of variables are computed by using observations with nonmissing row and column variable values. The "Sums of Squares and Crossproducts" table in Output 2.2.2 displays the crossproduct, sum of squares for the row variable, and sum of squares for the column variable for each pair of variables.

Output 2.2.2 Sums of Squares and Crossproducts

```
             Sums of Squares and Crossproducts
               SSCP / Row Var SS / Col Var SS

                              SepalLength         SepalWidth

    PetalLength               36214.00000         24756.00000
    Petal Length in mm.       10735.00000         10735.00000
                              123793.0000         58164.0000

    PetalWidth                 6113.00000          4191.00000
    Petal Width in mm.          355.00000           355.00000
                              121356.0000         56879.0000
```

The variances are computed by using observations with nonmissing row and column variable values. The "Variances and Covariances" table in Output 2.2.3 displays the covariance, variance for the row variable, variance for the column variable, and associated degrees of freedom for each pair of variables.

Output 2.2.3 Variances and Covariances

```
                  Variances and Covariances
       Covariance / Row Var Variance / Col Var Variance / DF

                              SepalLength         SepalWidth

    PetalLength                1.270833333         1.363095238
    Petal Length in mm.        2.625000000         2.625000000
                              12.33333333         14.60544218
                                       48                  48

    PetalWidth                 0.911347518         1.048315603
    Petal Width in mm.         1.063386525         1.063386525
                              11.80141844         13.62721631
                                       47                  47
```

When there are missing values in the analysis variables, the "Pearson Correlation Coefficients" table in Output 2.2.4 displays the correlation, the *p*-value under the null hypothesis of zero correlation, and the number of observations for each pair of variables. Only the correlation between PetalWidth and SepalLength and the correlation between PetalWidth and SepalWidth are slightly positive.

Output 2.2.4 Pearson Correlation Coefficients

```
             Pearson Correlation Coefficients
                  Prob > |r| under H0: Rho=0
                     Number of Observations

                                   Sepal         Sepal
                                   Length        Width

          PetalLength              0.22335       0.22014
          Petal Length in mm.      0.1229        0.1285
                                   49            49


          PetalWidth               0.25726       0.27539
          Petal Width in mm.       0.0775        0.0582
                                   48            48
```

When you specify the **ods graphics on** statement, the PROC CORR displays a scatter matrix plot by default. Output 2.2.5 displays a rectangular scatter plot matrix for the two sets of variables: the VAR variables SepalLength and SepalWidth are listed across the top of the matrix, and the WITH variables PetalLength and PetalWidth are listed down the side of the matrix. As measured in Output 2.2.4, the plot for PetalWidth and SepalLength and the plot for PetalWidth and SepalWidth also show slight positive correlations.

Output 2.2.5 Rectangular Matrix Plot

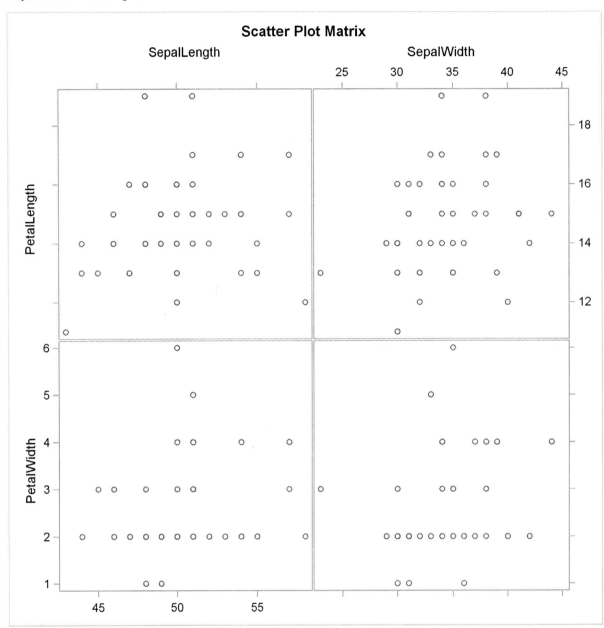

Note that this graphical display is requested by specifying the `ods graphics on` statement and the PLOTS option. For more information about the `ods graphics` statement, see Chapter 21, "Statistical Graphics Using ODS" (*SAS/STAT User's Guide*).

Example 2.3: Analysis Using Fisher's *z* Transformation

The following statements request Pearson correlation statistics by using Fisher's *z* transformation for the data set Fitness:

Example 2.3: Analysis Using Fisher's z Transformation

```
proc corr data=Fitness nosimple fisher;
   var weight oxygen runtime;
run;
```

The NOSIMPLE option suppresses the table of univariate descriptive statistics. By default, PROC CORR displays the "Pearson Correlation Coefficients" table in Output 2.3.1.

Output 2.3.1 Pearson Correlations

```
                       The CORR Procedure

                 Pearson Correlation Coefficients
                    Prob > |r| under H0: Rho=0
                      Number of Observations

                       Weight        Oxygen       RunTime

       Weight         1.00000       -0.15358       0.20072
                                     0.4264        0.2965
                           31            29            29

       Oxygen        -0.15358       1.00000       -0.86843
                      0.4264                       <.0001
                           29            29            28

       RunTime        0.20072      -0.86843        1.00000
                      0.2965        <.0001
                           29            28            29
```

Using the FISHER option, the CORR procedure displays correlation statistics by using Fisher's z transformation in Output 2.3.2.

Output 2.3.2 Correlation Statistics Using Fisher's z Transformation

```
           Pearson Correlation Statistics (Fisher's z Transformation)

           With                 Sample                     Bias    Correlation
Variable   Variable      N   Correlation   Fisher's z   Adjustment   Estimate

Weight     Oxygen       29      -0.15358     -0.15480    -0.00274    -0.15090
Weight     RunTime      29       0.20072      0.20348     0.00358     0.19727
Oxygen     RunTime      28      -0.86843     -1.32665    -0.01608    -0.86442

           Pearson Correlation Statistics (Fisher's z Transformation)

           With                                           p Value for
Variable   Variable       95% Confidence Limits            H0:Rho=0

Weight     Oxygen         -0.490289       0.228229          0.4299
Weight     RunTime        -0.182422       0.525765          0.2995
Oxygen     RunTime        -0.935728      -0.725221          <.0001
```

The table also displays confidence limits and a p-value for the default null hypothesis $H_0: \rho = \rho_0$. See the section "Fisher's z Transformation" on page 24 for details on Fisher's z transformation.

The following statements request one-sided hypothesis tests and confidence limits for the correlations using Fisher's z transformation:

```
proc corr data=Fitness nosimple nocorr fisher (type=lower);
   var weight oxygen runtime;
run;
```

The NOSIMPLE option suppresses the "Simple Statistics" table, and the NOCORR option suppresses the "Pearson Correlation Coefficients" table.

Output 2.3.3 displays correlation statistics by using Fisher's z transformation.

Output 2.3.3 One-Sided Correlation Analysis Using Fisher's z Transformation

```
                          The CORR Procedure

            Pearson Correlation Statistics (Fisher's z Transformation)

                       With              Sample                    Bias    Correlation
        Variable     Variable      N   Correlation   Fisher's z   Adjustment   Estimate

        Weight       Oxygen       29     -0.15358     -0.15480    -0.00274    -0.15090
        Weight       RunTime      29      0.20072      0.20348     0.00358     0.19727
        Oxygen       RunTime      28     -0.86843     -1.32665    -0.01608    -0.86442

            Pearson Correlation Statistics (Fisher's z Transformation)

                          With                        p Value for
                Variable  Variable   Lower 95% CL     H0:Rho<=0

                Weight    Oxygen       -0.441943       0.7850
                Weight    RunTime      -0.122077       0.1497
                Oxygen    RunTime      -0.927408       1.0000
```

The FISHER(TYPE=LOWER) option requests a lower confidence limit and a p-value for the test of the one-sided hypothesis $H_0: \rho \leq 0$ against the alternative hypothesis $H_1: \rho > 0$. Here Fisher's z, the bias adjustment, and the estimate of the correlation are the same as for the two-sided alternative. However, because TYPE=LOWER is specified, only a lower confidence limit is computed for each correlation, and one-sided p-values are computed.

Example 2.4: Applications of Fisher's z Transformation

This example illustrates some applications of Fisher's z transformation. For details, see the section "Fisher's z Transformation" on page 24.

The following statements simulate independent samples of variables X and Y from a bivariate normal distribution. The first batch of 150 observations is sampled using a known correlation of 0.3, the second batch of 150 observations is sampled using a known correlation of 0.25, and the third batch of 100 observations is sampled using a known correlation of 0.3.

```
data Sim (drop=i);
do i=1 to 400;
  X = rannor(135791);
  Batch = 1 + (i>150) + (i>300);
  if Batch = 1 then Y = 0.3*X + 0.9*rannor(246791);
  if Batch = 2 then Y = 0.25*X + sqrt(.8375)*rannor(246791);
  if Batch = 3 then Y = 0.3*X + 0.9*rannor(246791);
  output;
end;
run;
```

This data set will be used to illustrate the following applications of Fisher's z transformation:

- testing whether a population correlation is equal to a given value
- testing for equality of two population correlations
- combining correlation estimates from different samples

Testing Whether a Population Correlation Is Equal to a Given Value ρ_0

You can use the following statements to test the null hypothesis $H_0: \rho = 0.5$ against a two-sided alternative $H_1: \rho \neq 0.5$. The test is requested with the option FISHER(RHO0=0.5).

```
title 'Analysis for Batch 1';
proc corr data=Sim (where=(Batch=1)) fisher(rho0=.5);
   var X Y;
run;
```

Output 2.4.1 displays the results based on Fisher's transformation. The null hypothesis is rejected since the p-value is less than 0.0001.

Output 2.4.1 Fisher's Test for $H_0: \rho = \rho_0$

```
                           Analysis for Batch 1

                            The CORR Procedure

            Pearson Correlation Statistics (Fisher's z Transformation)

                      With                    Sample                          Bias    Correlation
      Variable    Variable         N     Correlation    Fisher's z      Adjustment       Estimate

      X           Y              150         0.22081       0.22451       0.0007410        0.22011

            Pearson Correlation Statistics (Fisher's z Transformation)

                     With                                              ------H0:Rho=Rho0-----
      Variable    Variable       95% Confidence Limits                    Rho0       p Value

      X           Y                0.062034       0.367409              0.50000         <.0001
```

Testing for Equality of Two Population Correlations

You can use the following statements to test for equality of two population correlations, ρ_1 and ρ_2. Here, the null hypothesis $H_0: \rho_1 = \rho_2$ is tested against the alternative $H_1: \rho_1 \neq \rho_2$.

```
ods output FisherPearsonCorr=SimCorr;
title 'Testing Equality of Population Correlations';
proc corr data=Sim (where=(Batch=1 or Batch=2)) fisher;
   var X Y;
   by Batch;
run;
```

The ODS OUTPUT statement saves the "FisherPearsonCorr" table into an output data set in the CORR procedure. The output data set SimCorr contains Fisher's z statistics for both batches.

The following statements display (in Figure 2.4.2) the output data set SimCorr:

```
proc print data=SimCorr;
run;
```

Output 2.4.2 Fisher's Correlation Statistics

```
                        With
     Obs    Batch   Var  Var       NObs         Corr         ZVal       BiasAdj

      1       1      X    Y         150       0.22081      0.22451     0.0007410
      2       2      X    Y         150       0.33694      0.35064     0.00113


     Obs        CorrEst            Lcl            Ucl       pValue

      1         0.22011         0.062034       0.367409     0.0065
      2         0.33594         0.185676       0.470853     <.0001
```

The *p*-value for testing H_0 is derived by treating the difference $z_1 - z_2$ as a normal random variable with mean zero and variance $1/(n_1 - 3) + 1/(n_2 - 3)$, where z_1 and z_2 are Fisher's *z* transformation of the sample correlations r_1 and r_2, respectively, and where n_1 and n_2 are the corresponding sample sizes.

The following statements compute the *p*-value in Output 2.4.3:

```
data SimTest (drop=Batch);
   merge SimCorr (where=(Batch=1) keep=Nobs ZVal Batch
                  rename=(Nobs=n1 ZVal=z1))
         SimCorr (where=(Batch=2) keep=Nobs ZVal Batch
                  rename=(Nobs=n2 ZVal=z2));
   variance = 1/(n1-3) + 1/(n2-3);
   z = (z1 - z2) / sqrt( variance );
   pval = probnorm(z);
   if (pval > 0.5) then pval = 1 - pval;
   pval = 2*pval;
run;

proc print data=SimTest noobs;
run;
```

Output 2.4.3 Test of Equality of Observed Correlations

n1	z1	n2	z2	variance	z	pval
150	0.22451	150	0.35064	0.013605	-1.08135	0.27954

In Output 2.4.3, the *p*-value of 0.2795 does not provide evidence to reject the null hypothesis that $\rho_1 = \rho_2$. The sample sizes $n_1 = 150$ and $n_2 = 150$ are not large enough to detect the difference $\rho_1 - \rho_2 = 0.05$ at a significance level of $\alpha = 0.05$.

Combining Correlation Estimates from Different Samples

Assume that sample correlations r_1 and r_2 are computed from two independent samples of n_1 and n_2 observations, respectively. A combined correlation estimate is given by $\bar{r} = \tanh(\bar{z})$, where $\bar{z}$ is the weighted average of the *z* transformations of r_1 and r_2:

$$\bar{z} = \frac{(n_1 - 3)z_1 + (n_2 - 3)z_2}{n_1 + n_2 - 6}$$

The following statements compute a combined estimate of ρ by using Batch 1 and Batch 3:

```
ods output FisherPearsonCorr=SimCorr2;
proc corr data=Sim (where=(Batch=1 or Batch=3)) fisher;
   var X Y;
   by Batch;
run;
```

```
   data SimComb (drop=Batch);
      merge SimCorr2 (where=(Batch=1) keep=Nobs ZVal Batch
                      rename=(Nobs=n1 ZVal=z1))
            SimCorr2 (where=(Batch=3) keep=Nobs ZVal Batch
                      rename=(Nobs=n2 ZVal=z2));
      z = ((n1-3)*z1 + (n2-3)*z2) / (n1+n2-6);
      corr = tanh(z);
      var = 1/(n1+n2-6);
      zlcl = z - probit(0.975)*sqrt(var);
      zucl = z + probit(0.975)*sqrt(var);
      lcl= tanh(zlcl);
      ucl= tanh(zucl);
      pval= probnorm( z/sqrt(var));
      if (pval > .5)  then pval= 1 - pval;
      pval= 2*pval;
   run;

   proc print data=SimComb noobs;
      var n1 z1 n2 z2 corr lcl ucl pval;
   run;
```

Output 2.4.4 displays the combined estimate of ρ. The table shows that a correlation estimate from the combined samples is $r = 0.2264$. The 95% confidence interval is $(0.10453, 0.34156)$, using the variance of the combined estimate. Note that this interval contains the population correlation 0.3.

Output 2.4.4 Combined Correlation Estimate

Obs	n1	z1	n2	z2	z	corr
1	150	0.22451	100	0.23929	0.23039	0.22640

Obs	var	zlcl	zucl	lcl	ucl	pval
1	.004098361	0.10491	0.35586	0.10453	0.34156	.000319748

Example 2.5: Computing Cronbach's Coefficient Alpha

The following statements create the data set Fish1 from the Fish data set used in Chapter 82, "The STEPDISC Procedure" (*SAS/STAT User's Guide*). The cubic root of the weight (Weight3) is computed as a one-dimensional measure of the size of a fish.

```
*------------------ Fish Measurement Data ---------------------*
| The data set contains 35 fish from the species Bream caught in  |
| Finland's lake Laengelmavesi with the following measurements:   |
| Weight    (in grams)                                            |
| Length3   (length from the nose to the end of its tail, in cm)  |
| HtPct     (max height, as percentage of Length3)                |
| WidthPct  (max width,  as percentage of Length3)                |
*-----------------------------------------------------------------*;
```

Example 2.5: Computing Cronbach's Coefficient Alpha ✦ 49

```
data Fish1 (drop=HtPct WidthPct);
   title 'Fish Measurement Data';
   input Weight Length3 HtPct WidthPct @@;
   Weight3= Weight**(1/3);
   Height=HtPct*Length3/100;
   Width=WidthPct*Length3/100;
   datalines;
242.0 30.0 38.4 13.4    290.0 31.2 40.0 13.8
340.0 31.1 39.8 15.1    363.0 33.5 38.0 13.3
430.0 34.0 36.6 15.1    450.0 34.7 39.2 14.2
500.0 34.5 41.1 15.3    390.0 35.0 36.2 13.4
450.0 35.1 39.9 13.8    500.0 36.2 39.3 13.7
475.0 36.2 39.4 14.1    500.0 36.2 39.7 13.3
500.0 36.4 37.8 12.0      .   37.3 37.3 13.6
600.0 37.2 40.2 13.9    600.0 37.2 41.5 15.0
700.0 38.3 38.8 13.8    700.0 38.5 38.8 13.5
610.0 38.6 40.5 13.3    650.0 38.7 37.4 14.8
575.0 39.5 38.3 14.1    685.0 39.2 40.8 13.7
620.0 39.7 39.1 13.3    680.0 40.6 38.1 15.1
700.0 40.5 40.1 13.8    725.0 40.9 40.0 14.8
720.0 40.6 40.3 15.0    714.0 41.5 39.8 14.1
850.0 41.6 40.6 14.9   1000.0 42.6 44.5 15.5
920.0 44.1 40.9 14.3    955.0 44.0 41.1 14.3
925.0 45.3 41.4 14.9    975.0 45.9 40.6 14.7
950.0 46.5 37.9 13.7
;
```

The following statements request a correlation analysis and compute Cronbach's coefficient alpha for the variables Weight3, Length3, Height, and Width:

```
ods graphics on;
title 'Fish Measurement Data';
proc corr data=fish1 nomiss alpha plots;
   var Weight3 Length3 Height Width;
 run;
ods graphics off;
```

The NOMISS option excludes observations with missing values, and the ALPHA option computes Cronbach's coefficient alpha for the analysis variables.

The "Simple Statistics" table in Output 2.5.1 displays univariate descriptive statistics for each analysis variable.

Output 2.5.1 Simple Statistics

```
                   Fish Measurement Data

                    The CORR Procedure

       4  Variables:    Weight3   Length3   Height   Width
```

Output 2.5.1 *continued*

```
                             Simple Statistics

 Variable          N          Mean       Std Dev           Sum       Minimum       Maximum

 Weight3          34       8.44751       0.97574      287.21524       6.23168      10.00000
 Length3          34      38.38529       4.21628           1305      30.00000      46.50000
 Height           34      15.22057       1.98159      517.49950      11.52000      18.95700
 Width            34       5.43805       0.72967      184.89370       4.02000       6.74970
```

The "Pearson Correlation Coefficients" table in Output 2.5.2 displays Pearson correlation statistics for pairs of analysis variables. When you specify the NOMISS option, the same set of 34 observations is used to compute the correlation for each pair of variables.

Output 2.5.2 Pearson Correlation Coefficients

```
                    Pearson Correlation Coefficients, N = 34
                          Prob > |r| under H0: Rho=0

                   Weight3         Length3          Height           Width

   Weight3         1.00000         0.96523         0.96261         0.92789
                                    <.0001          <.0001          <.0001

   Length3         0.96523         1.00000         0.95492         0.92171
                    <.0001                          <.0001          <.0001

   Height          0.96261         0.95492         1.00000         0.92632
                    <.0001          <.0001                          <.0001

   Width           0.92789         0.92171         0.92632         1.00000
                    <.0001          <.0001          <.0001
```

Since the data set contains only one species of fish, all the variables are highly correlated. Using the ALPHA option, the CORR procedure computes Cronbach's coefficient alpha in Output 2.5.3. The Cronbach's coefficient alpha is a lower bound for the reliability coefficient for the raw variables and the standardized variables. Positive correlation is needed for the alpha coefficient because variables measure a common entity.

Output 2.5.3 Cronbach's Coefficient Alpha

```
              Cronbach Coefficient Alpha

              Variables              Alpha
              ------------------------------
              Raw                  0.822134
              Standardized         0.985145
```

Because the variances of some variables vary widely, you should use the standardized score to estimate reliability. The overall standardized Cronbach's coefficient alpha of 0.985145 provides an acceptable lower bound for the reliability coefficient. This is much greater than the suggested value of 0.70 given by Nunnally and Bernstein (1994).

The standardized alpha coefficient provides information about how each variable reflects the reliability of the scale with standardized variables. If the standardized alpha decreases after removing a variable from the construct, then this variable is strongly correlated with other variables in the scale. On the other hand, if the standardized alpha increases after removing a variable from the construct, then removing this variable from the scale makes the construct more reliable. The "Cronbach Coefficient Alpha with Deleted Variables" table in Output 2.5.4 does not show significant increase or decrease in the standardized alpha coefficients. See the section "Cronbach's Coefficient Alpha" on page 27 for more information about Cronbach's alpha.

Output 2.5.4 Cronbach's Coefficient Alpha with Deleted Variables

	Cronbach Coefficient Alpha with Deleted Variable			
	Raw Variables		Standardized Variables	
Deleted Variable	Correlation with Total	Alpha	Correlation with Total	Alpha
Weight3	0.975379	0.783365	0.973464	0.977103
Length3	0.967602	0.881987	0.967177	0.978783
Height	0.964715	0.655098	0.968079	0.978542
Width	0.934635	0.824069	0.937599	0.986626

Example 2.6: Saving Correlations in an Output Data Set

The following statements compute Pearson correlations:

```
title 'Correlations for a Fitness and Exercise Study';
proc corr data=Fitness nomiss outp=CorrOutp;
   var weight oxygen runtime;
run;
```

The NOMISS option excludes observations with missing values of the VAR statement variables from the analysis—that is, the same set of 28 observations is used to compute the correlation for each pair of variables. The OUTP= option creates an output data set named CorrOutp that contains the Pearson correlation statistics.

"Pearson Correlation Coefficients" table in Output 2.6.1 displays the correlation and the p-value under the null hypothesis of zero correlation.

Output 2.6.1 Pearson Correlation Coefficients

```
              Correlations for a Fitness and Exercise Study

                            The CORR Procedure

                  Pearson Correlation Coefficients, N = 28
                        Prob > |r| under H0: Rho=0

                          Weight        Oxygen       RunTime

         Weight          1.00000      -0.18419       0.19505
                                        0.3481        0.3199

         Oxygen         -0.18419       1.00000      -0.86843
                          0.3481                     <.0001

         RunTime         0.19505      -0.86843       1.00000
                          0.3199        <.0001
```

The following statements display (in Output 2.6.2) the output data set:

```
title 'Output Data Set from PROC CORR';
proc print data=CorrOutp noobs;
run;
```

Output 2.6.2 OUTP= Data Set with Pearson Correlations

```
                   Output Data Set from PROC CORR

     _TYPE_      _NAME_       Weight      Oxygen      RunTime

     MEAN                    77.2168     47.1327      10.6954
     STD                      8.4495      5.5535       1.4127
     N                       28.0000     28.0000      28.0000
     CORR        Weight       1.0000     -0.1842       0.1950
     CORR        Oxygen      -0.1842      1.0000      -0.8684
     CORR        RunTime      0.1950     -0.8684       1.0000
```

The output data set has the default type CORR and can be used as an input data set for regression or other statistical procedures. For example, the following statements request a regression analysis using CorrOutp, without reading the original data in the REG procedure:

```
title 'Input Type CORR Data Set from PROC REG';
proc reg data=CorrOutp;
   model runtime= weight oxygen;
run;
```

The following statements generate the same results as the preceding statements:

```
proc reg data=Fitness;
   model runtime= weight oxygen;
run;
```

Example 2.7: Creating Scatter Plots

The following statements request a correlation analysis and a scatter plot matrix for the variables in the data set Fish1, which was created in Example 2.5. This data set contains 35 observations, one of which contains a missing value for the variable Weight3.

```
ods graphics on;
title 'Fish Measurement Data';
proc corr data=fish1 nomiss plots=matrix(histogram);
   var Height Width Length3 Weight3;
  run;
ods graphics off;
```

The "Simple Statistics" table in Output 2.7.1 displays univariate descriptive statistics for analysis variables.

Output 2.7.1 Simple Statistics

```
                        Fish Measurement Data

                         The CORR Procedure

             4   Variables:    Height    Width    Length3   Weight3

                          Simple Statistics

  Variable         N         Mean      Std Dev          Sum      Minimum      Maximum

  Height          34     15.22057      1.98159    517.49950     11.52000     18.95700
  Width           34      5.43805      0.72967    184.89370      4.02000      6.74970
  Length3         34     38.38529      4.21628         1305     30.00000     46.50000
  Weight3         34      8.44751      0.97574    287.21524      6.23168     10.00000
```

When you specify the NOMISS option, the same set of 34 observations is used to compute the correlation for each pair of variables. The "Pearson Correlation Coefficients" table in Output 2.7.2 displays Pearson correlation statistics for pairs of analysis variables.

Output 2.7.2 Pearson Correlation Coefficients

```
              Pearson Correlation Coefficients, N = 34
                     Prob > |r| under H0: Rho=0

                  Height        Width      Length3      Weight3

    Height       1.00000      0.92632      0.95492      0.96261
                              <.0001       <.0001       <.0001

    Width        0.92632      1.00000      0.92171      0.92789
                 <.0001                    <.0001       <.0001

    Length3      0.95492      0.92171      1.00000      0.96523
                 <.0001       <.0001                    <.0001

    Weight3      0.96261      0.92789      0.96523      1.00000
                 <.0001       <.0001       <.0001
```

The variables are highly correlated. For example, the correlation between Height and Width is 0.92632.

The PLOTS=MATRIX(HISTOGRAM) option requests a scatter plot matrix for the VAR statement variables in Output 2.7.3.

Output 2.7.3 Scatter Plot Matrix

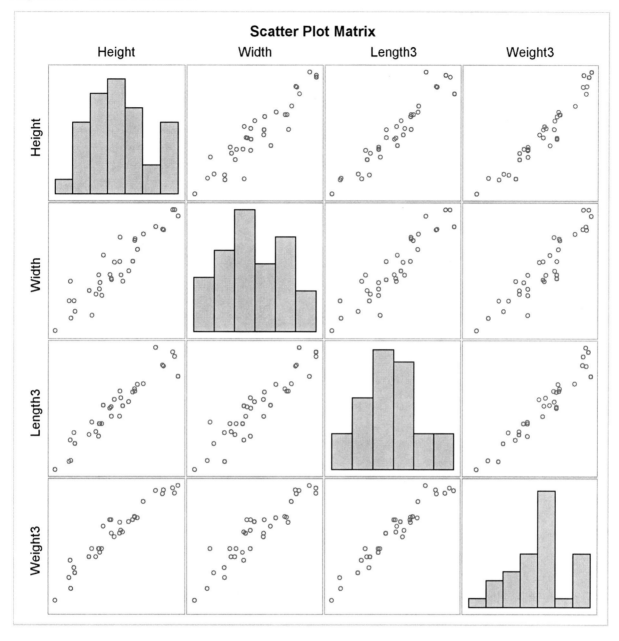

In order to create this display, you must specify the `ods graphics on` statement in addition to the PLOTS= option. For more information about the `ods graphics` statement, see Chapter 21, "Statistical Graphics Using ODS" (*SAS/STAT User's Guide*).

To explore the correlation between Height and Width, the following statements display (in Output 2.7.4) a scatter plot with prediction ellipses for the two variables:

```
ods graphics on;
proc corr data=fish1 nomiss
          plots=scatter(nvar=2 alpha=.20 .30);
   var Height Width Length3 Weight3;
 run;
ods graphics off;
```

The NOMISS option is specified with the original VAR statement to ensure that the same set of 34 observations is used for this analysis. The PLOTS=SCATTER(NVAR=2) option requests a scatter plot for the first two variables in the VAR list. The ALPHA=.20 .30 suboption requests 80% and 70% prediction ellipses, respectively.

Output 2.7.4 Scatter Plot with Prediction Ellipses

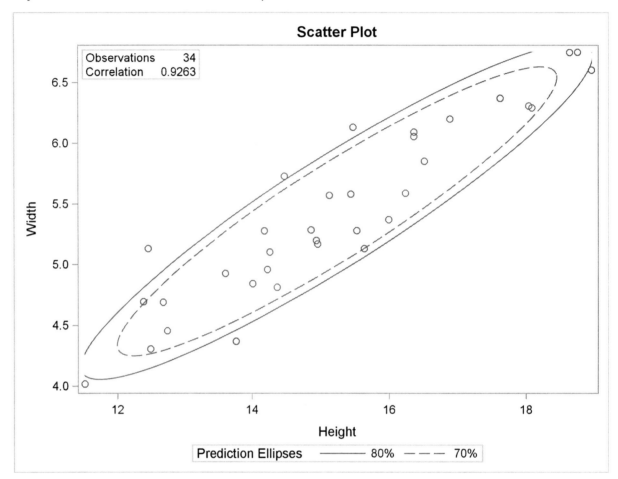

A prediction ellipse is a region for predicting a new observation from the population, assuming bivariate normality. It also approximates a region containing a specified percentage of the population. The displayed prediction ellipse is centered at the means $(\bar{x}, \bar{y})$. For further details, see the section "Confidence and Prediction Ellipses" on page 28.

Note that the following statements also display (in Output 2.7.5) a scatter plot for Height and Width:

```
ods graphics on;
proc corr data=fish1
          plots=scatter(alpha=.20 .30);
   var Height Width;
 run;
ods graphics off;
```

Output 2.7.5 Scatter Plot with Prediction Ellipses

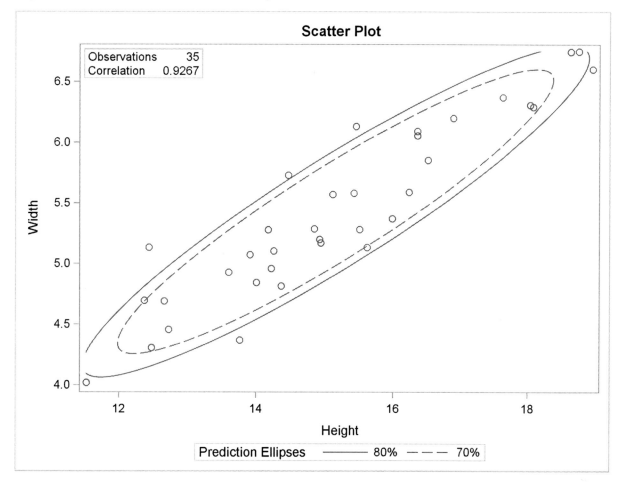

Output 2.7.5 includes the point (13.9, 5.1), which was excluded from Output 2.7.4 because the observation had a missing value for Weight3. The prediction ellipses in Output 2.7.5 also reflect the inclusion of this observation.

The following statements display (in Output 2.7.6) a scatter plot with confidence ellipses for the mean:

```
ods graphics on;
title 'Fish Measurement Data';
proc corr data=fish1 nomiss
          plots=scatter(ellipse=confidence nvar=2 alpha=.05 .01);
   var Height Width Length3 Weight3;
 run;
ods graphics off;
```

The NVAR=2 suboption within the PLOTS= option restricts the number of plots created to the first two variables in the VAR statement, and the ELLIPSE=CONFIDENCE suboption requests confidence ellipses for the mean. The ALPHA=.05 .01 suboption requests 95% and 99% confidence ellipses, respectively.

Output 2.7.6 Scatter Plot with Confidence Ellipses

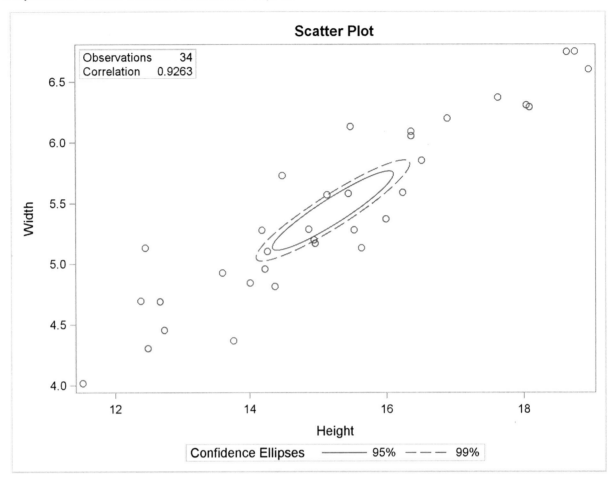

The confidence ellipse for the mean is centered at the means $(\bar{x}, \bar{y})$. For further details, see the section "Confidence and Prediction Ellipses" on page 28.

Example 2.8: Computing Partial Correlations

A partial correlation measures the strength of the linear relationship between two variables, while adjusting for the effect of other variables.

The following statements request a partial correlation analysis of variables Height and Width while adjusting for the variables Length3 and Weight. The latter variables, which are said to be "partialled out" of the analysis, are specified with the PARTIAL statement.

```
ods graphics on;
title 'Fish Measurement Data';
proc corr data=fish1 plots=scatter(alpha=.20 .30);
   var Height Width;
   partial Length3 Weight3;
 run;
ods graphics off;
```

Output 2.8.1 displays descriptive statistics for all the variables. The partial variance and partial standard deviation for the variables in the VAR statement are also displayed.

Output 2.8.1 Descriptive Statistics

```
                        Fish Measurement Data

                         The CORR Procedure

              2 Partial Variables:    Length3   Weight3
              2          Variables:   Height    Width

                          Simple Statistics

Variable         N        Mean      Std Dev         Sum     Minimum     Maximum

Length3         34    38.38529      4.21628        1305    30.00000    46.50000
Weight3         34     8.44751      0.97574   287.21524     6.23168    10.00000
Height          34    15.22057      1.98159   517.49950    11.52000    18.95700
Width           34     5.43805      0.72967   184.89370     4.02000     6.74970

                          Simple Statistics

                               Partial      Partial
                 Variable     Variance      Std Dev

                 Length3
                 Weight3
                 Height        0.26607      0.51582
                 Width         0.07315      0.27047
```

When you specify a PARTIAL statement, observations with missing values are excluded from the analysis. Output 2.8.2 displays partial correlations for the variables in the VAR statement.

Output 2.8.2 Pearson Partial Correlation Coefficients

```
            Pearson Partial Correlation Coefficients, N = 34
                   Prob > |r| under H0: Partial Rho=0

                               Height          Width

                Height        1.00000        0.25692
                                             0.1558

                Width         0.25692        1.00000
                              0.1558
```

The partial correlation between the variables Height and Width is 0.25692, which is much less than the unpartialled correlation, 0.92632 (in Output 2.8.2). The *p*-value for the partial correlation is 0.1558.

The PLOTS=SCATTER option displays (in Output 2.8.3) a scatter plot of the residuals for the variables Height and Width after controlling for the effect of variables Length3 and Weight. The ALPHA=.20 .30 suboption requests 80% and 70% prediction ellipses, respectively.

Output 2.8.3 Partial Residual Scatter Plot

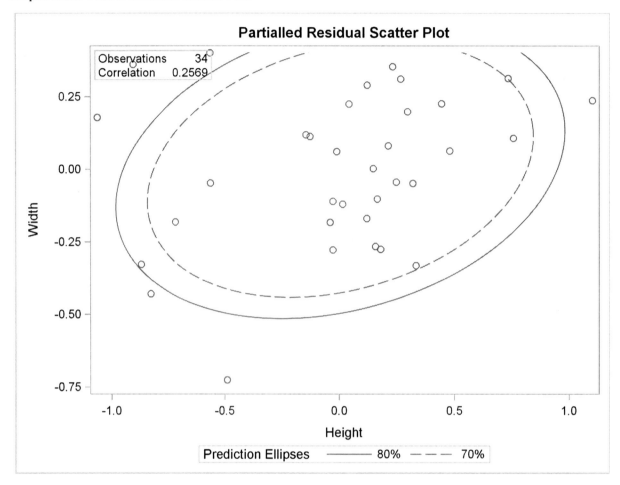

In Output 2.8.3, a standard deviation of Height has roughly the same length on the X axis as a standard deviation of Width on the Y axis. The major axis length is not significantly larger than the minor axis length, indicating a weak partial correlation between Height and Width.

References

Anderson, T. W. (1984), *An Introduction to Multivariate Statistical Analysis,* Second Edition, New York: John Wiley & Sons.

Blum, J. R., Kiefer, J., and Rosenblatt, M. (1961), "Distribution Free Tests of Independence Based on the Sample Distribution Function," *Annals of Mathematical Statistics,* 32, 485–498.

Conover, W. J. (1998), *Practical Nonparametric Statistics*, Third Edition, New York: John Wiley & Sons.

Cronbach, L. J. (1951), "Coefficient Alpha and the Internal Structure of Tests," *Psychometrika*, 16, 297–334.

Fisher, R. A. (1915), "Frequency Distribution of the Values of the Correlation Coefficient in Samples from an Indefinitely Large Population," *Biometrika*, 10, 507–521.

Fisher, R. A. (1921), "On the "Probable Error" of a Coefficient of Correlation Deduced from a Small Sample," *Metron*, 1, 3–32.

Fisher, R. A. (1936), "The Use of Multiple Measurements in Taxonomic Problems," *Annals of Eugenics*, 7, 179–188.

Fisher, R. A. (1970), *Statistical Methods for Research Workers,* Fourteenth Edition, Davien, CT: Hafner Publishing Company.

Hoeffding, W. (1948), "A Non-parametric Test of Independence," *Annals of Mathematical Statistics*, 19, 546–557.

Hollander, M. and Wolfe, D. (1999), *Nonparametric Statistical Methods*, Second Edition, New York: John Wiley & Sons.

Keeping, E. S. (1962), *Introduction to Statistical Inference,* New York: D. Van Nostrand.

Knight, W. E. (1966), "A Computer Method for Calculating Kendall's Tau with Ungrouped Data," *Journal of the American Statistical Association*, 61, 436–439.

Moore, D. S. (2000), *Statistics: Concepts and Controversies*, Fifth Edition, New York: W. H. Freeman.

Mudholkar, G. S. (1983), "Fisher's z-Transformation," *Encyclopedia of Statistical Sciences*, 3, 130–135.

Noether, G. E. (1967), *Elements of Nonparametric Statistics*, New York: John Wiley & Sons.

Novick, M. R. (1967), "Coefficient Alpha and the Reliability of Composite Measurements," *Psychometrika*, 32, 1–13.

Nunnally, J. C. and Bernstein, I. H. (1994), *Psychometric Theory*, Third Edition, New York: McGraw-Hill.

Ott, R. L. and Longnecker, M. T. (2000), *An Introduction to Statistical Methods and Data Analysis*, Fifth Edition, Belmont, CA: Wadsworth.

SAS Institute Inc. (1994), "Measuring the Internal Consistency of a Test, Using PROC CORR to Compute Cronbach's Coefficient Alpha," *SAS Communications*, 20:4, TT2–TT5.

Spector, P. E. (1992), *Summated Rating Scale Construction: An Introduction*, Newbury Park, CA: Sage.

Chapter 3
The FREQ Procedure

Contents

Overview: FREQ Procedure	**64**
Getting Started: FREQ Procedure	**66**
Frequency Tables and Statistics	66
Agreement Study	73
Syntax: FREQ Procedure	**75**
PROC FREQ Statement	76
BY Statement	78
EXACT Statement	79
OUTPUT Statement	82
TABLES Statement	85
TEST Statement	110
WEIGHT Statement	111
Details: FREQ Procedure	**112**
Inputting Frequency Counts	112
Grouping with Formats	113
Missing Values	114
In-Database Computation	116
Statistical Computations	118
Definitions and Notation	118
Chi-Square Tests and Statistics	119
Measures of Association	124
Binomial Proportion	133
Risks and Risk Differences	140
Odds Ratio and Relative Risks for 2 x 2 Tables	148
Cochran-Armitage Test for Trend	150
Jonckheere-Terpstra Test	151
Tests and Measures of Agreement	153
Cochran-Mantel-Haenszel Statistics	158
Exact Statistics	167
Computational Resources	171
Output Data Sets	172
Displayed Output	175
ODS Table Names	183
ODS Graphics	186

Examples: FREQ Procedure **187**
    Example 3.1: Output Data Set of Frequencies 187
    Example 3.2: Frequency Dot Plots 190
    Example 3.3: Chi-Square Goodness-of-Fit Tests 193
    Example 3.4: Binomial Proportions 197
    Example 3.5: Analysis of a 2x2 Contingency Table 200
    Example 3.6: Output Data Set of Chi-Square Statistics 203
    Example 3.7: Cochran-Mantel-Haenszel Statistics 205
    Example 3.8: Cochran-Armitage Trend Test 206
    Example 3.9: Friedman's Chi-Square Test 210
    Example 3.10: Cochran's Q Test 212
References **214**

Overview: FREQ Procedure

The FREQ procedure produces one-way to n-way frequency and contingency (crosstabulation) tables. For two-way tables, PROC FREQ computes tests and measures of association. For n-way tables, PROC FREQ provides stratified analysis by computing statistics across, as well as within, strata.

For one-way frequency tables, PROC FREQ computes goodness-of-fit tests for equal proportions or specified null proportions. For one-way tables, PROC FREQ also provides confidence limits and tests for binomial proportions, including tests for noninferiority and equivalence.

For contingency tables, PROC FREQ can compute various statistics to examine the relationships between two classification variables. For some pairs of variables, you might want to examine the existence or strength of any association between the variables. To determine if an association exists, chi-square tests are computed. To estimate the strength of an association, PROC FREQ computes measures of association that tend to be close to zero when there is no association and close to the maximum (or minimum) value when there is perfect association. The statistics for contingency tables include the following:

- chi-square tests and measures

- measures of association

- risks (binomial proportions) and risk differences for 2×2 tables

- odds ratios and relative risks for 2×2 tables

- tests for trend

- tests and measures of agreement

- Cochran-Mantel-Haenszel statistics

PROC FREQ computes asymptotic standard errors, confidence intervals, and tests for measures of association and measures of agreement. Exact p-values and confidence intervals are available for many test statistics and measures. PROC FREQ also performs analyses that adjust for any stratification variables by computing statistics across, as well as within, strata for n-way tables. These statistics include Cochran-Mantel-Haenszel statistics and measures of agreement.

In choosing measures of association to use in analyzing a two-way table, you should consider the study design (which indicates whether the row and column variables are dependent or independent), the measurement scale of the variables (nominal, ordinal, or interval), the type of association that each measure is designed to detect, and any assumptions required for valid interpretation of a measure. You should exercise care in selecting measures that are appropriate for your data.

Similar comments apply to the choice and interpretation of test statistics. For example, the Mantel-Haenszel chi-square statistic requires an ordinal scale for both variables and is designed to detect a linear association. The Pearson chi-square, on the other hand, is appropriate for all variables and can detect any kind of association, but it is less powerful for detecting a linear association because its power is dispersed over a greater number of degrees of freedom (except for 2×2 tables).

For more information about selecting the appropriate statistical analyses, see Agresti (2007) or Stokes, Davis, and Koch (2000).

Several SAS procedures produce frequency counts; only PROC FREQ computes chi-square tests for one-way to n-way tables and measures of association and agreement for contingency tables. Other procedures to consider for counting include the TABULATE and UNIVARIATE procedures. When you want to produce contingency tables and tests of association for sample survey data, use PROC SURVEYFREQ. See Chapter 14, "Introduction to Survey Procedures" (*SAS/STAT User's Guide*), for more information. When you want to fit models to categorical data, use a procedure such as CATMOD, GENMOD, GLIMMIX, LOGISTIC, PROBIT, or SURVEYLOGISTIC. See Chapter 8, "Introduction to Categorical Data Analysis Procedures" (*SAS/STAT User's Guide*), for more information.

PROC FREQ uses the Output Delivery System (ODS), a SAS subsystem that provides capabilities for displaying and controlling the output from SAS procedures. ODS enables you to convert any of the output from PROC FREQ into a SAS data set. See the section "ODS Table Names" on page 183 for more information.

PROC FREQ now uses ODS Graphics to create graphs as part of its output. For general information about ODS Graphics, see Chapter 21, "Statistical Graphics Using ODS" (*SAS/STAT User's Guide*). For specific information about the statistical graphics available with the FREQ procedure, see the PLOTS option in the TABLES statement and the section "ODS Graphics" on page 186.

Getting Started: FREQ Procedure

Frequency Tables and Statistics

The FREQ procedure provides easy access to statistics for testing for association in a crosstabulation table.

In this example, high school students applied for courses in a summer enrichment program; these courses included journalism, art history, statistics, graphic arts, and computer programming. The students accepted were randomly assigned to classes with and without internships in local companies. Table 3.1 contains counts of the students who enrolled in the summer program by gender and whether they were assigned an internship slot.

Table 3.1   Summer Enrichment Data

		Enrollment		
Gender	Internship	Yes	No	Total
boys	yes	35	29	64
boys	no	14	27	41
girls	yes	32	10	42
girls	no	53	23	76

The SAS data set SummerSchool is created by inputting the summer enrichment data as cell count data, or providing the frequency count for each combination of variable values. The following DATA step statements create the SAS data set SummerSchool:

```
data SummerSchool;
   input Gender $ Internship $ Enrollment $ Count @@;
   datalines;
boys   yes yes 35   boys   yes no 29
boys   no  yes 14   boys   no  no 27
girls  yes yes 32   girls  yes no 10
girls  no  yes 53   girls  no  no 23
;
```

The variable Gender takes the values 'boys' or 'girls,' the variable Internship takes the values 'yes' and 'no,' and the variable Enrollment takes the values 'yes' and 'no.' The variable Count contains the number of students that correspond to each combination of data values. The double at sign (@@) indicates that more than one observation is included on a single data line. In this DATA step, two observations are included on each line.

Researchers are interested in whether there is an association between internship status and summer program enrollment. The Pearson chi-square statistic is an appropriate statistic to assess the association in the corresponding 2×2 table. The following PROC FREQ statements specify this analysis.

You specify the table for which you want to compute statistics with the TABLES statement. You specify the statistics you want to compute with options after a slash (/) in the TABLES statement.

```
proc freq data=SummerSchool order=data;
   tables Internship*Enrollment / chisq;
   weight Count;
run;
```

The ORDER= option controls the order in which variable values are displayed in the rows and columns of the table. By default, the values are arranged according to the alphanumeric order of their unformatted values. If you specify ORDER=DATA, the data are displayed in the same order as they occur in the input data set. Here, because 'yes' appears before 'no' in the data, 'yes' appears first in any table. Other options for controlling order include ORDER=FORMATTED, which orders according to the formatted values, and ORDER=FREQUENCY, which orders by descending frequency count.

In the TABLES statement, Internship*Enrollment specifies a table where the rows are internship status and the columns are program enrollment. The CHISQ option requests chi-square statistics for assessing association between these two variables. Because the input data are in cell count form, the WEIGHT statement is required. The WEIGHT statement names the variable Count, which provides the frequency of each combination of data values.

Figure 3.1 presents the crosstabulation of Internship and Enrollment. In each cell, the values printed under the cell count are the table percentage, row percentage, and column percentage, respectively. For example, in the first cell, 63.21 percent of the students offered courses with internships accepted them and 36.79 percent did not.

Figure 3.1 Crosstabulation Table

```
                    The FREQ Procedure

              Table of Internship by Enrollment

         Internship     Enrollment

         Frequency|
         Percent  |
         Row Pct  |
         Col Pct  |yes     |no      | Total
         ---------+--------+--------+
         yes      |    67  |    39  |   106
                  |  30.04 |  17.49 |  47.53
                  |  63.21 |  36.79 |
                  |  50.00 |  43.82 |
         ---------+--------+--------+
         no       |    67  |    50  |   117
                  |  30.04 |  22.42 |  52.47
                  |  57.26 |  42.74 |
                  |  50.00 |  56.18 |
         ---------+--------+--------+
         Total         134       89      223
                     60.09    39.91   100.00
```

Figure 3.2 displays the statistics produced by the CHISQ option. The Pearson chi-square statistic is labeled 'Chi-Square' and has a value of 0.8189 with 1 degree of freedom. The associated *p*-value is 0.3655, which means that there is no significant evidence of an association between internship status and program enrollment. The other chi-square statistics have similar values and are asymptotically equivalent. The other statistics (phi coefficient, contingency coefficient, and Cramer's *V*) are measures of association derived from the Pearson chi-square. For Fisher's exact test, the two-sided *p*-value is 0.4122, which also shows no association between internship status and program enrollment.

Figure 3.2 Statistics Produced with the CHISQ Option

```
Statistic                      DF      Value      Prob
------------------------------------------------------
Chi-Square                      1      0.8189     0.3655
Likelihood Ratio Chi-Square     1      0.8202     0.3651
Continuity Adj. Chi-Square      1      0.5899     0.4425
Mantel-Haenszel Chi-Square      1      0.8153     0.3666
Phi Coefficient                        0.0606
Contingency Coefficient                0.0605
Cramer's V                             0.0606


              Fisher's Exact Test
         ---------------------------------
         Cell (1,1) Frequency  (F)      67
         Left-sided Pr <= F         0.8513
         Right-sided Pr >= F        0.2213

         Table Probability (P)      0.0726
         Two-sided Pr <= P          0.4122
```

The analysis, so far, has ignored gender. However, it might be of interest to ask whether program enrollment is associated with internship status after adjusting for gender. You can address this question by doing an analysis of a set of tables (in this case, by analyzing the set consisting of one for boys and one for girls). The Cochran-Mantel-Haenszel (CMH) statistic is appropriate for this situation: it addresses whether rows and columns are associated after controlling for the stratification variable. In this case, you would be stratifying by gender.

The PROC FREQ statements for this analysis are very similar to those for the first analysis, except that there is a third variable, Gender, in the TABLES statement. When you cross more than two variables, the two rightmost variables construct the rows and columns of the table, respectively, and the leftmost variables determine the stratification.

The following PROC FREQ statements also request frequency plots for the crosstabulation tables. PROC FREQ produces these plots by using ODS Graphics to create graphs as part of the procedure output. Before requesting graphs, you must enable ODS Graphics with the **ODS GRAPHICS ON** statement.

```
ods graphics on;
proc freq data=SummerSchool;
   tables Gender*Internship*Enrollment /
          chisq cmh plots(only)=freqplot;
   weight Count;
run;
ods graphics off;
```

This execution of PROC FREQ first produces two individual crosstabulation tables of Internship by Enrollment: one for boys and one for girls. Frequency plots and chi-square statistics are produced for each individual table. Figure 3.3, Figure 3.4, and Figure 3.5 show the results for boys. Note that the chi-square statistic for boys is significant at the $\alpha = 0.05$ level of significance. Boys offered a course with an internship are more likely to enroll than boys who are not.

Figure 3.4 displays the frequency plot of Internship by Enrollment for boys. By default, the frequency plot is displayed as a bar chart with vertical grouping by the row variable Internship. You can use PLOTS= options to request a dot plot instead of a bar chart or to change the orientation of the bars from vertical to horizontal. You can also use PLOTS= options to specify other two-way layouts such as stacked or horizontal grouping.

Figure 3.6, Figure 3.7, and Figure 3.8 display the crosstabulation table, frequency plot, and chi-square statistics for girls. You can see that there is no evidence of association between internship offers and program enrollment for girls.

Figure 3.3 Crosstabulation Table for Boys

```
                    The FREQ Procedure

            Table 1 of Internship by Enrollment
                  Controlling for Gender=boys

           Internship     Enrollment

           Frequency|
           Percent  |
           Row Pct  |
           Col Pct  |no       |yes      |  Total
           ---------+---------+---------+
           no       |      27 |      14 |     41
                    |   25.71 |   13.33 |  39.05
                    |   65.85 |   34.15 |
                    |   48.21 |   28.57 |
           ---------+---------+---------+
           yes      |      29 |      35 |     64
                    |   27.62 |   33.33 |  60.95
                    |   45.31 |   54.69 |
                    |   51.79 |   71.43 |
           ---------+---------+---------+
           Total           56        49       105
                        53.33     46.67    100.00
```

Figure 3.4 Frequency Plot for Boys

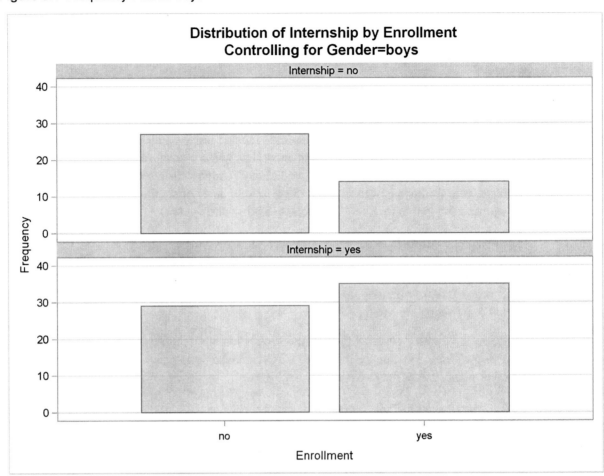

Figure 3.5 Chi-Square Statistics for Boys

```
Statistic                        DF       Value       Prob
-----------------------------------------------------------
Chi-Square                        1       4.2366      0.0396
Likelihood Ratio Chi-Square       1       4.2903      0.0383
Continuity Adj. Chi-Square        1       3.4515      0.0632
Mantel-Haenszel Chi-Square        1       4.1963      0.0405
Phi Coefficient                           0.2009
Contingency Coefficient                   0.1969
Cramer's V                                0.2009

             Fisher's Exact Test
             -----------------------------------
             Cell (1,1) Frequency (F)         27
             Left-sided Pr <= F           0.9885
             Right-sided Pr >= F          0.0311

             Table Probability (P)        0.0196
             Two-sided Pr <= P            0.0467
```

Figure 3.6 Crosstabulation Table for Girls

```
           Table 2 of Internship by Enrollment
                Controlling for Gender=girls

    Internship     Enrollment

    Frequency|
    Percent  |
    Row Pct  |
    Col Pct  |no       |yes      |   Total
    ---------+---------+---------+
    no       |    23   |    53   |      76
             | 19.49   | 44.92   |   64.41
             | 30.26   | 69.74   |
             | 69.70   | 62.35   |
    ---------+---------+---------+
    yes      |    10   |    32   |      42
             |  8.47   | 27.12   |   35.59
             | 23.81   | 76.19   |
             | 30.30   | 37.65   |
    ---------+---------+---------+
    Total          33        85        118
                27.97     72.03     100.00
```

Figure 3.7 Frequency Plot for Girls

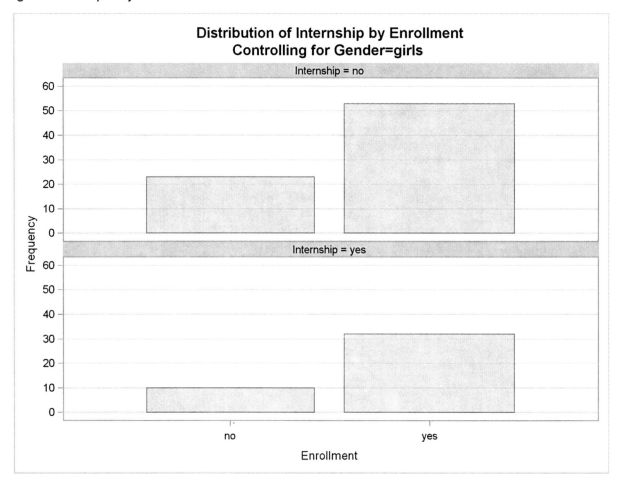

Figure 3.8 Chi-Square Statistics for Girls

```
Statistic                        DF       Value       Prob
-----------------------------------------------------------
Chi-Square                        1       0.5593      0.4546
Likelihood Ratio Chi-Square       1       0.5681      0.4510
Continuity Adj. Chi-Square        1       0.2848      0.5936
Mantel-Haenszel Chi-Square        1       0.5545      0.4565
Phi Coefficient                           0.0688
Contingency Coefficient                   0.0687
Cramer's V                                0.0688


             Fisher's Exact Test
          ---------------------------------
          Cell (1,1) Frequency (F)       23
          Left-sided Pr <= F         0.8317
          Right-sided Pr >= F        0.2994

          Table Probability (P)      0.1311
          Two-sided Pr <= P          0.5245
```

These individual table results demonstrate the occasional problems with combining information into one table and not accounting for information in other variables such as Gender. Figure 3.9 contains the CMH results. There are three summary (CMH) statistics; which one you use depends on whether your rows and/or columns have an order in $r \times c$ tables. However, in the case of 2×2 tables, ordering does not matter and all three statistics take the same value. The CMH statistic follows the chi-square distribution under the hypothesis of no association, and here, it takes the value 4.0186 with 1 degree of freedom. The associated *p*-value is 0.0450, which indicates a significant association at the $\alpha = 0.05$ level.

Thus, when you adjust for the effect of gender in these data, there is an association between internship and program enrollment. But, if you ignore gender, no association is found. Note that the CMH option also produces other statistics, including estimates and confidence limits for relative risk and odds ratios for 2×2 tables and the Breslow-Day Test. These results are not displayed here.

Figure 3.9 Test for the Hypothesis of No Association

```
        Cochran-Mantel-Haenszel Statistics (Based on Table Scores)

        Statistic    Alternative Hypothesis      DF      Value      Prob
        ---------------------------------------------------------------
            1        Nonzero Correlation          1      4.0186     0.0450
            2        Row Mean Scores Differ       1      4.0186     0.0450
            3        General Association          1      4.0186     0.0450
```

Agreement Study

Medical researchers are interested in evaluating the efficacy of a new treatment for a skin condition. Dermatologists from participating clinics were trained to conduct the study and to evaluate the condition. After the training, two dermatologists examined patients with the skin condition from a pilot study and rated the same patients. The possible evaluations are terrible, poor, marginal, and clear. Table 3.2 contains the data.

Table 3.2 Skin Condition Data

	Dermatologist 2			
Dermatologist 1	**Terrible**	**Poor**	**Marginal**	**Clear**
Terrible	10	4	1	0
Poor	5	10	12	2
Marginal	2	4	12	5
Clear	0	2	6	13

The following DATA step statements create the SAS dataset SkinCondition. The dermatologists' evaluations of the patients are contained in the variables Derm1 and Derm2; the variable Count is the number of patients given a particular pair of ratings.

```
data SkinCondition;
   input Derm1 $ Derm2 $ Count;
   datalines;
terrible terrible 10
terrible     poor  4
terrible marginal  1
terrible    clear  0
poor     terrible  5
poor         poor 10
poor     marginal 12
poor        clear  2
marginal terrible  2
marginal     poor  4
marginal marginal 12
marginal    clear  5
clear    terrible  0
clear        poor  2
clear    marginal  6
clear       clear 13
;
```

The following PROC FREQ statements request an agreement analysis of the skin condition data. In order to evaluate the agreement of the diagnoses (a possible contribution to measurement error in the study), the *kappa coefficient* is computed. The AGREE option in the TABLES statement requests the kappa coefficient, together with its standard error and confidence limits. The KAPPA option in the TEST statement requests a test for the null hypothesis that kappa equals zero, or that the agreement is purely by chance.

```
proc freq data=SkinCondition order=data;
   tables Derm1*Derm2 / agree noprint;
   test kappa;
   weight Count;
run;
```

Figure 3.10 shows the results. The kappa coefficient has the value 0.3449, which indicates slight agreement between the dermatologists, and the hypothesis test confirms that you can reject the null hypothesis of no agreement. This conclusion is further supported by the confidence interval of (0.2030, 0.4868), which suggests that the true kappa is greater than zero. The AGREE option also produces Bowker's test for symmetry and the weighted kappa coefficient, but that output is not shown here.

Figure 3.10 Agreement Study

```
                    The FREQ Procedure

           Statistics for Table of Derm1 by Derm2

                    Simple Kappa Coefficient
           ─────────────────────────────────────
           Kappa                         0.3449
           ASE                           0.0724
           95% Lower Conf Limit          0.2030
           95% Upper Conf Limit          0.4868

                    Test of H0: Kappa = 0

           ASE under H0                  0.0612
           Z                             5.6366
           One-sided Pr >  Z            <.0001
           Two-sided Pr > |Z|           <.0001
```

Syntax: FREQ Procedure

The following statements are available in PROC FREQ:

PROC FREQ < *options* > ;
    BY *variables* ;
    EXACT *statistic-options* < / *computation-options* > ;
    OUTPUT < **OUT=***SAS-data-set* > *options* ;
    TABLES *requests* < / *options* > ;
    TEST *options* ;
    WEIGHT *variable* < / *option* > ;

The PROC FREQ statement is the only required statement for the FREQ procedure. If you specify the following statements, PROC FREQ produces a one-way frequency table for each variable in the most recently created data set.

```
proc freq;
run;
```

The rest of this section gives detailed syntax information for the BY, EXACT, OUTPUT, TABLES, TEST, and WEIGHT statements in alphabetical order after the description of the PROC FREQ statement. Table 3.3 summarizes the basic function of each PROC FREQ statement.

Table 3.3 Summary of PROC FREQ Statements

Statement	Description
BY	provides separate analyses for each BY group
EXACT	requests exact tests
OUTPUT	requests an output data set
TABLES	specifies tables and requests analyses
TEST	requests tests for measures of association and agreement
WEIGHT	identifies a weight variable

PROC FREQ Statement

PROC FREQ < *options* > ;

The PROC FREQ statement invokes the procedure and optionally identifies the input data set. By default, the procedure uses the most recently created SAS data set.

Table 3.4 lists the options available in the PROC FREQ statement. Descriptions follow in alphabetical order.

Table 3.4 PROC FREQ Statement Options

Option	Description
COMPRESS	begins the next one-way table on the current page
DATA=	names the input data set
FORMCHAR=	specifies the outline and cell divider characters for crosstabulation tables
NLEVELS	displays the number of levels for all TABLES variables
NOPRINT	suppresses all displayed output
ORDER=	specifies the order for reporting variable values
PAGE	displays one table per page

You can specify the following options in the PROC FREQ statement.

COMPRESS
> begins display of the next one-way frequency table on the same page as the preceding one-way table if there is enough space to begin the table. By default, the next one-way table begins on the current page only if the entire table fits on that page. The COMPRESS option is not valid with the PAGE option.

DATA=*SAS-data-set*
> names the SAS data set to be analyzed by PROC FREQ. If you omit the DATA= option, the procedure uses the most recently created SAS data set.

FORMCHAR(1,2,7)='formchar-string'
defines the characters to be used for constructing the outlines and dividers for the cells of crosstabulation table displays. The *formchar-string* should be three characters long. The characters are used to draw the vertical separators (1), the horizontal separators (2), and the vertical-horizontal intersections (7). If you do not specify the FORMCHAR= option, PROC FREQ uses FORMCHAR(1,2,7)='|-+' by default. Table 3.5 summarizes the formatting characters used by PROC FREQ.

Table 3.5 Formatting Characters Used by PROC FREQ

Position	Default	Used to Draw
1	|	vertical separators
2	-	horizontal separators
7	+	intersections of vertical and horizontal separators

The FORMCHAR= option can specify 20 different SAS formatting characters used to display output; however, PROC FREQ uses only the first, second, and seventh formatting characters. Therefore, the proper specification for PROC FREQ is FORMCHAR(1,2,7)= '*formchar-string*'.

Specifying all blanks for *formchar-string* produces crosstabulation tables with no outlines or dividers—for example, FORMCHAR(1,2,7)=' '. You can use any character in *formchar-string*, including hexadecimal characters. If you use hexadecimal characters, you must put an *x* after the closing quote. For information about which hexadecimal codes to use for which characters, see the documentation for your hardware.

See the CALENDAR, PLOT, and TABULATE procedures in the *Base SAS Procedures Guide* for more information about form characters.

NLEVELS
displays the "Number of Variable Levels" table, which provides the number of levels for each variable named in the TABLES statements. See the section "Number of Variable Levels Table" on page 175 for details. PROC FREQ determines the variable levels from the formatted variable values, as described in the section "Grouping with Formats" on page 113.

NOPRINT
suppresses the display of all output. You can use the NOPRINT option when you only want to create an output data set. See the section "Output Data Sets" on page 172 for information about the output data sets produced by PROC FREQ. Note that the NOPRINT option temporarily disables the Output Delivery System (ODS). For more information, see Chapter 20, "Using the Output Delivery System" (*SAS/STAT User's Guide*).

NOTE: A NOPRINT option is also available in the TABLES statement. It suppresses display of the crosstabulation tables but allows display of the requested statistics.

ORDER=DATA | FORMATTED | FREQ | INTERNAL

specifies the order in which the values of the frequency and crosstabulation table variables are reported. PROC FREQ interprets the values of the ORDER= option as follows:

DATA
: orders values according to their order in the input data set

FORMATTED
: orders values by their formatted values (in ascending order). This order is dependent on the operating environment.

FREQ
: orders values by their descending frequency counts

INTERNAL
: orders values by their unformatted values, which yields the same order that the SORT procedure does. This order is dependent on the operating environment.

By default, ORDER=INTERNAL. The ORDER= option does not apply to missing values, which are always ordered first.

PAGE

displays only one table per page. Otherwise, PROC FREQ displays multiple tables per page as space permits. The PAGE option is not valid with the COMPRESS option.

BY Statement

BY *variables* ;

You can specify a BY statement with PROC FREQ to obtain separate analyses on observations in groups defined by the BY variables. When a BY statement appears, the procedure expects the input data set to be sorted in order of the BY variables.

If your input data set is not sorted in ascending order, use one of the following alternatives:

- Sort the data by using the SORT procedure with a similar BY statement.

- Specify the BY statement option NOTSORTED or DESCENDING in the BY statement for the FREQ procedure. The NOTSORTED option does not mean that the data are unsorted but rather that the data are arranged in groups (according to values of the BY variables) and that these groups are not necessarily in alphabetical or increasing numeric order.

- Create an index on the BY variables by using the DATASETS procedure.

For more information about the BY statement, see *SAS Language Reference: Concepts*. For more information about the DATASETS procedure, see the *Base SAS Procedures Guide*.

EXACT Statement

EXACT *statistic-options* < / *computation-options* > ;

The EXACT statement requests exact tests or confidence limits for the specified statistics. Optionally, PROC FREQ computes Monte Carlo estimates of the exact *p*-values. The *statistic-options* specify the statistics to provide exact tests or confidence limits for. The *computation-options* specify options for the computation of exact statistics. See the section "Exact Statistics" on page 167 for details.

CAUTION: PROC FREQ computes exact tests with fast and efficient algorithms that are superior to direct enumeration. Exact tests are appropriate when a data set is small, sparse, skewed, or heavily tied. For some large problems, computation of exact tests might require a considerable amount of time and memory. Consider using asymptotic tests for such problems. Alternatively, when asymptotic methods might not be sufficient for such large problems, consider using Monte Carlo estimation of exact *p*-values. See the section "Computational Resources" on page 169 for more information.

Statistic-Options

The *statistic-options* specify the statistics to provide exact tests or confidence limits for.

For one-way tables, exact *p*-values are available for the binomial proportion tests and the chi-square goodness-of-fit test. Exact confidence limits are available for the binomial proportion.

For two-way tables, exact *p*-values are available for the following tests: Pearson chi-square test, likelihood-ratio chi-square test, Mantel-Haenszel chi-square test, Fisher's exact test, Jonckheere-Terpstra test, and Cochran-Armitage test for trend. Exact *p*-values are also available for tests of the following statistics: Pearson correlation coefficient, Spearman correlation coefficient, simple kappa coefficient, and weighted kappa coefficient.

For 2 × 2 tables, PROC FREQ provides exact confidence limits for the odds ratio, exact unconditional confidence limits for the proportion difference, and McNemar's exact test. For stratified 2 × 2 tables, PROC FREQ provides Zelen's exact test for equal odds ratios, exact confidence limits for the common odds ratio, and an exact test for the common odds ratio.

Table 3.6 lists the available *statistic-options* and the exact statistics computed. Most of the option names are identical to the corresponding option names in the TABLES and OUTPUT statements. You can request exact computations for groups of statistics by using options that are identical to the following TABLES statement options: CHISQ, MEASURES, and AGREE. For example, when you specify the CHISQ option in the EXACT statement, PROC FREQ computes exact *p*-values for the Pearson chi-square, likelihood-ratio chi-square, and Mantel-Haenszel chi-square tests. You can request exact computations for an individual statistic by specifying the corresponding *statistic-option* from the list in Table 3.6.

Table 3.6 EXACT Statement *Statistic-Options*

Statistic-Option	Exact Statistics Computed
AGREE	McNemar's test (for 2×2 tables), simple kappa coefficient test, weighted kappa coefficient test
BINOMIAL	binomial proportion tests for one-way tables
CHISQ	chi-square goodness-of-fit test for one-way tables; Pearson chi-square, likelihood-ratio chi-square, and Mantel-Haenszel chi-square tests for two-way tables
COMOR	confidence limits for the common odds ratio and common odds ratio test (for $h \times 2 \times 2$ tables)
EQOR | ZELEN	Zelen's test for equal odds ratios (for $h \times 2 \times 2$ tables)
FISHER	Fisher's exact test
JT	Jonckheere-Terpstra test
KAPPA	test for the simple kappa coefficient
LRCHI	likelihood-ratio chi-square test
MCNEM	McNemar's test (for 2×2 tables)
MEASURES	tests for the Pearson correlation and Spearman correlation, confidence limits for the odds ratio (for 2×2 tables)
MHCHI	Mantel-Haenszel chi-square test
OR	confidence limits for the odds ratio (for 2×2 tables)
PCHI	Pearson chi-square test
PCORR	test for the Pearson correlation coefficient
RISKDIFF	confidence limits for the proportion differences (for 2×2 tables)
RISKDIFF1	confidence limits for the column 1 proportion difference
RISKDIFF2	confidence limits for the column 2 proportion difference
SCORR	test for the Spearman correlation coefficient
TREND	Cochran-Armitage test for trend
WTKAP	test for the weighted kappa coefficient

Using TABLES Statement Options with the EXACT Statement

If you use only one TABLES statement, you do not need to specify the same options in both the TABLES and EXACT statements; when you specify an option in the EXACT statement, PROC FREQ automatically invokes the corresponding TABLES statement option. However, when you use multiple TABLES statements and want exact computations, you must specify options in the TABLES statements to request the desired statistics. PROC FREQ then performs exact computations for all statistics that you also specify in the EXACT statement.

Note that the TABLES statement group option CHISQ includes tests that correspond to the following EXACT statement individual statistic-options: LRCHI, MHCHI, and PCHI. The MEASURES option in the TABLES statement includes statistics that correspond to the following EXACT statement statistic-options: OR, PCORR, and SCORR. The AGREE option in the TABLES statement produces analyses that correspond to the KAPPA, MCNEM, and WTKAP statistic-options in the EXACT statement. The CMH option in the TABLES statement produces analyses that correspond to the COMOR and EQOR (or ZELEN) statistic-options in the EXACT statement.

Computation-Options

The *computation-options* specify options for computation of exact statistics. You can specify the following *computation-options* in the EXACT statement after a slash (/).

ALPHA=α

specifies the level of the confidence limits for Monte Carlo *p*-value estimates. The value of α must be between 0 and 1, and the default is 0.01. A confidence level of α produces $100(1 - \alpha)\%$ confidence limits. The default of ALPHA=.01 produces 99% confidence limits for the Monte Carlo estimates.

The ALPHA= option invokes the MC option.

MAXTIME=*value*

specifies the maximum clock time (in seconds) that PROC FREQ can use to compute an exact *p*-value. If the procedure does not complete the computation within the specified time, the computation terminates. The value of MAXTIME= must be a positive number. The MAXTIME= option is valid for Monte Carlo estimation of exact *p*-values, as well as for direct exact *p*-value computation. See the section "Computational Resources" on page 169 for more information.

MC

requests Monte Carlo estimation of exact *p*-values instead of direct exact *p*-value computation. Monte Carlo estimation can be useful for large problems that require a considerable amount of time and memory for exact computations but for which asymptotic approximations might not be sufficient. See the section "Monte Carlo Estimation" on page 170 for more information.

The MC option is available for all EXACT *statistic-options* except the BINOMIAL option and the following options that apply only to 2×2 or $h \times 2 \times 2$ tables: COMOR, EQOR, MCNEM, RISKDIFF, and OR. PROC FREQ computes only exact tests or confidence limits for these statistics.

The ALPHA=, N=, and SEED= options also invoke the MC option.

N=*n*

specifies the number of samples for Monte Carlo estimation. The value of *n* must be a positive integer, and the default is 10,000. Larger values of *n* produce more precise estimates of exact *p*-values. Because larger values of *n* generate more samples, the computation time increases.

The N= option invokes the MC option.

POINT

requests exact point probabilities for the test statistics.

The POINT option is available for all the EXACT statement *statistic-options* except the OR and RISKDIFF options, which provide exact confidence limits. The POINT option is not available with the MC option.

SEED=_number_

specifies the initial seed for random number generation for Monte Carlo estimation. The value of the SEED= option must be an integer. If you do not specify the SEED= option or if the SEED= value is negative or zero, PROC FREQ uses the time of day from the computer's clock to obtain the initial seed.

The SEED= option invokes the MC option.

OUTPUT Statement

OUTPUT < OUT= SAS-data-set > options ;

The OUTPUT statement creates a SAS data set that contains statistics computed by PROC FREQ. You specify which statistics to store in the output data set with the OUTPUT statement *options*. The output data set contains one observation for each two-way table or stratum, and one observation for summary statistics across all strata. For more information about the contents of the output data set, see the section "Contents of the OUTPUT Statement Output Data Set" on page 174.

Only one OUTPUT statement is allowed for each execution of PROC FREQ. You must specify a TABLES statement with the OUTPUT statement. If you use multiple TABLES statements, the contents of the OUTPUT data set correspond to the last TABLES statement. If you use multiple table requests in a TABLES statement, the contents of the OUTPUT data set correspond to the last table request.

Note that you can use the Output Delivery System (ODS) to create a SAS data set from any piece of PROC FREQ output. For more information, see the section "ODS Table Names" on page 183.

Also note that the output data set created by the OUTPUT statement is not the same as the output data set created by the OUT= option in the TABLES statement. The OUTPUT statement creates a data set that contains statistics (such as the Pearson chi-square and its *p*-value), and the OUT= option in the TABLES statement creates a data set that contains frequency table counts and percentages. See the section "Output Data Sets" on page 172 for more information.

You can specify the following options in an OUTPUT statement:

OUT=_SAS-data-set_

names the output data set. If you omit the OUT= option, the data set is named DATA*n*, where *n* is the smallest integer that makes the name unique.

options

specify the statistics you want in the output data set. Table 3.7 lists the available options, together with the TABLES statement options needed to produce the statistics. You can output groups of statistics by using group options identical to those available in the TABLES statement, which include the AGREE, ALL, CHISQ, CMH, and MEASURES options. Or you can request statistics individually.

When you specify an option in the OUTPUT statement, the output data set contains all statistics from that analysis—the estimate or test statistic plus any associated standard error, confidence limits, *p*-values, and degrees of freedom. See the section "Contents of the OUTPUT Statement Output Data Set" on page 174 for details.

If you want to store a statistic in the output data set, you must also request computation of that statistic with the appropriate TABLES or EXACT statement option. For example, you cannot specify the option PCHI (Pearson chi-square) in the OUTPUT statement without also specifying a TABLES or EXACT statement option to compute the Pearson chi-square test. The TABLES statement option ALL or CHISQ requests the Pearson chi-square test. If you have only one TABLES statement, the EXACT statement option CHISQ or PCHI also requests the Pearson chi-square test. Table 3.7 lists the TABLES statement options required to produce the OUTPUT data set statistics. Note that the ALL option in the TABLES statement invokes the CHISQ, MEASURES, and CMH options.

Table 3.7 OUTPUT Statement Options

Option	Output Data Set Statistics	Required TABLES Statement Option
AGREE	McNemar's test, Bowker's test, simple and weighted kappas; for multiple strata, overall simple and weighted kappas, tests for equal kappas, and Cochran's Q	AGREE
AJCHI	continuity-adjusted chi-square (2 × 2 tables)	CHISQ
ALL	CHISQ, MEASURES, and CMH statistics; N	ALL
BDCHI	Breslow-Day test ($h \times 2 \times 2$ tables)	CMH, CMH1, or CMH2
BINOMIAL	binomial statistics for one-way tables	BINOMIAL
CHISQ	for one-way tables, goodness-of-fit test; for two-way tables, Pearson, likelihood-ratio, continuity-adjusted, and Mantel-Haenszel chi-squares, Fisher's exact test (2 × 2 tables), phi and contingency coefficients, Cramer's V	CHISQ
CMH	Cochran-Mantel-Haenszel (CMH) correlation, row mean scores (ANOVA), and general association statistics; for 2 × 2 tables, logit and Mantel-Haenszel adjusted odds ratios and relative risks, Breslow-Day test	CMH
CMH1	CMH output, except row mean scores (ANOVA) and general association statistics	CMH or CMH1
CMH2	CMH output, except general association statistic	CMH or CMH2
CMHCOR	CMH correlation statistic	CMH, CMH1, or CMH2
CMHGA	CMH general association statistic	CMH
CMHRMS	CMH row mean scores (ANOVA) statistic	CMH or CMH2
COCHQ	Cochran's Q ($h \times 2 \times 2$ tables)	AGREE
CONTGY	contingency coefficient	CHISQ
CRAMV	Cramer's V	CHISQ
EQKAP	test for equal simple kappas	AGREE
EQOR | ZELEN	Zelen's test for equal odds ratios ($h \times 2 \times 2$ tables)	CMH and EXACT EQOR
EQWKP	test for equal weighted kappas	AGREE
FISHER	Fisher's exact test	CHISQ or FISHER [1]
GAMMA	gamma	MEASURES

[1] CHISQ computes Fisher's exact test for 2 × 2 tables. Use the FISHER option to compute Fisher's exact test for general rxc tables.

Table 3.7 *continued*

Option	Output Data Set Statistics	Required TABLES Statement Option
JT	Jonckheere-Terpstra test	JT
KAPPA	simple kappa coefficient	AGREE
KENTB	Kendall's tau-b	MEASURES
LAMCR	lambda asymmetric $(C\|R)$	MEASURES
LAMDAS	lambda symmetric	MEASURES
LAMRC	lambda asymmetric $(R\|C)$	MEASURES
LGOR	adjusted logit odds ratio ($h \times 2 \times 2$ tables)	CMH, CMH1, or CMH2
LGRRC1	adjusted column 1 logit relative risk	CMH, CMH1, or CMH2
LGRRC2	adjusted column 2 logit relative risk	CMH, CMH1, or CMH2
LRCHI	likelihood-ratio chi-square	CHISQ
MCNEM	McNemar's test (2×2 tables)	AGREE
MEASURES	gamma, Kendall's tau-b, Stuart's tau-c, Somers' $D(C\|R)$ and $D(R\|C)$, Pearson and Spearman correlations, lambda asymmetric $(C\|R)$ and $(R\|C)$, lambda symmetric, uncertainty coefficients $(C\|R)$ and $(R\|C)$, symmetric uncertainty coefficient; odds ratio and relative risks (2×2 tables)	MEASURES
MHCHI	Mantel-Haenszel chi-square	CHISQ
MHOR | COMOR	adjusted Mantel-Haenszel odds ratio ($h \times 2 \times 2$ tables)	CMH, CMH1, or CMH2
MHRRC1	adjusted column 1 Mantel-Haenszel relative risk	CMH, CMH1, or CMH2
MHRRC2	adjusted column 2 Mantel-Haenszel relative risk	CMH, CMH1, or CMH2
N	number of nonmissing observations	
NMISS	number of missing observations	
OR	odds ratio (2×2 tables)	MEASURES or RELRISK
PCHI	chi-square goodness-of-fit test for one-way tables, Pearson chi-square for two-way tables	CHISQ
PCORR	Pearson correlation coefficient	MEASURES
PHI	phi coefficient	CHISQ
PLCORR	polychoric correlation coefficient	PLCORR
RDIF1	column 1 risk difference (row 1 - row 2)	RISKDIFF
RDIF2	column 2 risk difference (row 1 - row 2)	RISKDIFF
RELRISK	odds ratio and relative risks (2×2 tables)	MEASURES or RELRISK
RISKDIFF	risks and risk differences (2×2 tables)	RISKDIFF
RISKDIFF1	column 1 risks and risk difference	RISKDIFF
RISKDIFF2	column 2 risks and risk difference	RISKDIFF
RRC1	column 1 relative risk	MEASURES or RELRISK
RRC2	column 2 relative risk	MEASURES or RELRISK
RSK1	column 1 risk, overall	RISKDIFF
RSK11	column 1 risk, for row 1	RISKDIFF
RSK12	column 2 risk, for row 1	RISKDIFF
RSK2	column 2 risk, overall	RISKDIFF
RSK21	column 1 risk, for row 2	RISKDIFF

Table 3.7   continued

Option	Output Data Set Statistics	Required TABLES Statement Option
RSK22	column 2 risk, for row 2	RISKDIFF
SCORR	Spearman correlation coefficient	MEASURES
SMDCR	Somers' $D(C\|R)$	MEASURES
SMDRC	Somers' $D(R\|C)$	MEASURES
STUTC	Stuart's tau-c	MEASURES
TREND	Cochran-Armitage test for trend	TREND
TSYMM	Bowker's test of symmetry	AGREE
U	symmetric uncertainty coefficient	MEASURES
UCR	uncertainty coefficient $(C\|R)$	MEASURES
URC	uncertainty coefficient $(R\|C)$	MEASURES
WTKAP	weighted kappa coefficient	AGREE

TABLES Statement

TABLES requests < / options > ;

The TABLES statement requests one-way to n-way frequency and crosstabulation tables and statistics for those tables.

If you omit the TABLES statement, PROC FREQ generates one-way frequency tables for all data set variables that are not listed in the other statements.

The following argument is required in the TABLES statement.

requests

specify the frequency and crosstabulation tables to produce. A request is composed of one variable name or several variable names separated by asterisks. To request a one-way frequency table, use a single variable. To request a two-way crosstabulation table, use an asterisk between two variables. To request a multiway table (an n-way table, where $n>2$), separate the desired variables with asterisks. The unique values of these variables form the rows, columns, and strata of the table. You can include up to 50 variables in a single multiway table request.

For two-way to multiway tables, the values of the last variable form the crosstabulation table columns, while the values of the next-to-last variable form the rows. Each level (or combination of levels) of the other variables forms one stratum. PROC FREQ produces a separate crosstabulation table for each stratum. For example, a specification of A*B*C*D in a TABLES statement produces k tables, where k is the number of different combinations of values for A and B. Each table lists the values for C down the side and the values for D across the top.

You can use multiple TABLES statements in the PROC FREQ step. PROC FREQ builds all the table requests in one pass of the data, so that there is essentially no loss of efficiency. You can also specify any number of table requests in a single TABLES statement. To specify multiple table requests quickly, use a grouping syntax by placing parentheses around several variables and joining other variables or variable combinations. For example, the statements shown in Table 3.8 illustrate grouping syntax.

Table 3.8 Grouping Syntax

Request	Equivalent to
tables A*(B C);	tables A*B A*C;
tables (A B)*(C D);	tables A*C B*C A*D B*D;
tables (A B C)*D;	tables A*D B*D C*D;
tables A - - C;	tables A B C;
tables (A - - C)*D;	tables A*D B*D C*D;

The TABLES statement variables are one or more variables from the DATA= input data set. These variables can be either character or numeric, but the procedure treats them as categorical variables. PROC FREQ uses the formatted values of the TABLES variable to determine the categorical variable levels. So if you assign a format to a variable with a FORMAT statement, PROC FREQ formats the values before dividing observations into the levels of a frequency or crosstabulation table. See the discussion of the FORMAT procedure in the *Base SAS Procedures Guide* and the discussions of the FORMAT statement and SAS formats in *SAS Language Reference: Dictionary*.

If you use PROC FORMAT to create a user-written format that combines missing and nonmissing values into one category, PROC FREQ treats the entire category of formatted values as missing. See the discussion in the section "Grouping with Formats" on page 113 for more information.

The frequency or crosstabulation table lists the values of both character and numeric variables in ascending order based on internal (unformatted) variable values unless you change the order with the ORDER= option. To list the values in ascending order by formatted value, use ORDER=FORMATTED in the PROC FREQ statement.

Without Options

If you request a one-way frequency table for a variable without specifying options, PROC FREQ produces frequencies, cumulative frequencies, percentages of the total frequency, and cumulative percentages for each value of the variable. If you request a two-way or an *n*-way crosstabulation table without specifying options, PROC FREQ produces crosstabulation tables that include cell frequencies, cell percentages of the total frequency, cell percentages of row frequencies, and cell percentages of column frequencies. The procedure excludes observations with missing values from the table but displays the total frequency of missing observations below each table.

Options

Table 3.9 lists the options available in the TABLES statement. Descriptions follow in alphabetical order.

Table 3.9 TABLES Statement Options

Option	Description
Control Statistical Analysis	
AGREE	requests tests and measures of classification agreement
ALL	requests tests and measures of association produced by CHISQ, MEASURES, and CMH
ALPHA=	sets the confidence level for confidence limits
BDT	requests Tarone's adjustment for the Breslow-Day test
BINOMIAL	requests binomial proportion, confidence limits, and tests for one-way tables
BINOMIALC	requests BINOMIAL statistics with a continuity correction
CHISQ	requests chi-square tests and measures based on chi-square
CL	requests confidence limits for the MEASURES statistics
CMH	requests all Cochran-Mantel-Haenszel statistics
CMH1	requests CMH correlation statistic, adjusted odds ratios, and adjusted relative risks
CMH2	requests CMH correlation and row mean scores (ANOVA) statistics, adjusted odds ratios, and adjusted relative risks
CONVERGE=	specifies convergence criterion for polychoric correlation
FISHER	requests Fisher's exact test for tables larger than 2×2
JT	requests Jonckheere-Terpstra test
MAXITER=	specifies maximum number of iterations for polychoric correlation
MEASURES	requests measures of association
MISSING	treats missing values as nonmissing
PLCORR	requests polychoric correlation
RELRISK	requests relative risk measures for 2×2 tables
RISKDIFF	requests risks and risk differences for 2×2 tables
RISKDIFFC	requests RISKDIFF statistics with a continuity correction
SCORES=	specifies the type of row and column scores
TESTF=	specifies expected frequencies for one-way chi-square test
TESTP=	specifies expected proportions for one-way chi-square test
TREND	requests Cochran-Armitage test for trend
Control Additional Table Information	
CELLCHI2	displays cell contributions to the Pearson chi-square statistic
CUMCOL	displays cumulative column percentages
DEVIATION	displays deviations of cell frequencies from expected values
EXPECTED	displays expected cell frequencies
MISSPRINT	displays missing value frequencies
SPARSE	includes all possible combinations of variable levels in LIST and OUT=
TOTPCT	displays percentages of total frequency for n-way tables ($n > 2$)

Table 3.9 *continued*

Option	Description
Control Displayed Output	
CONTENTS=	specifies the contents label for crosstabulation tables
CROSSLIST	displays crosstabulation tables in ODS column format
FORMAT=	formats the frequencies in crosstabulation tables
LIST	displays two-way to *n*-way tables in list format
NOCOL	suppresses display of column percentages
NOCUM	suppresses display of cumulative frequencies and percentages
NOFREQ	suppresses display of frequencies
NOPERCENT	suppresses display of percentages
NOPRINT	suppresses display of crosstabulation tables but displays statistics
NOROW	suppresses display of row percentages
NOSPARSE	suppresses zero frequency levels in CROSSLIST, LIST and OUT=
NOWARN	suppresses log warning message for the chi-square test
PRINTKWT	displays kappa coefficient weights
SCOROUT	displays row and column scores
Produce Statistical Graphics	
PLOTS=	requests plots from ODS Graphics
Create an Output Data Set	
OUT=	names an output data set to contain frequency counts
OUTCUM	includes cumulative frequencies and percentages in the output data set for one-way tables
OUTEXPECT	includes expected frequencies in the output data set
OUTPCT	includes row, column, and two-way table percentages in the output data set

You can specify the following options in a TABLES statement.

AGREE < (WT=FC) >

requests tests and measures of classification agreement for square tables. The AGREE option provides McNemar's test for 2 × 2 tables and Bowker's test of symmetry for square tables with more than two response categories. The AGREE option also produces the simple kappa coefficient, the weighted kappa coefficient, their asymptotic standard errors, and their confidence limits. When there are multiple strata, the AGREE option provides overall simple and weighted kappas as well as tests for equal kappas among strata. When there are multiple strata and two response categories, PROC FREQ computes Cochran's Q test. See the section "Tests and Measures of Agreement" on page 153 for details about these statistics.

If you specify the WT=FC option in parentheses following the AGREE option, PROC FREQ uses Fleiss-Cohen weights to compute the weighted kappa coefficient. By default, PROC FREQ uses Cicchetti-Allison weights. See the section "Weighted Kappa Coefficient" on page 155 for details. You can specify the PRINTKWT option to display the kappa coefficient weights.

AGREE statistics are computed only for square tables, where the number of rows equals the number of columns. If your table is not square due to observations with zero weights, you can specify the ZEROS option in the WEIGHT statement to include these observations. For more details, see the section "Tables with Zero Rows and Columns" on page 158.

You can use the TEST statement to request asymptotic tests for the simple and weighted kappa coefficients. You can request exact *p*-values for the simple and weighted kappa coefficient tests, as well as for McNemar's test, by specifying the corresponding options in the EXACT statement. See the section "Exact Statistics" on page 167 for more information.

ALL

requests all of the tests and measures that are computed by the CHISQ, MEASURES, and CMH options. The number of CMH statistics computed can be controlled by the CMH1 and CMH2 options.

ALPHA=α

specifies the level of confidence limits. The value of α must be between 0 and 1, and the default is 0.05. A confidence level of α produces $100(1-\alpha)\%$ confidence limits. The default of ALPHA=0.05 produces 95% confidence limits.

ALPHA= applies to confidence limits requested by TABLES statement options. There is a separate ALPHA= option in the EXACT statement that sets the level of confidence limits for Monte Carlo estimates of exact *p*-values, which are requested in the EXACT statement.

BDT

requests Tarone's adjustment in the Breslow-Day test for homogeneity of odds ratios. (You must specify the CMH option to compute the Breslow-Day test.) See the section "Breslow-Day Test for Homogeneity of the Odds Ratios" on page 164 for more information.

BINOMIAL < (*binomial-options***) >**

requests the binomial proportion for one-way tables. When you specify the BINOMIAL option, PROC FREQ also provides the asymptotic standard error, asymptotic (Wald) and exact (Clopper-Pearson) confidence limits, and the asymptotic equality test for the binomial proportion by default.

You can specify *binomial-options* inside the parentheses following the BINOMIAL option. The LEVEL= binomial-option identifies the variable level for which to compute the proportion. If you do not specify LEVEL=, PROC FREQ computes the proportion for the first level that appears in the output. The P= binomial-option specifies the null proportion for the binomial tests. If you do not specify P=, PROC FREQ uses P=0.5 by default.

You can also specify *binomial-options* to request confidence limits and tests of noninferiority, superiority, and equivalence for the binomial proportion. Table 3.10 summarizes the *binomial-options*.

Available confidence limits for the binomial proportion include Agresti-Coull, exact (Clopper-Pearson), Jeffreys, Wald, and Wilson (score) confidence limits. You can specify more than one type of binomial confidence limits in the same analysis. If you do not specify any confidence limit requests with *binomial-options*, PROC FREQ computes Wald asymptotic confidence limits and exact (Clopper-Pearson) confidence limits by default. The

ALPHA= option determines the confidence level, and the default of ALPHA=0.05 produces 95% confidence limits for the binomial proportion.

To request exact tests for the binomial proportion, specify the BINOMIAL option in the EXACT statement. PROC FREQ then computes exact *p*-values for all binomial tests that you request with *binomial-options*, which can include tests of noninferiority, superiority, and equivalence, as well as the test of equality.

See the section "Binomial Proportion" on page 133 for details.

Table 3.10   BINOMIAL Options

Task	Binomial-Option
Specify the variable level	LEVEL=
Specify the null proportion	P=
Request a continuity correction	CORRECT
Request confidence limits	AGRESTICOULL | AC ALL EXACT | CLOPPERPEARSON JEFFREYS | J WILSON | W WALD
Request tests	EQUIV | EQUIVALENCE NONINF | NONINFERIORITY SUP | SUPERIORITY
Specify the test margin	MARGIN=
Specify the test variance	VAR=SAMPLE | NULL

You can specify the following *binomial-options* inside parentheses following the BINOMIAL option.

AGRESTICOULL | AC
> requests Agresti-Coull confidence limits for the binomial proportion. See the section "Agresti-Coull Confidence Limits" on page 134 for details.

ALL
> requests all available types of confidence limits for the binomial proportion. These include the following: Agresti-Coull, exact (Clopper-Pearson), Jeffreys, Wald, and Wilson (score) confidence limits.

CORRECT
> includes a continuity correction in the asymptotic Wald confidence limits and tests. The CORRECT binomial-option has the same effect as the BINOMIALC option.

EQUIV | EQUIVALENCE
> requests a test of equivalence for the binomial proportion. See the section "Equivalence Test" on page 139 for details. You can specify the equivalence test margins, the

null proportion, and the variance type with the MARGIN=, P=, and VAR= binomial-options, respectively.

EXACT

CLOPPERPEARSON

requests exact (Clopper-Pearson) confidence limits for the binomial proportion. See the section "Exact (Clopper-Pearson) Confidence Limits" on page 135 for details. If you do not request any binomial confidence limits by specifying binomial-options, PROC FREQ produces Wald and exact (Clopper-Pearson) confidence limits by default.

JEFFREYS | J

requests Jeffreys confidence limits for the binomial proportion. See the section "Jeffreys Confidence Limits" on page 134 for details.

LEVEL=*level-number* | '*level-value*'

specifies the variable level for the binomial proportion. By default, PROC FREQ computes the proportion of observations for the first variable level that appears in the output. To request a different level, use LEVEL=*level-number* or LEVEL='*level-value*', where *level-number* is the variable level's number or order in the output, and *level-value* is the formatted value of the variable level. The value of *level-number* must be a positive integer. You must enclose *level-value* in single quotes.

MARGIN=*value* | (*lower,upper*)

specifies the margin for the noninferiority, superiority, and equivalence tests, which you request with the NONINF, SUP, and EQUIV binomial-options, respectively. If you do not specify MARGIN=, PROC FREQ uses a margin of 0.2 by default.

For noninferiority and superiority tests, specify a single *value* for the MARGIN= option. The MARGIN= *value* must be a positive number. You can specify *value* as a number between 0 and 1. Or you can specify *value* in percentage form as a number between 1 and 100, and PROC FREQ converts that number to a proportion. The procedure treats the value 1 as 1%.

For noninferiority and superiority tests, the test limits must be between 0 and 1. The limits are determined by the null proportion value (which you can specify with the P= binomial-option) and by the margin value. The noninferiority limit equals the null proportion minus the margin. By default, the null proportion equals 0.5 and the margin equals 0.2, which gives a noninferiority limit of 0.3. The superiority limit equals the null proportion plus the margin, which is 0.7 by default.

For an equivalence test, you can specify a single MARGIN= *value*, or you can specify both *lower* and *upper* values. If you specify a single MARGIN= *value*, it must be a positive number, as described previously. If you specify a single MARGIN= *value* for an equivalence test, PROC FREQ uses –*value* as the lower margin and *value* as the upper margin for the test. If you specify both *lower* and *upper* values for an equivalence test, you can specify them in proportion form as numbers between –1 or 1. Or you can specify them in percentage form as numbers between –100 and 100, and PROC FREQ converts the numbers to proportions. The value of *lower* must be less than the value of *upper*.

The equivalence limits must be between 0 and 1. The equivalence limits are determined by the null proportion value (which you can specify with the P= binomial-option) and by the margin values. The lower equivalence limit equals the null proportion plus the lower margin. By default, the null proportion equals 0.5 and the lower margin equals –0.2, which gives a lower equivalence limit of 0.3. The upper equivalence limit equals the null proportion plus the upper margin, which is 0.7 by default.

See the sections "Noninferiority Test" on page 137 and "Equivalence Test" on page 139 for details.

NONINF | NONINFERIORITY

requests a test of noninferiority for the binomial proportion. See the section "Noninferiority Test" on page 137 for details. You can specify the noninferiority test margin, the null proportion, and the variance type with the MARGIN=, P=, and VAR= binomial-options, respectively.

P=*value*

specifies the null hypothesis proportion for the binomial tests. If you omit the P= option, PROC FREQ uses 0.5 as the null proportion. The null proportion *value* must be a positive number. You can specify *value* as a number between 0 and 1. Or you can specify *value* in percentage form as a number between 1 and 100, and PROC FREQ converts that number to a proportion. The procedure treats the value 1 as 1%.

SUP | SUPERIORITY

requests a test of superiority for the binomial proportion. See the section "Superiority Test" on page 138 for details. You can specify the superiority test margin, the null proportion, and the variance type with the MARGIN=, P=, and VAR= binomial-options, respectively.

VAR=SAMPLE | NULL

specifies the type of variance estimate to use for the tests of noninferiority, superiority, and equivalence. The default is VAR=SAMPLE, which estimates the variance from the sample proportion. VAR=NULL uses the null proportion to compute the variance. See the sections "Noninferiority Test" on page 137 and "Equivalence Test" on page 139 for details.

WALD

requests Wald confidence limits for the binomial proportion. See the section "Wald Confidence Limits" on page 134 for details. If you do not request any binomial confidence limits by specifying binomial-options, PROC FREQ produces Wald and exact (Clopper-Pearson) confidence limits by default.

WILSON | W
SCORE

requests Wilson confidence limits for the binomial proportion. These are also known as *score* confidence limits. See the section "Wilson (Score) Confidence Limits" on page 135 for details.

BINOMIALC < (*binomial-options*) >

requests the BINOMIAL option statistics for one-way tables, and includes a continuity correction in the asymptotic Wald confidence limits and tests. The BINOMIAL option statistics include the binomial proportion, its asymptotic standard error, asymptotic (Wald) and exact (Clopper-Pearson) confidence limits, and the asymptotic equality test for the binomial proportion by default. The *binomial-options* available with the BINOMIALC option are the same as those available with BINOMIAL. See the description of the BINOMIAL option for details.

CELLCHI2

displays each crosstabulation table cell's contribution to the total Pearson chi-square statistic. The cell contribution is computed as

$$\frac{(frequency - expected)^2}{expected}$$

where *frequency* is the table cell frequency or count and *expected* is the expected cell frequency, which is computed under the null hypothesis that the row and column variables are independent. See the section "Pearson Chi-Square Test for Two-Way Tables" on page 120 for details.

The CELLCHI2 option has no effect for one-way tables or for tables that are displayed with the LIST option.

CHISQ

requests chi-square tests of homogeneity or independence and measures of association based on the chi-square statistic. The tests include the Pearson chi-square, likelihood-ratio chi-square, and Mantel-Haenszel chi-square. The measures include the phi coefficient, the contingency coefficient, and Cramer's V. For 2×2 tables, the CHISQ option also provides Fisher's exact test and the continuity-adjusted chi-square. See the section "Chi-Square Tests and Statistics" on page 119 for details.

For one-way tables, the CHISQ option provides a chi-square goodness-of-fit test for equal proportions. If you specify the null hypothesis proportions with the TESTP= option, PROC FREQ computes a chi-square goodness-of-fit test for the specified proportions. If you specify null hypothesis frequencies with the TESTF= option, PROC FREQ computes a chi-square goodness-of-fit test for the specified frequencies. See the section "Chi-Square Test for One-Way Tables" on page 120 for more information.

To request Fisher's exact test for tables larger than 2×2, use the FISHER option in the EXACT statement. Exact tests are also available for other CHISQ statistics, including the Pearson, likelihood-ratio, and Mantel-Haenszel chi-square, and the chi-square goodness-of-fit test for one-way tables. You can use the EXACT statement to request these tests. See the section "Exact Statistics" on page 167 for details.

CL

requests confidence limits for the MEASURES statistics. If you omit the MEASURES option, the CL option invokes MEASURES. You can set the level of the confidence limits by using the ALPHA= option. The default of ALPHA=0.5 produces 95% confidence limits. See the sections "Measures of Association" on page 124 and "Confidence Limits" on page 125 for more information.

CMH

requests Cochran-Mantel-Haenszel statistics, which test for association between the row and column variables after adjusting for the remaining variables in a multiway table. The Cochran-Mantel-Haenszel statistics include the nonzero correlation statistic, the row mean scores differ (ANOVA) statistic, and the general association statistic. In addition, for 2×2 tables, the CMH option provides the adjusted Mantel-Haenszel and logit estimates of the odds ratio and relative risks, together with their confidence limits. For stratified 2×2 tables, the CMH option provides the Breslow-Day test for homogeneity of odds ratios. (To request Tarone's adjustment for the Breslow-Day test, specify the BDT option.) See the section "Cochran-Mantel-Haenszel Statistics" on page 158 for details.

You can use the CMH1 or CMH2 option to control the number of CMH statistics that PROC FREQ computes.

For stratified 2×2 tables, you can request Zelen's exact test for equal odds ratios by specifying the EQOR option in the EXACT statement. See the section "Zelen's Exact Test for Equal Odds Ratios" on page 164 for details. You can request exact confidence limits for the common odds ratio by specifying the COMOR option in the EXACT statement. This option also provides a common odds ratio test. See the section "Exact Confidence Limits for the Common Odds Ratio" on page 165 for details.

CMH1

requests the Cochran-Mantel-Haenszel correlation statistic. It does not provide the CMH row mean scores differ (ANOVA) statistic or the general association statistic, which are provided by the CMH option. For tables larger than 2×2, the CMH1 option requires less memory than the CMH option, which can require an enormous amount of memory for large tables.

For 2×2 tables, the CMH1 option also provides the adjusted Mantel-Haenszel and logit estimates of the common odds ratio and relative risks and the Breslow-Day test.

CMH2

requests the Cochran-Mantel-Haenszel correlation statistic and the row mean scores (ANOVA) statistic. It does not provide the CMH general association statistic, which is provided by the CMH option. For tables larger than 2×2, the CMH2 option requires less memory than the CMH option, which can require an enormous amount of memory for large tables.

For 2×2 tables, the CMH2 option also provides the adjusted Mantel-Haenszel and logit estimates of the common odds ratio and relative risks and the Breslow-Day test.

CONTENTS='*string*'

specifies the label to use for crosstabulation tables in the contents file, the Results window, and the trace record. For information about output presentation, see the *SAS Output Delivery System: User's Guide*.

If you omit the CONTENTS= option, the contents label for crosstabulation tables is "Cross-Tabular Freq Table" by default.

Note that contents labels for all crosstabulation tables that are produced by a single TABLES statement use the same text. To specify different contents labels for different crosstabulation tables, request the tables in separate TABLES statements and use the CONTENTS= option in each TABLES statement.

To remove the crosstabulation table entry from the contents file, you can specify a null label with CONTENTS=''.

The CONTENTS= option affects only contents labels for crosstabulation tables. It does not affect contents labels for other PROC FREQ tables.

To specify the contents label for any PROC FREQ table, you can use PROC TEMPLATE to create a customized table definition. The CONTENTS_LABEL attribute in the DEFINE TABLE statement of PROC TEMPLATE specifies the contents label for the table. See the chapter "The TEMPLATE Procedure" in the *SAS Output Delivery System: User's Guide* for more information.

CONVERGE=*value*

specifies the convergence criterion for computing the polychoric correlation, which you request with the PLCORR option. The CONVERGE= *value* must be a positive number. By default CONVERGE=0.0001. Iterative computation of the polychoric correlation stops when the convergence measure falls below the value of CONVERGE= or when the number of iterations exceeds the value specified in the MAXITER= option, whichever happens first. See the section "Polychoric Correlation" on page 130 for details.

CROSSLIST

displays crosstabulation tables in ODS column format instead of the default crosstabulation cell format. In a CROSSLIST table display, the rows correspond to the crosstabulation table cells, and the columns correspond to descriptive statistics such as Frequency and Percent. The CROSSLIST table displays the same information as the default crosstabulation table, but uses an ODS column format instead of the table cell format. See the section "Multiway Tables" on page 177 for details about the contents of the CROSSLIST table.

You can control the contents of a CROSSLIST table with the same options available for the default crosstabulation table. These include the NOFREQ, NOPERCENT, NOROW, and NOCOL options. You can request additional information in a CROSSLIST table with the CELLCHI2, DEVIATION, EXPECTED, MISSPRINT, and TOTPCT options.

The FORMAT= option and the CUMCOL option have no effect for CROSSLIST tables. You cannot specify both the LIST option and the CROSSLIST option in the same TABLES statement.

You can use the NOSPARSE option to suppress display of variable levels with zero frequency in CROSSLIST tables. By default for CROSSLIST tables, PROC FREQ displays all levels of the column variable within each level of the row variable, including any column variable levels with zero frequency for that row. And for multiway tables displayed with the CROSSLIST option, the procedure displays all levels of the row variable for each stratum of the table by default, including any row variable levels with zero frequency for the stratum.

CUMCOL

displays the cumulative column percentages in the cells of the crosstabulation table. The CUMCOL option does not apply to crosstabulation tables produced with the LIST or CROSSLIST option.

DEVIATION

displays the deviation of the frequency from the expected frequency for each cell of the crosstabulation table. See the section "Pearson Chi-Square Test for Two-Way Tables" on page 120 for details. The DEVIATION option does not apply to crosstabulation tables produced with the LIST option.

EXPECTED

displays the expected cell frequencies under the hypothesis of independence (or homogeneity) for crosstabulation tables. See the section "Pearson Chi-Square Test for Two-Way Tables" on page 120 for details. The EXPECTED option does not apply to tables produced with the LIST option.

FISHER | EXACT

requests Fisher's exact test for tables that are larger than 2×2. (For 2×2 tables, the CHISQ option provides Fisher's exact test.) This test is also known as the Freeman-Halton test. See the sections "Fisher's Exact Test" on page 122 and "Exact Statistics" on page 167 for more information.

If you omit the CHISQ option in the TABLES statement, the FISHER option invokes CHISQ. You can also request Fisher's exact test by specifying the FISHER option in the EXACT statement.

CAUTION: PROC FREQ computes exact tests with fast and efficient algorithms that are superior to direct enumeration. Exact tests are appropriate when a data set is small, sparse, skewed, or heavily tied. For some large problems, computation of exact tests might require a considerable amount of time and memory. Consider using asymptotic tests for such problems. Alternatively, when asymptotic methods might not be sufficient for such large problems, consider using Monte Carlo estimation of exact *p*-values. See the section "Computational Resources" on page 169 for more information.

FORMAT=*format-name*

specifies a format for the following crosstabulation table cell values: frequency, expected frequency, and deviation. PROC FREQ also uses the specified format to display the row and column total frequencies and the overall total frequency in crosstabulation tables.

You can specify any standard SAS numeric format or a numeric format defined with the FORMAT procedure. The format length must not exceed 24. If you omit the FORMAT= option, by default PROC FREQ uses the BEST6. format to display frequencies less than 1E6, and the BEST7. format otherwise.

The FORMAT= option applies only to crosstabulation tables displayed in the default format. It does not apply to crosstabulation tables produced with the LIST or CROSSLIST option.

To change display formats in any FREQ table, you can use PROC TEMPLATE. See the chapter "The TEMPLATE Procedure" in the *SAS Output Delivery System: User's Guide* for more information.

JT

requests the Jonckheere-Terpstra test. See the section "Jonckheere-Terpstra Test" on page 151 for details.

LIST

displays two-way to n-way crosstabulation tables in a list format instead of the default crosstabulation cell format. The LIST option displays the entire multiway table in one table, instead of displaying a separate two-way table for each stratum. See the section "Multiway Tables" on page 177 for details.

The LIST option is not available when you also specify statistical options. You must use the standard crosstabulation table display or the CROSSLIST display when you request statistical tests or measures.

MAXITER=number

specifies the maximum number of iterations for computing the polychoric correlation, which you request with the PLCORR option. The value of the MAXITER= option must be a positive integer. By default MAXITER=20. Iterative computation of the polychoric correlation stops when the number of iterations exceeds the MAXITER= value or when the convergence measures falls below the value of the CONVERGE= option, whichever happens first. See the section "Polychoric Correlation" on page 130 for details.

MEASURES

requests several measures of association and their asymptotic standard errors. The MEASURES option provides the following statistics: gamma, Kendall's tau-b, Stuart's tau-c, Somers' $D(C|R)$, Somers' $D(R|C)$, the Pearson and Spearman correlation coefficients, lambda (symmetric and asymmetric), and uncertainty coefficients (symmetric and asymmetric). To request confidence limits for these measures of association, you can specify the CL option.

For 2×2 tables, the MEASURES option also provides the odds ratio, column 1 relative risk, column 2 relative risk, and the corresponding confidence limits. Alternatively, you can obtain the odds ratio and relative risks, without the other measures of association, by specifying the RELRISK option.

See the section "Measures of Association" on page 124 for details.

You can use the TEST statement to request asymptotic tests for the following measures of association: gamma, Kendall's tau-b, Stuart's tau-c, Somers' $D(C|R)$, Somers' $D(R|C)$, and the Pearson and Spearman correlation coefficients. You can use the EXACT statement to request exact tests for the Pearson and Spearman correlation coefficients and exact confidence limits for the odds ratio. See the section "Exact Statistics" on page 167 for more information.

MISSING

treats missing values as a valid nonmissing level for all TABLES variables. The MISSING option displays the missing levels in frequency and crosstabulation tables and includes them in all calculations of percentages, tests, and measures.

By default, if you do not specify the MISSING or MISSPRINT option, an observation is excluded from a table if it has a missing value for any of the variables in the TABLES request. When PROC FREQ excludes observations with missing values, it displays the total frequency of missing observations below the table. See the section "Missing Values" on page 114 for more information.

MISSPRINT

displays missing value frequencies in frequency and crosstabulation tables but does not include the missing value frequencies in any computations of percentages, tests, or measures.

By default, if you do not specify the MISSING or MISSPRINT option, an observation is excluded from a table if it has a missing value for any of the variables in the TABLES request. When PROC FREQ excludes observations with missing values, it displays the total frequency of missing observations below the table. See the section "Missing Values" on page 114 for more information.

NOCOL

suppresses the display of column percentages in crosstabulation table cells.

NOCUM

suppresses the display of cumulative frequencies and percentages in one-way frequency tables. The NOCUM option also suppresses the display of cumulative frequencies and percentages in crosstabulation tables in list format, which you request with the LIST option.

NOFREQ

suppresses the display of cell frequencies in crosstabulation tables. The NOFREQ option also suppresses row total frequencies. This option has no effect for one-way tables or for crosstabulation tables in list format, which you request with the LIST option.

NOPERCENT

suppresses the display of overall percentages in crosstabulation tables. These percentages include the cell percentages of the total (two-way) table frequency, as well as the row and column percentages of the total table frequency. To suppress the display of cell percentages of row or column totals, use the NOROW or NOCOL option, respectively.

For one-way frequency tables and crosstabulation tables in list format, the NOPERCENT option suppresses the display of percentages and cumulative percentages.

NOPRINT

suppresses the display of frequency and crosstabulation tables but displays all requested tests and statistics. To suppress the display of all output, including tests and statistics, use the NOPRINT option in the PROC FREQ statement.

NOROW

suppresses the display of row percentages in crosstabulation table cells.

NOSPARSE

suppresses the display of cells with a zero frequency count in LIST output and omits them from the OUT= data set. The NOSPARSE option applies when you specify the ZEROS option in the WEIGHT statement to include observations with zero weights. By default, the ZEROS option invokes the SPARSE option, which displays table cells with a zero frequency count in the LIST output and includes them in the OUT= data set. See the description of the ZEROS option for more information.

The NOSPARSE option also suppresses the display of variable levels with zero frequency in CROSSLIST tables. By default for CROSSLIST tables, PROC FREQ displays all levels of the column variable within each level of the row variable, including any column variable levels with zero frequency for that row. For multiway tables displayed with the CROSSLIST option, the procedure displays all levels of the row variable for each stratum of the table by default, including any row variable levels with zero frequency for the stratum.

NOWARN

suppresses the log warning message that the asymptotic chi-square test might not be valid. By default, PROC FREQ displays this log message when you request the CHISQ option and more than 20 percent of the table cells have expected frequencies less than five.

OUT=SAS-data-set

names an output data set that contains frequency or crosstabulation table counts and percentages. If more than one table request appears in the TABLES statement, the contents of the OUT= data set correspond to the last table request in the TABLES statement. The OUT= data set variable COUNT contains the frequencies and the variable PERCENT contains the percentages. See the section "Output Data Sets" on page 172 for details. You can specify the following options to include additional information in the OUT= data set: OUTCUM, OUTEXPECT, and OUTPCT.

OUTCUM

includes cumulative frequencies and cumulative percentages in the OUT= data set for one-way tables. The variable CUM_FREQ contains the cumulative frequencies, and the variable CUM_PCT contains the cumulative percentages. See the section "Output Data Sets" on page 172 for details. The OUTCUM option has no effect for two-way or multiway tables.

OUTEXPECT

includes expected cell frequencies in the OUT= data set for crosstabulation tables. The variable EXPECTED contains the expected cell frequencies. See the section "Output Data Sets" on page 172 for details. The EXPECTED option has no effect for one-way tables.

OUTPCT

includes the following additional variables in the OUT= data set for crosstabulation tables:

PCT_COL	percentage of column frequency
PCT_ROW	percentage of row frequency
PCT_TABL	percentage of stratum (two-way table) frequency, for n-way tables where $n > 2$

See the section "Output Data Sets" on page 172 for details. The OUTPCT option has no effect for one-way tables.

PLCORR

requests the polychoric correlation coefficient. For 2×2 tables, this statistic is more commonly known as the tetrachoric correlation coefficient, and it is labeled as such in the displayed output. See the section "Polychoric Correlation" on page 130 for details. Also see the descriptions of the CONVERGE= and MAXITER= options, which you can specify to control the iterative computation of the polychoric correlation coefficient.

If you omit the MEASURES option, the PLCORR option invokes MEASURES.

PLOTS < (*global-plot-options*) > < = *plot-request* < (*plot-options*) > >
PLOTS < (*global-plot-options*) > < = (*plot-request* < (*plot-options*) > <... *plot-request* < (*plot-options*) >>

requests plots for PROC FREQ to produce by using ODS Graphics. When you specify only one *plot-request*, you can omit the parentheses around the request. For example:

```
plots=all
plots=freqplot
plots=(freqplot oddsratioplot)
plots(only)=(cumfreqplot deviationplot)
```

For information about ODS Graphics, see Chapter 21, "Statistical Graphics Using ODS" (*SAS/STAT User's Guide*). You must enable ODS Graphics before requesting plots, as shown in the following statements:

```
ods graphics on;
proc freq;
   tables treatment*response / chisq plots=freqplot;
   weight wt;
run;
ods graphics off;
```

The PLOTS= option has no effect when you specify the NOPRINT option in the PROC FREQ statement.

If you do not specify the PLOTS= option but have enabled ODS Graphics, then PROC FREQ produces all plots associated with the analyses you request in the current TABLES statement.

Table 3.11 lists the available *plot-requests*, together with their *plot-options* and required TABLES statement options.

Table 3.11   PLOTS= Options

Plot-Request	*Plot-Options*	**Required TABLES Statement Option**
CUMFREQPLOT	ORIENT= SCALE= TYPE=	one-way table request
DEVIATIONPLOT	NOSTATS ORIENT= TYPE=	CHISQ (one-way table)
FREQPLOT	ORIENT= SCALE= TYPE=	any table request
FREQPLOT	TWOWAY=	two-way or multiway table
KAPPAPLOT	NPANELPOS= ORDER= RANGE= STATS	AGREE ($h \times r \times r$ table)
ODDSRATIOPLOT	LOGBASE= NPANELPOS= ORDER= RANGE= STATS	MEASURES or RELRISK ($h \times 2 \times 2$ table)
WTKAPPAPLOT	NPANELPOS= ORDER= RANGE= STATS	AGREE ($h \times r \times r$ table, $r > 2$)

Global-Plot-Options

A *global-plot-option* applies to all plots for which the option is available, unless it is altered by a specific *plot-option*. You can specify the following *global-plot-options* in parentheses following the PLOTS option:

NPANELPOS=n

>applies to ODDSRATIOPLOT, KAPPAPLOT, and WTKAPPAPLOT. The NPANEL-POS= plot-option breaks the plot into multiple graphics that have at most $|n|$ odds ratios or kappa statistics per graphic. If n is positive, the number of statistics per graphic is balanced; but if n is negative, the number of statistics per graphic is not balanced. By default, $n = 0$ and all statistics are displayed in a single plot. For example, suppose you want to display 21 odds ratios. Then NPANELPOS=20 displays two plots, the first with 11 odds ratios and the second with 10; NPANELPOS=–20 displays 20 odds ratios in the first plot but only 1 in the second.

ONLY

>suppresses the default plots and requests only the plots specified as *plot-requests*.

ORDER=ASCENDING | DESCENDING

applies to ODDSRATIOPLOT, KAPPPAPLOT, and WTKAPPAPLOT. The ORDER= plot-option displays the odds ratios or kappa statistics in sorted order. By default, the statistics are displayed in the order that the corresponding strata appear in the multiway table display.

ORIENT=HORIZONTAL | VERTICAL

applies to FREQPLOT, CUMFREQPLOT, and DEVIATIONPLOT. The ORIENT= plot-option controls the orientation of the plot. ORIENT=HORIZONTAL places the variable levels on the y-axis and the frequencies or statistic-values on the x-axis. ORIENT=VERTICAL places the variable levels on the x-axis. The default orientation is ORIENT=VERTICAL for bar charts and ORIENT=HORIZONTAL for dot plots.

RANGE=(< min ><, max >) | CLIP

applies to ODDSRATIOPLOT, KAPPAPLOT, and WTKAPPAPLOT. The RANGE= plot-option specifies the range of values to display. If you specify RANGE=CLIP, the confidence intervals are clipped and the display range is determined by the minimum and maximum values of the estimates. By default, the display range includes all confidence limits.

SCALE=FREQ | LOG | PERCENT | SQRT

applies to FREQPLOT and CUMFREQPLOT. The SCALE= plot-option specifies the scale of the frequencies to display. The default is SCALE=FREQ, which displays unscaled frequencies. SCALE=LOG displays log (base 10) frequencies. SCALE=PERCENT displays percentages or relative frequencies. SCALE=SQRT displays square roots of the frequencies, which produces a plot known as a *rootogram*.

STATS

applies to ODDSRATIOPLOT, KAPPAPLOT, and WTKAPPAPLOT. The STATS plot-option displays the values of the statistics and their confidence limits on the right side of the plot. If you do not request the STATS option, the statistic values are not displayed.

TYPE=BARCHART | DOTPLOT

applies to FREQPLOT, CUMFREQPLOT, and DEVIATIONPLOT. The TYPE= plot-option specifies the plot type. TYPE=BARCHART produces a bar chart, and TYPE=DOTPLOT produces a dot plot. The default is TYPE=BARCHART.

Plot-Requests

The following *plot-requests* are available with the PLOTS= option:

ALL

requests all plots associated with the specified analyses. This is the default if you do not specify the PLOTS(ONLY) option.

CUMFREQPLOT <(*plot-options*)>

requests a plot of cumulative frequencies for a one-way frequency table. The following *plot-options* are available for CUMFREQPLOT: ORIENT=, SCALE=, and TYPE=.

DEVIATIONPLOT < (*plot-options*) >

requests a plot of relative deviations from expected frequencies for a one-way table. The DEVIATIONPLOT is associated with the CHISQ option for a one-way table request. The following *plot-options* are available for DEVIATIONPLOT: ORIENT= and TYPE=.

FREQPLOT < (*plot-options*) >

requests a frequency plot. Frequency plots are available for frequency and crosstabulation tables. For multiway tables, PROC FREQ provides a two-way frequency plot for each stratum. The following *plot-options* are available for FREQPLOT for all tables: ORIENT=, SCALE=, and TYPE=. For two-way and multiway tables, you can use the TWOWAY= plot-option to specify the layout of the two-way frequency plot.

KAPPAPLOT < (*plot-options*) >

requests a plot of kappa statistics and confidence limits for a multiway table. The KAPPAPLOT is associated with the AGREE option for multiway square tables. The following *plot-options* are available for KAPPAPLOT: NPANELPOS=, ORDER=, RANGE=, and STATS.

NONE

suppresses all plots.

ODDSRATIOPLOT < (*plot-options*) >

requests a plot of odds ratios and confidence limits for a multiway table. The ODDSRATIOPLOT is associated with the MEASURES or RELRISK option for multiway 2×2 tables. The following *plot-options* are available for ODDSRATIOPLOT: LOGBASE=, NPANELPOS=, ORDER=, RANGE=, and STATS.

WTKAPPAPLOT < (*plot-options*) >

requests a plot of weighted kappa statistics and confidence limits for a multiway table. The WTKAPPAPLOT is associated with the AGREE option for multiway square tables with more than two rows. (For 2×2 tables, the simple kappa and weighted kappa statistics are the same, so weighted kappas are not presented for 2×2 tables.) The following *plot-options* are available for WTKAPPAPLOT: NPANELPOS=, ORDER=, RANGE=, and STATS.

Plot-Options

You can specify the following *plot-options* in parentheses after a *plot-request*.

LOGBASE=2 | E | 10

applies only to ODDSRATIOPLOT. The LOGBASE= plot-option displays the odds ratio axis on the specified log scale.

NOSTATS

applies only to DEVIATIONPLOT. The NOSTATS plot-option suppresses the chi-square *p*-value that is displayed by default in the deviation plot.

NPANELPOS=n

applies to ODDSRATIOPLOT, KAPPAPLOT, and WTKAPPAPLOT. The NPANELPOS= plot-option breaks the plot into multiple graphics that have at most $|n|$ odds ratios or kappa statistics per graphic. If n is positive, the number of statistics per graphic is balanced; but if n is negative, the number of statistics per graphic is not balanced. By default, $n = 0$ and all statistics are displayed in a single plot. For example, suppose you want to display 21 odds ratios. Then NPANELPOS=20 displays two plots, the first with 11 odds ratios and the second with 10; NPANELPOS=–20 displays 20 odds ratios in the first plot but only 1 in the second.

ORDER=ASCENDING | DESCENDING

applies to ODDSRATIOPLOT, KAPPPAPLOT, and WTKAPPAPLOT. The ORDER= plot-option displays the odds ratios or kappa statistics in sorted order. By default, the statistics are displayed in the order that the corresponding strata appear in the multiway table display.

ORIENT=HORIZONTAL | VERTICAL

applies to FREQPLOT, CUMFREQPLOT, and DEVIATIONPLOT. The ORIENT= plot-option controls the orientation of the plot. ORIENT=HORIZONTAL places the variable levels on the y-axis and the frequencies or statistic-values on the x-axis. ORIENT=VERTICAL places the variable levels on the x-axis. The default orientation is ORIENT=VERTICAL for bar charts and ORIENT=HORIZONTAL for dot plots.

RANGE=(< *min* ><, *max* >) | CLIP

applies to ODDSRATIOPLOT, KAPPAPLOT, and WTKAPPAPLOT. The RANGE= plot-option specifies the range of values to display. If you specify RANGE=CLIP, the confidence intervals are clipped and the display range is determined by the minimum and maximum values of the estimates. By default, the display range includes all confidence limits.

SCALE=FREQ | LOG | PERCENT| SQRT

applies to FREQPLOT and CUMFREQPLOT. The SCALE= option specifies the scale of the frequencies to display. The default is SCALE=FREQ, which displays unscaled frequencies. SCALE=LOG displays log (base 10) frequencies. SCALE=PERCENT displays percentages or relative frequencies. SCALE=SQRT displays square roots of the frequencies, which produces a plot known as a *rootogram*.

STATS

applies to ODDSRATIOPLOT, KAPPAPLOT, and WTKAPPAPLOT. The STATS plot-option displays the values of the statistics and their confidence limits on the right side of the plot. If you do not request the STATS option, the statistic values are not displayed.

TWOWAY=GROUPVERTICAL | GROUPHORIZONTAL | STACKED

applies to FREQPLOT for two-way and multiway tables. For multiway tables, PROC FREQ provides a two-way frequency plot for each stratum. The TWOWAY= plot-option specifies the layout for two-way frequency plots. The default is TWOWAY=GROUPVERTICAL, which produces a grouped plot with a vertical common baseline. The plots are grouped by the row variable, which is the first variable you specify in a two-way table request. TWOWAY=GROUPHORIZONTAL produces a grouped plot with a horizontal common baseline.

TWOWAY=STACKED produces stacked frequency plots for two-way tables. In a stacked bar chart, the bars correspond to the column variable values, and the row frequencies are stacked within each column. For dot plots, the dotted lines correspond to the columns, and the row frequencies within columns are plotted as data dots on the same column line.

The TYPE= and ORIENT= plot-options are available for each TWOWAY= layout option.

TYPE=BARCHART | DOTPLOT

applies to FREQPLOT, CUMFREQPLOT, and DEVIATIONPLOT. The TYPE= plot-option specifies the plot type. TYPE=BARCHART produces a bar chart, and TYPE=DOTPLOT produces a dot plot. The default is TYPE=BARCHART.

PRINTKWT

displays the weights that PROC FREQ uses to compute the weighted kappa coefficient. You must also specify the AGREE option to request the weighted kappa coefficient. You can specify (WT=FC) with the AGREE option to request Fleiss-Cohen weights. By default, PROC FREQ uses Cicchetti-Allison weights to compute the weighted kappa coefficient. See the section "Weighted Kappa Coefficient" on page 155 for details.

RELRISK

requests relative risk measures and their confidence limits for 2 × 2 tables. These measures include the odds ratio and the column 1 and 2 relative risks. See the section "Odds Ratio and Relative Risks for 2 x 2 Tables" on page 148 for details.

You can also obtain the RELRISK measures by specifying the MEASURES option, which produces other measures of association in addition to the relative risks.

You can request exact confidence limits for the odds ratio by specifying the OR option in the EXACT statement.

RISKDIFF < (*riskdiff-options***) >**

requests risks, or binomial proportions, for 2 × 2 tables. For column 1 and column 2, PROC FREQ computes the row 1 risk, row 2 risk, total risk, and risk difference (row 1 – row 2), together with their asymptotic standard errors and asymptotic (Wald) confidence limits. PROC FREQ also provides exact (Clopper-Pearson) confidence limits for the row 1, row 2, and total risks. The ALPHA= option determines the confidence level, and the default of ALPHA=0.05 produces 95% confidence limits. See the section "Risks and Risk Differences" on page 140 for details.

You can specify *riskdiff-options* inside the parentheses following the RISKDIFF option to request tests of noninferiority, superiority, and equivalence for the risk difference. Available test methods include Farrington-Manning, Hauck-Anderson, and Newcombe score (Wilson), in addition to the Wald test. Table 3.12 summarizes the *riskdiff-options*.

You can request exact unconditional confidence limits for the risk difference by specifying the RISKDIFF option in the EXACT statement. See the section "Exact Unconditional Confidence Limits for the Risk Difference" on page 147 for more information.

Table 3.12   RISKDIFF (Proportion Difference) Options

Task	Riskdiff-Option
Specify the column	COLUMN=1 | 2
Request a continuity correction	CORRECT
Request tests	EQUAL
	EQUIV | EQUIVALENCE
	NONINF | NONINFERIORITY
	SUP | SUPERIORITY
Specify the test method	METHOD=
Specify the test margin	MARGIN=
Specify the test variance	VAR=SAMPLE | NULL

You can specify the following *riskdiff-options* inside parentheses following the RISKDIFF option.

COLUMN=1 | 2 | BOTH

specifies the table column for which to compute the risk difference tests of noninferiority, superiority, or equivalence, which you request with the NONINF, SUP, and EQUIV riskdiff-options, respectively. You can specify COLUMN=1, COLUMN=2, or COLUMN=BOTH. If you do not specify the COLUMN= option, PROC FREQ computes the risk difference tests for column 1. The COLUMN= option has no effect on the table of risk estimates and confidence limits or on the equality test; PROC FREQ computes these statistics for both column 1 and column 2.

CORRECT

includes a continuity correction in the asymptotic Wald confidence limits and tests. The CORRECT riskdiff-option also includes a continuity correction in the Newcombe score confidence limits, which you request with the METHOD=SCORE riskdiff-option. METHOD=HA and METHOD=FM do not use continuity corrections. The CORRECT riskdiff-option has the same effect as the RISKDIFFC option.

EQUAL

requests a test of the null hypothesis that the risk difference equals zero. PROC FREQ provides an asymptotic Wald test of equality. See the section "Equality Test" on page 143 for details. You can specify the test variance type with the VAR= riskdiff-option.

EQUIV | EQUIVALENCE

requests a test of equivalence for the risk difference. See the section "Equivalence Tests" on page 146 for details. You can specify the equivalence test margins with the MARGIN= riskdiff-option and the test method with the METHOD= riskdiff-option. PROC FREQ uses METHOD=WALD by default.

MARGIN=*value* | (*lower,upper*)

specifies the margin for the noninferiority, superiority, and equivalence tests, which you request with the NONINF, SUP, and EQUIV riskdiff-options, respectively. If you do not specify MARGIN=, PROC FREQ uses a margin of 0.2 by default.

For noninferiority and superiority tests, specify a single *value* for the MARGIN= riskdiff-option. The MARGIN= *value* must be a positive number. You can specify *value* as a number between 0 and 1. Or you can specify *value* in percentage form as a number between 1 and 100, and PROC FREQ converts that number to a proportion. The procedure treats the value 1 as 1%.

For an equivalence test, you can specify a single MARGIN= *value*, or you can specify both *lower* and *upper* values. If you specify a single MARGIN= *value*, it must be a positive number, as described previously. If you specify a single MARGIN= *value* for an equivalence test, PROC FREQ uses –*value* as the lower margin and *value* as the upper margin for the test. If you specify both *lower* and *upper* values for an equivalence test, you can specify them in proportion form as numbers between –1 or 1. Or you can specify them in percentage form as numbers between –100 and 100, and PROC FREQ converts the numbers to proportions. The value of *lower* must be less than the value of *upper*.

METHOD=*method*

specifies the method for the noninferiority, superiority, and equivalence tests, which you request with the NONINF, SUP, and EQUIV riskdiff-options, respectively. The following methods are available:

FM	Farrington-Manning
HA	Hauck-Anderson
SCORE | NEWCOMBE | WILSON	Newcombe score (Wilson)
WALD	Wald

The default is METHOD=WALD. See the section "Noninferiority Test" on page 143 for descriptions of these methods.

For METHOD=SCORE and METHOD=WALD, you can request a continuity correction with the CORRECT riskdiff-option. For METHOD=WALD, you can specify the variance type with the VAR= riskdiff-option.

NONINF | NONINFERIORITY

requests a test of noninferiority for the risk difference. See the section "Noninferiority Test" on page 143 for details. You can specify the test margin with the MARGIN= riskdiff-option and the test method with the METHOD= riskdiff-option. PROC FREQ uses METHOD=WALD by default.

SUP | SUPERIORITY

requests a test of superiority for the binomial proportion. See the section "Superiority Test" on page 146 for details. You can specify the test margin with the MARGIN= riskdiff-option and the test method with the METHOD= riskdiff-option. PROC FREQ uses METHOD=WALD by default.

VAR=SAMPLE | NULL

specifies the type of variance estimate to use for the Wald tests of noninferiority, superiority, equivalence, and equality. The default is VAR=SAMPLE, which estimates the variance from the sample proportions. VAR=NULL uses the null hypothesis values to compute the variance. See the sections "Equality Test" on page 143 and "Noninferiority Test" on page 143 for details.

RISKDIFFC < (*riskdiff-options*) >

requests the RISKDIFF option statistics for 2 × 2 tables and includes a continuity correction in the asymptotic Wald confidence limits and tests.

The RISKDIFF option statistics include risks, or binomial proportions, for 2 × 2 tables. For column 1 and column 2, PROC FREQ computes the row 1 risk, row 2 risk, total risk, and risk difference (row 1 − row 2), together with their asymptotic standard errors and asymptotic (Wald) confidence limits. PROC FREQ also provides exact (Clopper-Pearson) confidence limits for the row 1, row 2, and total risks. See the section "Risks and Risk Differences" on page 140 for details.

You can request additional tests and statistics for the risk difference by specifying *riskdiff-options* in parentheses after RISKDIFFC. The *riskdiff-options* are the same as those available with RISKDIFF. See the description of the RISKDIFF option for details.

You can request exact unconditional confidence limits for the risk difference by specifying the RISKDIFF option in the EXACT statement. See the section "Exact Unconditional Confidence Limits for the Risk Difference" on page 147 for more information.

SCORES=*type*

specifies the type of row and column scores that PROC FREQ uses to compute the following statistics: Mantel-Haenszel chi-square, Pearson correlation, Cochran-Armitage test for trend, weighted kappa coefficient, and Cochran-Mantel-Haenszel statistics. The value of *type* can be one of the following:

- MODRIDIT
- RANK
- RIDIT
- TABLE

See the section "Scores" on page 118 for descriptions of these score types.

If you do not specify the SCORES= option, PROC FREQ uses SCORES=TABLE by default. For character variables, the row and column TABLE scores are the row and column numbers. That is, the TABLE score is 1 for row 1, 2 for row 2, and so on. For numeric variables, the row and column TABLE scores equal the variable values. See the section "Scores" on page 118 for details. Using MODRIDIT, RANK, or RIDIT scores yields nonparametric analyses.

You can use the SCOROUT option to display the row and column scores.

SCOROUT

displays the row and column scores that PROC FREQ uses to compute score-based tests and statistics. You can specify the score type with the SCORES= option. See the section "Scores" on page 118 for details.

The scores are computed and displayed only when PROC FREQ computes statistics for two-way tables. You can use ODS to store the scores in an output data set. See the section "ODS Table Names" on page 183 for more information.

SPARSE

reports all possible combinations of the variable values for an *n*-way table when $n > 1$, even if a combination does not occur in the data. The SPARSE option applies only to crosstabulation tables displayed in LIST format and to the OUT= output data set. If you do not use the LIST or OUT= option, the SPARSE option has no effect.

When you specify the SPARSE and LIST options, PROC FREQ displays all combinations of variable values in the table listing, including those with a frequency count of zero. By default, without the SPARSE option, PROC FREQ does not display zero-frequency levels in LIST output. When you use the SPARSE and OUT= options, PROC FREQ includes empty crosstabulation table cells in the output data set. By default, PROC FREQ does not include zero-frequency table cells in the output data set.

See the section "Missing Values" on page 114 for more information.

TESTF=(*values*)

specifies the null hypothesis frequencies for a one-way chi-square goodness-of-fit test, which you request with the CHISQ option. See the section "Chi-Square Test for One-Way Tables" on page 120 for details.

You can separate the TESTF= *values* with blanks or commas. The number of *values* must equal the number of variable levels in the one-way table. The sum of the *values* must equal the total frequency for the one-way table. List the *values* in the order in which the corresponding variable levels appear in the output. If you omit the CHISQ option, the TESTF= option invokes CHISQ.

TESTP=(*values*)

specifies the null hypothesis proportions for a one-way chi-square goodness-of-fit test, which you request with the CHISQ option. See the section "Chi-Square Test for One-Way Tables" on page 120 for details.

You can separate the TESTP= *values* with blanks or commas. The number of *values* must equal the number of variable levels in the one-way table. List the *values* in the order in which the corresponding variable levels appear in the output. You can specify *values* in probability form as numbers between 0 and 1, where the proportions sum to 1. Or you can specify *values* in percentage form as numbers between 0 and 100, where the percentages sum to 100. If you omit the CHISQ option, the TESTP= option invokes CHISQ.

TOTPCT

displays the percentage of the total multiway table frequency in crosstabulation tables for *n*-way tables, where $n > 2$. By default, PROC FREQ displays the percentage of the individual two-way table frequency but does not display the percentage of the total frequency for multiway crosstabulation tables. See the section "Multiway Tables" on page 177 for more information.

The percentage of total multiway table frequency is displayed by default when you specify the LIST option. It is also provided by default in the PERCENT variable in the OUT= output data set.

TREND

requests the Cochran-Armitage test for trend. The table must be $2 \times C$ or $R \times 2$ to compute the trend test. See the section "Cochran-Armitage Test for Trend" on page 150 for details.

TEST Statement

TEST *options* ;

The TEST statement requests asymptotic tests for measures of association and measures of agreement. You must use a TABLES statement with the TEST statement.

options

specify the statistics for which to provide asymptotic tests. Table 3.13 lists the available statistics, which include measures of association and agreement. The option names are identical to those in the TABLES and OUTPUT statements. You can request all tests for groups of statistics by using group options MEASURES or AGREE. Or you can request tests individually by using the options shown in Table 3.13.

For each measure of association or agreement that you specify, PROC FREQ provides an asymptotic test that the measure equals zero. PROC FREQ displays the asymptotic standard error under the null hypothesis, the test statistic, and the *p*-values. Additionally, PROC FREQ reports the confidence limits for the measure. The ALPHA= option in the TABLES statement determines the confidence level, which by default equals 0.05 and provides 95% confidence limits. See the sections "Asymptotic Tests" on page 125 and "Confidence Limits" on page 125 for details. Also see the section "Statistical Computations" on page 118 for information about individual measures.

You can request exact tests for selected measures of association and agreement by using the EXACT statement. See the section "Exact Statistics" on page 167 for more information.

If you use only one TABLES statement, you do not need to specify the same options in both the TABLES and TEST statements; when you specify an option in the TEST statement, PROC FREQ automatically invokes the corresponding TABLES statement option. However, when you use the TEST statement with multiple TABLES statements, you must specify options in the TABLES statements to request the desired statistics. PROC FREQ then provides asymptotic tests for those statistics that you also specify in the TEST statement.

Table 3.13 TEST Statement Options

Option	Asymptotic Tests Computed	Required TABLES Statement Option
AGREE	simple and weighted kappa coefficients	AGREE
GAMMA	gamma	ALL or MEASURES
KAPPA	simple kappa coefficient	AGREE
KENTB	Kendall's tau-b	ALL or MEASURES
MEASURES	gamma, Kendall's tau-b, Stuart's tau-c, Somers' $D(C\|R)$, Somers' $D(R\|C)$, Pearson and Spearman correlations	ALL or MEASURES
PCORR	Pearson correlation coefficient	ALL or MEASURES
SCORR	Spearman correlation coefficient	ALL or MEASURES
SMDCR	Somers' $D(C\|R)$	ALL or MEASURES
SMDRC	Somers' $D(R\|C)$	ALL or MEASURES
STUTC	Stuart's tau-c	ALL or MEASURES
WTKAP	weighted kappa coefficient	AGREE

WEIGHT Statement

WEIGHT variable < / option > ;

The WEIGHT statement names a numeric variable that provides a weight for each observation in the input data set. The WEIGHT statement is most commonly used to input cell count data. See the section "Inputting Frequency Counts" on page 112 for more information. If you use a WEIGHT statement, PROC FREQ assumes that an observation represents n observations, where n is the value of *variable*. The value of the WEIGHT *variable* is not required to be an integer.

If the value of the WEIGHT *variable* is missing, PROC FREQ does not use that observation in the analysis. If the value of the WEIGHT *variable* is zero, PROC FREQ ignores the observation unless you specify the ZEROS option, which includes observations with zero weights. If you do not specify a WEIGHT statement, each observation has a default weight of 1. The sum of the WEIGHT *variable* values represents the total number of observations.

If any value of the WEIGHT *variable* is negative, PROC FREQ displays the frequencies computed from the weighted values but does not compute percentages and statistics. If you create an output data set by using the OUT= option in the TABLES statement, PROC FREQ assigns missing values to the PERCENT variable. PROC FREQ also assigns missing values to the variables that the OUTEXPECT and OUTPCT options provide. If any value of the WEIGHT *variable* is negative, you cannot create an output data set by using the OUTPUT statement because statistics are not computed when there are negative weights.

You can specify the following option in the WEIGHT statement:

ZEROS

> includes observations with zero weight values. By default, PROC FREQ ignores observations with zero weights.
>
> If you specify the ZEROS option, frequency and and crosstabulation tables display any levels corresponding to observations with zero weights. Without the ZEROS option, PROC FREQ does not process observations with zero weights, and so does not display levels that contain only observations with zero weights.
>
> With the ZEROS option, PROC FREQ includes levels with zero weights in the chi-square goodness-of-fit test for one-way tables. Also, PROC FREQ includes any levels with zero weights in binomial computations for one-way tables. This makes it possible to compute binomial tests and estimates when the specified level contains no observations with positive weights.
>
> For two-way tables, the ZEROS option enables computation of kappa statistics when there are levels that contain no observations with positive weight. For more information, see the section "Tables with Zero Rows and Columns" on page 158.
>
> Note that even with the ZEROS option, PROC FREQ does not compute the CHISQ or MEASURES statistics for two-way tables when the table has a zero row or zero column because most of these statistics are undefined in this case.
>
> The ZEROS option invokes the SPARSE option in the TABLES statement, which includes table cells with a zero frequency count in the LIST output and in the OUT= data set. By default, without the SPARSE option, PROC FREQ does not include zero frequency cells in the LIST output or in the OUT= data set. If you specify the ZEROS option in the WEIGHT statement but do not want the SPARSE option, you can specify the NOSPARSE option in the TABLES statement.

Details: FREQ Procedure

Inputting Frequency Counts

PROC FREQ can use either raw data or cell count data to produce frequency and crosstabulation tables. *Raw data*, also known as case-record data, report the data as one record for each subject or sample member. *Cell count data* report the data as a table, listing all possible combinations of data values along with the frequency counts. This way of presenting data often appears in published results.

The following DATA step statements store raw data in a SAS data set:

```
data Raw;
   input Subject $ R C @@;
   datalines;
01 1 1   02 1 1   03 1 1   04 1 1   05 1 1
06 1 2   07 1 2   08 1 2   09 2 1   10 2 1
11 2 1   12 2 1   13 2 2   14 2 2   14 2 2
;
```

You can store the same data as cell counts by using the following DATA step statements:

```
data CellCounts;
   input R C Count @@;
   datalines;
1 1 5    1 2 3
2 1 4    2 2 3
;
```

The variable R contains the values for the rows, and the variable C contains the values for the columns. The variable Count contains the cell count for each row and column combination.

Both the Raw data set and the CellCounts data set produce identical frequency counts, two-way tables, and statistics. When using the CellCounts data set, you must include a WEIGHT statement to specify that the variable Count contains cell counts. For example, the following PROC FREQ statements create a two-way crosstabulation table by using the CellCounts data set:

```
proc freq data=CellCounts;
   tables R*C;
   weight Count;
run;
```

Grouping with Formats

PROC FREQ groups a variable's values according to its formatted values. If you assign a format to a variable with a FORMAT statement, PROC FREQ formats the variable values before dividing observations into the levels of a frequency or crosstabulation table.

For example, suppose that variable X has the values 1.1, 1.4, 1.7, 2.1, and 2.3. Each of these values appears as a level in the frequency table. If you decide to round each value to a single digit, include the following statement in the PROC FREQ step:

```
format X 1.;
```

Now the table lists the frequency count for formatted level 1 as two and for formatted level 2 as three.

PROC FREQ treats formatted character variables in the same way. The formatted values are used to group the observations into the levels of a frequency table or crosstabulation table. PROC FREQ uses the entire value of a character format to classify an observation.

You can also use the FORMAT statement to assign formats that were created with the FORMAT procedure to the variables. User-written formats determine the number of levels for a variable and provide labels for a table. If you use the same data with different formats, then you can produce frequency counts and statistics for different classifications of the variable values.

When you use PROC FORMAT to create a user-written format that combines missing and nonmissing values into one category, PROC FREQ treats the entire category of formatted values as missing. For example, a questionnaire codes 1 as yes, 2 as no, and 8 as a no answer. The following PROC FORMAT statements create a user-written format:

```
proc format;
   value Questfmt 1   ='Yes'
                  2   ='No'
                  8,. ='Missing';
run;
```

When you use a FORMAT statement to assign Questfmt. to a variable, the variable's frequency table no longer includes a frequency count for the response of 8. You must use the MISSING or MISSPRINT option in the TABLES statement to list the frequency for no answer. The frequency count for this level includes observations with either a value of 8 or a missing value (.).

The frequency or crosstabulation table lists the values of both character and numeric variables in ascending order based on internal (unformatted) variable values unless you change the order with the ORDER= option. To list the values in ascending order by formatted values, use ORDER=FORMATTED in the PROC FREQ statement.

For more information about the FORMAT statement, see *SAS Language Reference: Concepts*.

Missing Values

When the value of the WEIGHT variable is missing, PROC FREQ does not include that observation in the analysis.

PROC FREQ treats missing BY variable values like any other BY variable value. The missing values form a separate BY group.

If an observation has a missing value for a variable in a TABLES request, by default PROC FREQ does not include that observation in the frequency or crosstabulation table. Also by default, PROC FREQ does not include observations with missing values in the computation of percentages and statistics. The procedure displays the number of missing observations below each table.

PROC FREQ also reports the number of missing values in output data sets. The TABLES statement OUT= data set includes an observation that contains the missing value frequency. The NMISS option in the OUTPUT statement provides an output data set variable that contains the missing value frequency.

The following options change the way in which PROC FREQ handles missing values of TABLES variables:

MISSPRINT   displays missing value frequencies in frequency or crosstabulation tables but does not include them in computations of percentages or statistics.

MISSING   treats missing values as a valid nonmissing level for all TABLES variables. Displays missing levels in frequency and crosstabulation tables and includes them in computations of percentages and statistics.

This example shows the three ways that PROC FREQ can handle missing values of TABLES variables. The following DATA step statements create a data set with a missing value for the variable A.

```
data one;
   input A Freq;
   datalines;
1 2
2 2
. 2
;
```

The following PROC FREQ statements request a one-way frequency table for the variable A. The first request does not specify a missing value option. The second request specifies the MISSPRINT option in the TABLES statement. The third request specifies the MISSING option in the TABLES statement.

```
proc freq data=one;
   tables A;
   weight Freq;
   title 'Default';
run;
proc freq data=one;
   tables A / missprint;
   weight Freq;
   title 'MISSPRINT Option';
run;
proc freq data=one;
   tables A / missing;
   weight Freq;
   title 'MISSING Option';
run;
```

Figure 3.11 displays the frequency tables produced by this example. The first table shows PROC FREQ's default behavior for handling missing values. The observation with a missing value of the TABLES variable A is not included in the table, and the frequency of missing values is displayed below the table. The second table, for which the MISSPRINT option is specified, displays the missing observation but does not include its frequency when computing the total frequency and percentages. The third table shows that PROC FREQ treats the missing level as a valid nonmissing level when the MISSING option is specified. The table displays the missing level, and PROC FREQ includes this level when computing frequencies and percentages.

Figure 3.11 Missing Values in Frequency Tables

```
                          Default

                     The FREQ Procedure

                                    Cumulative    Cumulative
      A       Frequency    Percent   Frequency     Percent
      ---------------------------------------------------------
      1           2         50.00        2           50.00
      2           2         50.00        4          100.00

                     Frequency Missing = 2

                      MISSPRINT Option

                     The FREQ Procedure

                                    Cumulative    Cumulative
      A       Frequency    Percent   Frequency     Percent
      ---------------------------------------------------------
      .           2           .          .             .
      1           2         50.00        2           50.00
      2           2         50.00        4          100.00

                     Frequency Missing = 2

                       MISSING Option

                     The FREQ Procedure

                                    Cumulative    Cumulative
      A       Frequency    Percent   Frequency     Percent
      ---------------------------------------------------------
      .           2         33.33        2           33.33
      1           2         33.33        4           66.67
      2           2         33.33        6          100.00
```

When a combination of variable values for a two-way table is missing, PROC FREQ assigns zero to the frequency count for the table cell. By default, PROC FREQ does not display missing combinations in LIST format. Also, PROC FREQ does not include missing combinations in the OUT= output data set by default. To include missing combinations, you can specify the SPARSE option with the LIST or OUT= option in the TABLES statement.

In-Database Computation

The FREQ procedure can use in-database computation to construct frequency and crosstabulation tables when the DATA= input data set is stored as a table in a Teradata database management system (DBMS). In-database computation can provide the advantages of faster processing and reduced data transfer between the database and SAS software. For information about in-database computation, see the section "In-Database Procedures in Teradata" in *SAS/ACCESS 9.2 for Relational Databases: Reference*.

PROC FREQ performs in-database computation by using SQL implicit pass-through. The procedure generates SQL queries that are based on the tables you request in the TABLES statement. The database executes these SQL queries to construct initial summary tables, which are then transmitted to PROC FREQ. The procedure uses this summary information to perform the remaining analyses and tasks in the usual way (out of the database). So instead of transferring the entire data set over the network between the database and SAS software, the in-database method transfers only the summary tables. This can substantially reduce processing time when the dimensions of the summary tables (in terms of rows and columns) are much smaller than the dimensions of the entire database table (in terms of individual observations). Additionally, in-database summarization uses efficient parallel processing, which can also provide performance advantages.

In-database computation is controlled by the SQLGENERATION option, which you can specify in either a LIBNAME statement or an OPTIONS statement. See the section "In-Database Computing in Teradata" in *SAS/ACCESS 9.2 for Relational Databases: Reference* for details about the SQLGENERATION option and other options that affect in-database computation. By default, PROC FREQ uses in-database computation when possible. There are no FREQ procedure options that control in-database computation.

PROC FREQ uses formatted values to group observations into the levels of frequency and crosstabulation tables. See the section "Grouping with Formats" on page 113 for more information. If formats are available in the Teradata database, then in-database summarization uses the formats. If formats are not available in the Teradata database, then in-database summarization is based on the raw data values, and PROC FREQ performs the final, formatted classification (out of the database). For more information, see the section "Deploying and Using SAS Formats in Teradata" in *SAS/ACCESS 9.2 for Relational Databases: Reference*.

The order of observations is not inherently defined for DBMS tables. The following options relate to the order of observations and therefore should not be specified for PROC FREQ in-database computation:

- If you specify the FIRSTOBS= or OBS= data set option, PROC FREQ does not perform in-database computation.

- If you specify the NOTSORTED option in the BY statement, PROC FREQ in-database computation ignores it and uses the default ASCENDING order for BY variables.

- If you specify the ORDER=DATA option for input data in a DBMS table, PROC FREQ computation might produce different results for separate runs of the same analysis. In addition to determining the order of variable levels in crosstabulation table displays, the ORDER= option can also affect the values of many of the test statistics and measures that PROC FREQ computes.

Statistical Computations

Definitions and Notation

A two-way table represents the crosstabulation of row variable X and column variable Y. Let the table row values or levels be denoted by X_i, $i = 1, 2, \ldots, R$, and the column values by Y_j, $j = 1, 2, \ldots, C$. Let n_{ij} denote the frequency of the table cell in the ith row and jth column and define the following notation:

$$n_{i\cdot} = \sum_j n_{ij} \quad \text{(row totals)}$$

$$n_{\cdot j} = \sum_i n_{ij} \quad \text{(column totals)}$$

$$n = \sum_i \sum_j n_{ij} \quad \text{(overall total)}$$

$$p_{ij} = n_{ij}/n \quad \text{(cell percentages)}$$
$$p_{i\cdot} = n_{i\cdot}/n \quad \text{(row percentages of total)}$$
$$p_{\cdot j} = n_{\cdot j}/n \quad \text{(column percentages of total)}$$

$$R_i = \text{score for row } i$$
$$C_j = \text{score for column } j$$

$$\bar{R} = \sum_i n_{i\cdot} R_i / n \quad \text{(average row score)}$$

$$\bar{C} = \sum_j n_{\cdot j} C_j / n \quad \text{(average column score)}$$

$$A_{ij} = \sum_{k>i} \sum_{l>j} n_{kl} + \sum_{k<i} \sum_{l<j} n_{kl}$$

$$D_{ij} = \sum_{k>i} \sum_{l<j} n_{kl} + \sum_{k<i} \sum_{l>j} n_{kl}$$

$$P = \sum_i \sum_j n_{ij} A_{ij} \quad \text{(twice the number of concordances)}$$

$$Q = \sum_i \sum_j n_{ij} D_{ij} \quad \text{(twice the number of discordances)}$$

Scores

PROC FREQ uses scores of the variable values to compute the Mantel-Haenszel chi-square, Pearson correlation, Cochran-Armitage test for trend, weighted kappa coefficient, and Cochran-Mantel-Haenszel statistics. The SCORES= option in the TABLES statement specifies the score type that PROC FREQ uses. The available score types are TABLE, RANK, RIDIT, and MODRIDIT scores. The default score type is TABLE. Using MODRIDIT, RANK, or RIDIT scores yields nonparametric analyses.

For numeric variables, table scores are the values of the row and column levels. If the row or column variable is formatted, then the table score is the internal numeric value corresponding to that level. If two or more numeric values are classified into the same formatted level, then the internal numeric value for that level is the smallest of these values. For character variables, table scores are defined as the row numbers and column numbers (that is, 1 for the first row, 2 for the second row, and so on).

Rank scores, which you request with the SCORES=RANK option, are defined as

$$R1_i = \sum_{k<i} n_{k.} + (n_{i.} + 1)/2 \quad i = 1, 2, \ldots, R$$

$$C1_j = \sum_{l<j} n_{.l} + (n_{.j} + 1)/2 \quad j = 1, 2, \ldots, C$$

where $R1_i$ is the rank score of row i, and $C1_j$ is the rank score of column j. Note that rank scores yield midranks for tied values.

Ridit scores, which you request with the SCORES=RIDIT option, are defined as rank scores standardized by the sample size (Bross 1958, Mack and Skillings 1980). Ridit scores are derived from the rank scores as

$$R2_i = R1_i/n \quad i = 1, 2, \ldots, R$$
$$C2_j = C1_j/n \quad j = 1, 2, \ldots, C$$

Modified ridit scores (SCORES=MODRIDIT) represent the expected values of the order statistics of the uniform distribution on (0,1) (van Elteren 1960, Lehmann 1975). Modified ridit scores are derived from rank scores as

$$R3_i = R1_i/(n+1) \quad i = 1, 2, \ldots, R$$
$$C3_j = C1_j/(n+1) \quad j = 1, 2, \ldots, C$$

Chi-Square Tests and Statistics

The CHISQ option provides chi-square tests of homogeneity or independence and measures of association based on the chi-square statistic. When you specify the CHISQ option in the TABLES statement, PROC FREQ computes the following chi-square tests for each two-way table: the Pearson chi-square, likelihood-ratio chi-square, and Mantel-Haenszel chi-square. PROC FREQ provides the following measures of association based on the Pearson chi-square statistic: the phi coefficient, contingency coefficient, and Cramer's V. For 2×2 tables, the CHISQ option also provides Fisher's exact test and the continuity-adjusted chi-square. You can request Fisher's exact test for general $R \times C$ tables by specifying the FISHER option in the TABLES or EXACT statement.

For one-way frequency tables, the CHISQ option provides a chi-square goodness-of-fit test. The other chi-square tests and statistics described in this section are computed only for two-way tables.

All of the two-way test statistics described in this section test the null hypothesis of no association between the row variable and the column variable. When the sample size n is large, these test statistics have an asymptotic chi-square distribution when the null hypothesis is true. When the sample size is not large, exact tests might be useful. PROC FREQ provides exact tests for the

Pearson chi-square, the likelihood-ratio chi-square, and the Mantel-Haenszel chi-square (in addition to Fisher's exact test). PROC FREQ also provides an exact chi-square goodness-of-fit test for one-way tables. You can request these exact tests by specifying the corresponding options in the EXACT statement. See the section "Exact Statistics" on page 167 for more information.

Note that the Mantel-Haenszel chi-square statistic is appropriate only when both variables lie on an ordinal scale. The other chi-square tests and statistics in this section are appropriate for either nominal or ordinal variables. The following sections give the formulas that PROC FREQ uses to compute the chi-square tests and statistics. See Agresti (2007), Stokes, Davis, and Koch (2000), and the other references cited for each statistic for more information.

Chi-Square Test for One-Way Tables

For one-way frequency tables, the CHISQ option in the TABLES statement provides a chi-square goodness-of-fit test. Let C denote the number of classes, or levels, in the one-way table. Let f_i denote the frequency of class i (or the number of observations in class i) for $i = 1, 2, \ldots, C$. Then PROC FREQ computes the one-way chi-square statistic as

$$Q_P = \sum_{i=1}^{C} \frac{(f_i - e_i)^2}{e_i}$$

where e_i is the expected frequency for class i under the null hypothesis.

In the test for equal proportions, which is the default for the CHISQ option, the null hypothesis specifies equal proportions of the total sample size for each class. Under this null hypothesis, the expected frequency for each class equals the total sample size divided by the number of classes,

$$e_i = n/C \quad \text{for} \quad i = 1, 2, \ldots, C$$

In the test for specified frequencies, which PROC FREQ computes when you input null hypothesis frequencies by using the TESTF= option, the expected frequencies are the TESTF= values that you specify. In the test for specified proportions, which PROC FREQ computes when you input null hypothesis proportions by using the TESTP= option, the expected frequencies are determined from the specified TESTP= proportions p_i as

$$e_i = p_i \times n \quad \text{for} \quad i = 1, 2, \ldots, C$$

Under the null hypothesis (of equal proportions, specified frequencies, or specified proportions), Q_P has an asymptotic chi-square distribution with $C - 1$ degrees of freedom.

In addition to the asymptotic test, you can request an exact one-way chi-square test by specifying the CHISQ option in the EXACT statement. See the section "Exact Statistics" on page 167 for more information.

Pearson Chi-Square Test for Two-Way Tables

The Pearson chi-square for two-way tables involves the differences between the observed and expected frequencies, where the expected frequencies are computed under the null hypothesis of independence. The Pearson chi-square statistic is computed as

$$Q_P = \sum_i \sum_j \frac{(n_{ij} - e_{ij})^2}{e_{ij}}$$

where n_{ij} is the observed frequency in table cell (i, j) and e_{ij} is the expected frequency for table cell (i, j). The expected frequency is computed under the null hypothesis that the row and column variables are independent,

$$e_{ij} = \frac{n_{i.} \cdot n_{.j}}{n}$$

When the row and column variables are independent, Q_P has an asymptotic chi-square distribution with $(R-1)(C-1)$ degrees of freedom. For large values of Q_P, this test rejects the null hypothesis in favor of the alternative hypothesis of general association.

In addition to the asymptotic test, you can request an exact Pearson chi-square test by specifying the PCHI or CHISQ option in the EXACT statement. See the section "Exact Statistics" on page 167 for more information.

For 2 × 2 tables, the Pearson chi-square is also appropriate for testing the equality of two binomial proportions. For $R \times 2$ and $2 \times C$ tables, the Pearson chi-square tests the homogeneity of proportions. See Fienberg (1980) for details.

Likelihood-Ratio Chi-Square Test

The likelihood-ratio chi-square involves the ratios between the observed and expected frequencies. The likelihood-ratio chi-square statistic is computed as

$$G^2 = 2 \sum_i \sum_j n_{ij} \ln \left(\frac{n_{ij}}{e_{ij}} \right)$$

where n_{ij} is the observed frequency in table cell (i, j) and e_{ij} is the expected frequency for table cell (i, j).

When the row and column variables are independent, G^2 has an asymptotic chi-square distribution with $(R-1)(C-1)$ degrees of freedom.

In addition to the asymptotic test, you can request an exact likelihood-ratio chi-square test by specifying the LRCHI or CHISQ option in the EXACT statement. See the section "Exact Statistics" on page 167 for more information.

Continuity-Adjusted Chi-Square Test

The continuity-adjusted chi-square for 2 × 2 tables is similar to the Pearson chi-square, but it is adjusted for the continuity of the chi-square distribution. The continuity-adjusted chi-square is most useful for small sample sizes. The use of the continuity adjustment is somewhat controversial; this chi-square test is more conservative (and more like Fisher's exact test) when the sample size is small. As the sample size increases, the continuity-adjusted chi-square becomes more like the Pearson chi-square.

The continuity-adjusted chi-square statistic is computed as

$$Q_C = \sum_i \sum_j \frac{\left(\max(0, |n_{ij} - e_{ij}| - 0.5)\right)^2}{e_{ij}}$$

Under the null hypothesis of independence, Q_C has an asymptotic chi-square distribution with $(R-1)(C-1)$ degrees of freedom.

Mantel-Haenszel Chi-Square Test

The Mantel-Haenszel chi-square statistic tests the alternative hypothesis that there is a linear association between the row variable and the column variable. Both variables must lie on an ordinal scale. The Mantel-Haenszel chi-square statistic is computed as

$$Q_{MH} = (n-1)r^2$$

where r^2 is the Pearson correlation between the row variable and the column variable. For a description of the Pearson correlation, see the "Pearson Correlation Coefficient" on page 128. The Pearson correlation and thus the Mantel-Haenszel chi-square statistic use the scores that you specify in the SCORES= option in the TABLES statement. See Mantel and Haenszel (1959) and Landis, Heyman, and Koch (1978) for more information.

Under the null hypothesis of no association, Q_{MH} has an asymptotic chi-square distribution with one degree of freedom.

In addition to the asymptotic test, you can request an exact Mantel-Haenszel chi-square test by specifying the MHCHI or CHISQ option in the EXACT statement. See the section "Exact Statistics" on page 167 for more information.

Fisher's Exact Test

Fisher's exact test is another test of association between the row and column variables. This test assumes that the row and column totals are fixed, and then uses the hypergeometric distribution to compute probabilities of possible tables conditional on the observed row and column totals. Fisher's exact test does not depend on any large-sample distribution assumptions, and so it is appropriate even for small sample sizes and for sparse tables.

2 × 2 Tables For 2 × 2 tables, PROC FREQ gives the following information for Fisher's exact test: table probability, two-sided p-value, left-sided p-value, and right-sided p-value. The table probability equals the hypergeometric probability of the observed table, and is in fact the value of the test statistic for Fisher's exact test.

Where p is the hypergeometric probability of a specific table with the observed row and column totals, Fisher's exact p-values are computed by summing probabilities p over defined sets of tables,

$$PROB = \sum_A p$$

The two-sided p-value is the sum of all possible table probabilties (conditional on the observed row and column totals) that are less than or equal to the observed table probability. For the two-sided

p-value, the set A includes all possible tables with hypergeometric probabilities less than or equal to the probability of the observed table. A small two-sided p-value supports the alternative hypothesis of association between the row and column variables.

For 2×2 tables, one-sided p-values for Fisher's exact test are defined in terms of the frequency of the cell in the first row and first column of the table, the (1,1) cell. Denoting the observed (1,1) cell frequency by n_{11}, the left-sided p-value for Fisher's exact test is the probability that the (1,1) cell frequency is less than or equal to n_{11}. For the left-sided p-value, the set A includes those tables with a (1,1) cell frequency less than or equal to n_{11}. A small left-sided p-value supports the alternative hypothesis that the probability of an observation being in the first cell is actually less than expected under the null hypothesis of independent row and column variables.

Similarly, for a right-sided alternative hypothesis, A is the set of tables where the frequency of the (1,1) cell is greater than or equal to that in the observed table. A small right-sided p-value supports the alternative that the probability of the first cell is actually greater than that expected under the null hypothesis.

Because the (1,1) cell frequency completely determines the 2×2 table when the marginal row and column sums are fixed, these one-sided alternatives can be stated equivalently in terms of other cell probabilities or ratios of cell probabilities. The left-sided alternative is equivalent to an odds ratio less than 1, where the odds ratio equals $(n_{11}n_{22}/n_{12}n_{21})$. Additionally, the left-sided alternative is equivalent to the column 1 risk for row 1 being less than the column 1 risk for row 2, $p_{1|1} < p_{1|2}$. Similarly, the right-sided alternative is equivalent to the column 1 risk for row 1 being greater than the column 1 risk for row 2, $p_{1|1} > p_{1|2}$. See Agresti (2007) for details.

R × C Tables Fisher's exact test was extended to general $R \times C$ tables by Freeman and Halton (1951), and this test is also known as the Freeman-Halton test. For $R \times C$ tables, the two-sided p-value definition is the same as for 2×2 tables. The set A contains all tables with p less than or equal to the probability of the observed table. A small p-value supports the alternative hypothesis of association between the row and column variables. For $R \times C$ tables, Fisher's exact test is inherently two-sided. The alternative hypothesis is defined only in terms of general, and not linear, association. Therefore, Fisher's exact test does not have right-sided or left-sided p-values for general $R \times C$ tables.

For $R \times C$ tables, PROC FREQ computes Fisher's exact test by using the network algorithm of Mehta and Patel (1983), which provides a faster and more efficient solution than direct enumeration. See the section "Exact Statistics" on page 167 for more details.

Phi Coefficient

The phi coefficient is a measure of association derived from the Pearson chi-square. The range of the phi coefficient is $-1 \leq \phi \leq 1$ for 2×2 tables. For tables larger than 2×2, the range is $0 \leq \phi \leq \min(\sqrt{R-1}, \sqrt{C-1})$ (Liebetrau 1983). The phi coefficient is computed as

$$\phi = (n_{11}n_{22} - n_{12}n_{21}) / \sqrt{n_{1\cdot}n_{2\cdot}n_{\cdot 1}n_{\cdot 2}} \quad \text{for } 2 \times 2 \text{ tables}$$

$$\phi = \sqrt{Q_P/n} \quad \text{otherwise}$$

See Fleiss, Levin, and Paik (2003, pp. 98–99) for more information.

Contingency Coefficient

The contingency coefficient is a measure of association derived from the Pearson chi-square. The range of the contingency coefficient is $0 \leq P \leq \sqrt{(m-1)/m}$, where $m = \min(R, C)$ (Liebetrau 1983). The contingency coefficient is computed as

$$P = \sqrt{Q_P / (Q_P + n)}$$

See Kendall and Stuart (1979, pp. 587–588) for more information.

Cramer's V

Cramer's V is a measure of association derived from the Pearson chi-square. It is designed so that the attainable upper bound is always 1. The range of Cramer's V is $-1 \leq V \leq 1$ for 2×2 tables; for tables larger than 2×2, the range is $0 \leq V \leq 1$. Cramer's V is computed as

$$V = \phi \quad \text{for } 2 \times 2 \text{ tables}$$

$$V = \sqrt{\frac{Q_P/n}{\min(R-1, C-1)}} \quad \text{otherwise}$$

See Kendall and Stuart (1979, p. 588) for more information.

Measures of Association

When you specify the MEASURES option in the TABLES statement, PROC FREQ computes several statistics that describe the association between the row and column variables of the contingency table. The following are measures of ordinal association that consider whether the column variable Y tends to increase as the row variable X increases: gamma, Kendall's tau-*b*, Stuart's tau-*c*, and Somers' D. These measures are appropriate for ordinal variables, and they classify pairs of observations as *concordant* or *discordant*. A pair is concordant if the observation with the larger value of X also has the larger value of Y. A pair is discordant if the observation with the larger value of X has the smaller value of Y. See Agresti (2007) and the other references cited for the individual measures of association.

The Pearson correlation coefficient and the Spearman rank correlation coefficient are also appropriate for ordinal variables. The Pearson correlation describes the strength of the linear association between the row and column variables, and it is computed by using the row and column scores specified by the SCORES= option in the TABLES statement. The Spearman correlation is computed with rank scores. The polychoric correlation (requested by the PLCORR option) also requires ordinal variables and assumes that the variables have an underlying bivariate normal distribution. The following measures of association do not require ordinal variables and are appropriate for nominal variables: lambda asymmetric, lambda symmetric, and the uncertainty coefficients.

PROC FREQ computes estimates of the measures according to the formulas given in the following sections. For each measure, PROC FREQ computes an asymptotic standard error (*ASE*), which is the square root of the asymptotic variance denoted by var in the following sections.

Confidence Limits

If you specify the CL option in the TABLES statement, PROC FREQ computes asymptotic confidence limits for all MEASURES statistics. The confidence coefficient is determined according to the value of the ALPHA= option, which, by default, equals 0.05 and produces 95% confidence limits.

The confidence limits are computed as

$$est \pm (z_{\alpha/2} \times ASE)$$

where *est* is the estimate of the measure, $z_{\alpha/2}$ is the $100(1 - \alpha/2)$th percentile of the standard normal distribution, and *ASE* is the asymptotic standard error of the estimate.

Asymptotic Tests

For each measure that you specify in the TEST statement, PROC FREQ computes an asymptotic test of the null hypothesis that the measure equals zero. Asymptotic tests are available for the following measures of association: gamma, Kendall's tau-*b*, Stuart's tau-*c*, Somers' $D(R|C)$, Somers' $D(C|R)$, the Pearson correlation coefficient, and the Spearman rank correlation coefficient. To compute an asymptotic test, PROC FREQ uses a standardized test statistic z, which has an asymptotic standard normal distribution under the null hypothesis. The test statistic is computed as

$$z = est / \sqrt{var_0(est)}$$

where *est* is the estimate of the measure and $var_0(est)$ is the variance of the estimate under the null hypothesis. Formulas for $var_0(est)$ for the individual measures of association are given in the following sections.

Note that the ratio of *est* to $\sqrt{var_0(est)}$ is the same for the following measures: gamma, Kendall's tau-*b*, Stuart's tau-*c*, Somers' $D(R|C)$, and Somers' $D(C|R)$. Therefore, the tests for these measures are identical. For example, the *p*-values for the test of H_0: gamma $= 0$ equal the *p*-values for the test of H_0: tau-$b = 0$.

PROC FREQ computes one-sided and two-sided *p*-values for each of these tests. When the test statistic z is greater than its null hypothesis expected value of zero, PROC FREQ displays the right-sided *p*-value, which is the probability of a larger value of the statistic occurring under the null hypothesis. A small right-sided *p*-value supports the alternative hypothesis that the true value of the measure is greater than zero. When the test statistic is less than or equal to zero, PROC FREQ displays the left-sided *p*-value, which is the probability of a smaller value of the statistic occurring under the null hypothesis. A small left-sided *p*-value supports the alternative hypothesis that the true value of the measure is less than zero. The one-sided *p*-value P_1 can be expressed as

$$P_1 = \begin{cases} \text{Prob}(Z > z) & \text{if } z > 0 \\ \text{Prob}(Z < z) & \text{if } z \leq 0 \end{cases}$$

where Z has a standard normal distribution. The two-sided *p*-value P_2 is computed as

$$P_2 = \text{Prob}(|Z| > |z|)$$

Exact Tests

Exact tests are available for two measures of association: the Pearson correlation coefficient and the Spearman rank correlation coefficient. If you specify the PCORR option in the EXACT statement, PROC FREQ computes the exact test of the hypothesis that the Pearson correlation equals zero. If you specify the SCORR option in the EXACT statement, PROC FREQ computes the exact test of the hypothesis that the Spearman correlation equals zero. See the section "Exact Statistics" on page 167 for more information.

Gamma

The gamma (Γ) statistic is based only on the number of concordant and discordant pairs of observations. It ignores tied pairs (that is, pairs of observations that have equal values of X or equal values of Y). Gamma is appropriate only when both variables lie on an ordinal scale. The range of gamma is $-1 \leq \Gamma \leq 1$. If the row and column variables are independent, then gamma tends to be close to zero. Gamma is estimated by

$$G = (P - Q) / (P + Q)$$

and the asymptotic variance is

$$\text{var}(G) = \frac{16}{(P+Q)^4} \sum_i \sum_j n_{ij}(QA_{ij} - PD_{ij})^2$$

For 2×2 tables, gamma is equivalent to Yule's Q. See Goodman and Kruskal (1979) and Agresti (2002) for more information.

The variance under the null hypothesis that gamma equals zero is computed as

$$\text{var}_0(G) = \frac{4}{(P+Q)^2} \left(\sum_i \sum_j n_{ij}(A_{ij} - D_{ij})^2 - (P-Q)^2/n \right)$$

See Brown and Benedetti (1977) for details.

Kendall's Tau-b

Kendall's tau-b (τ_b) is similar to gamma except that tau-b uses a correction for ties. Tau-b is appropriate only when both variables lie on an ordinal scale. The range of tau-b is $-1 \leq \tau_b \leq 1$. Kendall's tau-b is estimated by

$$t_b = (P - Q) / \sqrt{w_r w_c}$$

and the asymptotic variance is

$$\text{var}(t_b) = \frac{1}{w^4} \left(\sum_i \sum_j n_{ij}(2w d_{ij} + t_b v_{ij})^2 - n^3 t_b^2 (w_r + w_c)^2 \right)$$

where

$$w = \sqrt{w_r w_c}$$
$$w_r = n^2 - \sum_i n_{i\cdot}^2$$
$$w_c = n^2 - \sum_j n_{\cdot j}^2$$
$$d_{ij} = A_{ij} - D_{ij}$$
$$v_{ij} = n_{i\cdot} w_c + n_{\cdot j} w_r$$

See Kendall (1955) for more information.

The variance under the null hypothesis that tau-b equals zero is computed as

$$\text{var}_0(t_b) = \frac{4}{w_r w_c} \left(\sum_i \sum_j n_{ij}(A_{ij} - D_{ij})^2 - (P-Q)^2/n \right)$$

See Brown and Benedetti (1977) for details.

Stuart's Tau-c

Stuart's tau-c (τ_c) makes an adjustment for table size in addition to a correction for ties. Tau-c is appropriate only when both variables lie on an ordinal scale. The range of tau-c is $-1 \leq \tau_c \leq 1$. Stuart's tau-c is estimated by

$$t_c = m(P-Q) / n^2(m-1)$$

and the asymptotic variance is

$$\text{var}(t_c) = \frac{4m^2}{(m-1)^2 n^4} \left(\sum_i \sum_j n_{ij} d_{ij}^2 - (P-Q)^2/n \right)$$

where $m = \min(R,C)$ and $d_{ij} = A_{ij} - D_{ij}$. The variance under the null hypothesis that tau-c equals zero is the same as the asymptotic variance var,

$$\text{var}_0(t_c) = \text{var}(t_c)$$

See Brown and Benedetti (1977) for details.

Somers' D

Somers' $D(C|R)$ and Somers' $D(R|C)$ are asymmetric modifications of tau-b. $C|R$ indicates that the row variable X is regarded as the independent variable and the column variable Y is regarded as dependent. Similarly, $R|C$ indicates that the column variable Y is regarded as the independent variable and the row variable X is regarded as dependent. Somers' D differs from tau-b in that it uses a correction only for pairs that are tied on the independent variable. Somers' D is appropriate only when both variables lie on an ordinal scale. The range of Somers' D is $-1 \leq D \leq 1$. Somers' $D(C|R)$ is computed as

$$D(C|R) = (P - Q) / w_r$$

and its asymptotic variance is

$$\text{var}(D(C|R)) = \frac{4}{w_r^4} \sum_i \sum_j n_{ij} (w_r d_{ij} - (P - Q)(n - n_{i.}))^2$$

where $d_{ij} = A_{ij} - D_{ij}$ and

$$w_r = n^2 - \sum_i n_{i.}^2$$

See Somers (1962), Goodman and Kruskal (1979), and Liebetrau (1983) for more information.

The variance under the null hypothesis that $D(C|R)$ equals zero is computed as

$$\text{var}_0(D(C|R)) = \frac{4}{w_r^2} \left(\sum_i \sum_j n_{ij} (A_{ij} - D_{ij})^2 - (P - Q)^2/n \right)$$

See Brown and Benedetti (1977) for details.

Formulas for Somers' $D(R|C)$ are obtained by interchanging the indices.

Pearson Correlation Coefficient

The Pearson correlation coefficient (ρ) is computed by using the scores specified in the SCORES= option. This measure is appropriate only when both variables lie on an ordinal scale. The range of the Pearson correlation is $-1 \leq \rho \leq 1$. The Pearson correlation coefficient is estimated by

$$r = v/w = ss_{rc}/\sqrt{ss_r ss_c}$$

and its asymptotic variance is

$$\text{var}(r) = \frac{1}{w^4} \sum_i \sum_j n_{ij} \left(w(R_i - \bar{R})(C_j - \bar{C}) - \frac{b_{ij} v}{2w} \right)^2$$

where R_i and C_j are the row and column scores and

$$ss_r = \sum_i \sum_j n_{ij} (R_i - \bar{R})^2$$

$$ss_c = \sum_i \sum_j n_{ij} (C_j - \bar{C})^2$$

$$ss_{rc} = \sum_i \sum_j n_{ij} (R_i - \bar{R})(C_j - \bar{C})$$

$$b_{ij} = (R_i - \bar{R})^2 ss_c + (C_j - \bar{C})^2 ss_r$$

$$v = ss_{rc}$$

$$w = \sqrt{ss_r ss_c}$$

See Snedecor and Cochran (1989) for more information.

The SCORES= option in the TABLES statement determines the type of row and column scores used to compute the Pearson correlation (and other score-based statistics). The default is SCORES=TABLE. See the section "Scores" on page 118 for details about the available score types and how they are computed.

The variance under the null hypothesis that the correlation equals zero is computed as

$$\text{var}_0(r) = \left(\sum_i \sum_j n_{ij}(R_i - \bar{R})^2 (C_j - \bar{C})^2 - ss_{rc}^2/n \right) / ss_r ss_c$$

Note that this expression for the variance is derived for multinomial sampling in a contingency table framework, and it differs from the form obtained under the assumption that both variables are continuous and normally distributed. See Brown and Benedetti (1977) for details.

PROC FREQ also provides an exact test for the Pearson correlation coefficient. You can request this test by specifying the PCORR option in the EXACT statement. See the section "Exact Statistics" on page 167 for more information.

Spearman Rank Correlation Coefficient

The Spearman correlation coefficient (ρ_s) is computed by using rank scores, which are defined in the section "Scores" on page 118. This measure is appropriate only when both variables lie on an ordinal scale. The range of the Spearman correlation is $-1 \leq \rho_s \leq 1$. The Spearman correlation coefficient is estimated by

$$r_s = v / w$$

and its asymptotic variance is

$$\text{var}(r_s) = \frac{1}{n^2 w^4} \sum_i \sum_j n_{ij} (z_{ij} - \bar{z})^2$$

where $R1_i$ and $C1_j$ are the row and column rank scores and

$$v = \sum_i \sum_j n_{ij} R(i) C(j)$$

$$w = \frac{1}{12} \sqrt{FG}$$

$$F = n^3 - \sum_i n_{i\cdot}^3$$

$$G = n^3 - \sum_j n_{\cdot j}^3$$

$$R(i) = R1_i - n/2$$

$$C(j) = C1_j - n/2$$

$$\bar{z} = \frac{1}{n}\sum_i\sum_j n_{ij} z_{ij}$$

$$z_{ij} = wv_{ij} - vw_{ij}$$

$$v_{ij} = n\left(R(i)C(j) + \frac{1}{2}\sum_l n_{il}C(l) + \frac{1}{2}\sum_k n_{kj}R(k) + \sum_l\sum_{k>i} n_{kl}C(l) + \sum_k\sum_{l>j} n_{kl}R(k)\right)$$

$$w_{ij} = \frac{-n}{96w}\left(Fn_{\cdot j}^2 + Gn_{i\cdot}^2\right)$$

See Snedecor and Cochran (1989) for more information.

The variance under the null hypothesis that the correlation equals zero is computed as

$$\text{var}_0(r_s) = \frac{1}{n^2 w^2}\sum_i\sum_j n_{ij}(v_{ij} - \bar{v})^2$$

where

$$\bar{v} = \sum_i\sum_j n_{ij} v_{ij}/n$$

Note that the asymptotic variance is derived for multinomial sampling in a contingency table framework, and it differs from the form obtained under the assumption that both variables are continuous and normally distributed. See Brown and Benedetti (1977) for details.

PROC FREQ also provides an exact test for the Spearman correlation coefficient. You can request this test by specifying the SCORR option in the EXACT statement. See the section "Exact Statistics" on page 167 for more information.

Polychoric Correlation

When you specify the PLCORR option in the TABLES statement, PROC FREQ computes the polychoric correlation. This measure of association is based on the assumption that the ordered, categorical variables of the frequency table have an underlying bivariate normal distribution. For 2×2 tables, the polychoric correlation is also known as the tetrachoric correlation. See Drasgow (1986) for an overview of polychoric correlation. The polychoric correlation coefficient is the maximum likelihood estimate of the product-moment correlation between the normal variables, estimating thresholds from the observed table frequencies. The range of the polychoric correlation is from −1 to 1. Olsson (1979) gives the likelihood equations and an asymptotic covariance matrix for the estimates.

To estimate the polychoric correlation, PROC FREQ iteratively solves the likelihood equations by a Newton-Raphson algorithm that uses the Pearson correlation coefficient as the initial approximation. Iteration stops when the convergence measure falls below the convergence criterion or when the maximum number of iterations is reached, whichever occurs first. The CONVERGE= option sets the convergence criterion, and the default value is 0.0001. The MAXITER= option sets the maximum number of iterations, and the default value is 20.

Lambda (Asymmetric)

Asymmetric lambda, $\lambda(C|R)$, is interpreted as the probable improvement in predicting the column variable Y given knowledge of the row variable X. The range of asymmetric lambda is $0 \leq \lambda(C|R) \leq 1$. Asymmetric lambda $(C|R)$ is computed as

$$\lambda(C|R) = \frac{\sum_i r_i - r}{n - r}$$

and its asymptotic variance is

$$\text{var}(\lambda(C|R)) = \frac{n - \sum_i r_i}{(n-r)^3} \left(\sum_i r_i + r - 2 \sum_i (r_i \mid l_i = l) \right)$$

where

$$r_i = \max_j (n_{ij})$$

$$r = \max_j (n_{.j})$$

$$c_j = \max_i (n_{ij})$$

$$c = \max_i (n_{i.})$$

The values of l_i and l are determined as follows. Denote by l_i the unique value of j such that $r_i = n_{ij}$, and let l be the unique value of j such that $r = n_{.j}$. Because of the uniqueness assumptions, ties in the frequencies or in the marginal totals must be broken in an arbitrary but consistent manner. In case of ties, l is defined as the smallest value of j such that $r = n_{.j}$.

For those columns containing a cell (i, j) for which $n_{ij} = r_i = c_j$, cs_j records the row in which c_j is assumed to occur. Initially cs_j is set equal to -1 for all j. Beginning with $i = 1$, if there is at least one value j such that $n_{ij} = r_i = c_j$, and if $cs_j = -1$, then l_i is defined to be the smallest such value of j, and cs_j is set equal to i. Otherwise, if $n_{il} = r_i$, then l_i is defined to be equal to l. If neither condition is true, then l_i is taken to be the smallest value of j such that $n_{ij} = r_i$.

The formulas for lambda asymmetric $(R|C)$ can be obtained by interchanging the indices.

See Goodman and Kruskal (1979) for more information.

Lambda (Symmetric)

The nondirectional lambda is the average of the two asymmetric lambdas, $\lambda(C|R)$ and $\lambda(R|C)$. Its range is $0 \leq \lambda \leq 1$. Lambda symmetric is computed as

$$\lambda = \frac{\sum_i r_i + \sum_j c_j - r - c}{2n - r - c} = \frac{w - v}{w}$$

and its asymptotic variance is computed as

$$\mathrm{var}(\lambda) = \frac{1}{w^4}\left(wvy - 2w^2\left(n - \sum_i\sum_j(n_{ij} \mid j = l_i, i = k_j)\right) - 2v^2(n - n_{kl})\right)$$

where

$$r_i = \max_j(n_{ij})$$

$$r = \max_j(n_{.j})$$

$$c_j = \max_i(n_{ij})$$

$$c = \max_i(n_{i.})$$

$$w = 2n - r - c$$

$$v = 2n - \sum_i r_i - \sum_j c_j$$

$$x = \sum_i (r_i \mid l_i = l) + \sum_j (c_j \mid k_j = k) + r_k + c_l$$

$$y = 8n - w - v - 2x$$

The definitions of l_i and l are given in the previous section. The values k_j and k are defined in a similar way for lambda asymmetric $(R|C)$.

See Goodman and Kruskal (1979) for more information.

Uncertainty Coefficients (Asymmetric)

The uncertainty coefficient $U(C|R)$ measures the proportion of uncertainty (entropy) in the column variable Y that is explained by the row variable X. Its range is $0 \leq U(C|R) \leq 1$. The uncertainty coefficient is computed as

$$U(C|R) = (H(X) + H(Y) - H(XY)) / H(Y) = v/w$$

and its asymptotic variance is

$$\mathrm{var}(U(C|R)) = \frac{1}{n^2 w^4}\sum_i\sum_j n_{ij}\left(H(Y)\ln\left(\frac{n_{ij}}{n_{i.}}\right) + (H(X) - H(XY))\ln\left(\frac{n_{.j}}{n}\right)\right)^2$$

where

$$v = H(X) + H(Y) - H(XY)$$

$$w = H(Y)$$

$$H(X) = -\sum_i \left(\frac{n_{i\cdot}}{n}\right) \ln\left(\frac{n_{i\cdot}}{n}\right)$$

$$H(Y) = -\sum_j \left(\frac{n_{\cdot j}}{n}\right) \ln\left(\frac{n_{\cdot j}}{n}\right)$$

$$H(XY) = -\sum_i \sum_j \left(\frac{n_{ij}}{n}\right) \ln\left(\frac{n_{ij}}{n}\right)$$

The formulas for the uncertainty coefficient $U(R|C)$ can be obtained by interchanging the indices. See Theil (1972, pp. 115–120) and Goodman and Kruskal (1979) for more information.

Uncertainty Coefficient (Symmetric)

The uncertainty coefficient U is the symmetric version of the two asymmetric uncertainty coefficients. Its range is $0 \leq U \leq 1$. The uncertainty coefficient is computed as

$$U = 2\left(H(X) + H(Y) - H(XY)\right) / (H(X) + H(Y))$$

and its asymptotic variance is

$$\text{var}(U) = 4 \sum_i \sum_j \frac{n_{ij} \left(H(XY) \ln\left(\frac{n_{i\cdot} n_{\cdot j}}{n^2}\right) - (H(X) + H(Y)) \ln\left(\frac{n_{ij}}{n}\right)\right)^2}{n^2 (H(X) + H(Y))^4}$$

where $H(X)$, $H(Y)$, and $H(XY)$ are defined in the previous section. See Goodman and Kruskal (1979) for more information.

Binomial Proportion

If you specify the BINOMIAL option in the TABLES statement, PROC FREQ computes the binomial proportion for one-way tables. By default, this is the proportion of observations in the first variable level that appears in the output. (You can use the LEVEL= option to specify a different level for the proportion.) The binomial proportion is computed as

$$\hat{p} = n_1 / n$$

where n_1 is the frequency of the first (or designated) level and n is the total frequency of the one-way table. The standard error of the binomial proportion is computed as

$$se(\hat{p}) = \sqrt{\hat{p}(1-\hat{p})/n}$$

Confidence Limits

By default, PROC FREQ provides asymptotic and exact (Clopper-Pearson) confidence limits for the binomial proportion. If you do not specify any confidence limit requests with *binomial-options*, PROC FREQ computes the standard Wald asymptotic confidence limits. You can also request Agresti-Coull, Jeffreys, and Wilson (score) confidence limits for the binomial proportion. See Brown, Cai, and DasGupta (2001), Agresti and Coull (1998), and Newcombe (1998) for details about these binomial confidence limits, including comparisons of their performance.

Wald Confidence Limits The standard Wald asymptotic confidence limits are based on the normal approximation to the binomial distribution. PROC FREQ computes the Wald confidence limits for the binomial proportion as

$$\hat{p} \pm (z_{\alpha/2} \times \text{se}(\hat{p}))$$

where $z_{\alpha/2}$ is the $100(1 - \alpha/2)$th percentile of the standard normal distribution. The confidence level α is determined by the ALPHA= option, which, by default, equals 0.05 and produces 95% confidence limits.

If you specify the CORRECT binomial-option or the BINOMIALC option, PROC FREQ includes a continuity correction of $1/2n$ in the Wald asymptotic confidence limits. The purpose of this correction is to adjust for the difference between the normal approximation and the binomial distribution, which is a discrete distribution. See Fleiss, Levin, and Paik (2003) for more information. With the continuity correction, the asymptotic confidence limits for the binomial proportion are computed as

$$\hat{p} \pm (z_{\alpha/2} \times \text{se}(\hat{p}) + (1/2n))$$

Agresti-Coull Confidence Limits If you specify the AGRESTICOULL binomial-option, PROC FREQ computes Agresti-Coull confidence limits for the binomial proportion as

$$\tilde{p} \pm (z_{\alpha/2} \times \sqrt{\tilde{p}(1-\tilde{p})/\tilde{n}})$$

where

$$\tilde{n}_1 = n_1 + (z_{\alpha/2})/2$$
$$\tilde{n} = n + z_{\alpha/2}^2$$
$$\tilde{p} = \tilde{n}_1 / \tilde{n}$$

The Agresti-Coull confidence interval has the same basic form as the standard Wald interval but uses $\tilde{p}$ in place of $\hat{p}$. For $\alpha = 0.05$, the value of $z_{\alpha/2}$ is close to 2, and this interval is the "add 2 successes and 2 failures" adjusted Wald interval in Agresti and Coull (1998).

Jeffreys Confidence Limits If you specify the JEFFREYS binomial-option, PROC FREQ computes the Jeffreys confidence limits for the binomial proportion as

$$(\beta(\alpha/2, n_1 + 1/2, n - n_1 + 1/2), \beta(1 - \alpha/2, n_1 + 1/2, n - n_1 + 1/2))$$

where $\beta(\alpha, b, c)$ is the αth percentile of the beta distribution with shape parameters b and c. The lower confidence limit is set to 0 when $n_1 = 0$, and the upper confidence limit is set to 1 when $n_1 = n$. This is an equal-tailed interval based on the noninformative Jeffreys prior for a binomial proportion. See Brown, Cai, and DasGupta (2001) for details. See Berger (1985) for information about using beta priors for inference on the binomial proportion.

Wilson (Score) Confidence Limits If you specify the WILSON binomial-option, PROC FREQ computes Wilson confidence limits for the binomial proportion. These are also known as score confidence limits and are attributed to Wilson (1927). The confidence limits are based on inverting the normal test that uses the null proportion in the variance (the score test). Wilson confidence limits are the roots of

$$|p - \hat{p}| = z_{\alpha/2} \sqrt{p(1-p)/n}$$

and are computed as

$$\left(\hat{p} + z_{\alpha/2}^2/2n\right) \pm \left(z_{\alpha/2} \sqrt{\left(\hat{p}(1-\hat{p}) + z_{\alpha/2}^2\right)/4n} \Big/ \left(1 + z_{\alpha/2}^2/n\right)\right)$$

The Wilson interval has been shown to have better performance than the Wald interval and the exact (Clopper-Pearson) interval. See Agresti and Coull (1998), Brown, Cai, and DasGupta (2001), and Newcombe (1998) for more information.

Exact (Clopper-Pearson) Confidence Limits The exact or Clopper-Pearson confidence limits for the binomial proportion are constructed by inverting the equal-tailed test based on the binomial distribution. This method is attributed to Clopper and Pearson (1934). The exact confidence limits p_L and p_U satisfy the following equations, for $n_1 = 1, 2, \ldots n - 1$:

$$\sum_{x=n_1}^{n} \binom{n}{x} p_L^x (1 - p_L)^{n-x} = \alpha/2$$

$$\sum_{x=0}^{n_1} \binom{n}{x} p_U^x (1 - p_U)^{n-x} = \alpha/2$$

The lower confidence limit equals 0 when $n_1 = 0$, and the upper confidence limit equals 1 when $n_1 = n$.

PROC FREQ computes the exact (Clopper-Pearson) confidence limits by using the F distribution as

$$p_L = \left(1 + \frac{n - n_1 + 1}{n_1 \, F(1 - \alpha/2, \, 2n_1, \, 2(n - n_1 + 1))}\right)^{-1}$$

$$p_U = \left(1 + \frac{n - n_1}{(n_1 + 1) \, F(\alpha/2, \, 2(n_1 + 1), \, 2(n - n_1))}\right)^{-1}$$

where $F(\alpha, b, c)$ is the αth percentile of the F distribution with b and c degrees of freedom. See Leemis and Trivedi (1996) for a derivation of this expression. Also see Collett (1991) for more information about exact binomial confidence limits.

Because this is a discrete problem, the confidence coefficient (or coverage probability) of the exact (Clopper-Pearson) interval is not exactly $(1-\alpha)$ but is at least $(1-\alpha)$. Thus, this confidence interval is conservative. Unless the sample size is large, the actual coverage probability can be much larger than the target value. See Agresti and Coull (1998), Brown, Cai, and DasGupta (2001), and Leemis and Trivedi (1996) for more information about the performance of these confidence limits.

Tests

The BINOMIAL option provides an asymptotic equality test for the binomial proportion by default. You can also specify *binomial-options* to request tests of noninferiority, superiority, and equivalence for the binomial proportion. If you specify the BINOMIAL option in the EXACT statement, PROC FREQ also computes exact *p*-values for the tests that you request with the *binomial-options*.

Equality Test PROC FREQ computes an asymptotic test of the hypothesis that the binomial proportion equals p_0, where you can specify the value of p_0 with the P= binomial-option. If you do not specify a null value with P=, PROC FREQ uses $p_0 = 0.5$ by default. The binomial test statistic is computed as

$$z = (\hat{p} - p_0)/se$$

By default, the standard error is based on the null hypothesis proportion as

$$se = \sqrt{p_0(1 - p_0)/n}$$

If you specify the VAR=SAMPLE binomial-option, the standard error is computed from the sample proportion as

$$se = \sqrt{\hat{p}(1 - \hat{p})/n}$$

If you specify the CORRECT binomial-option or the BINOMIALC option, PROC FREQ includes a continuity correction in the asymptotic test statistic, towards adjusting for the difference between the normal approximation and the discrete binomial distribution. See Fleiss, Levin, and Paik (2003) for details. The continuity correction of $(1/2n)$ is subtracted from the numerator of the test statistic if $(\hat{p} - p_0)$ is positive; otherwise, the continuity correction is added to the numerator.

PROC FREQ computes one-sided and two-sided *p*-values for this test. When the test statistic z is greater than zero (its expected value under the null hypothesis), PROC FREQ computes the right-sided *p*-value, which is the probability of a larger value of the statistic occurring under the null hypothesis. A small right-sided *p*-value supports the alternative hypothesis that the true value of the proportion is greater than p_0. When the test statistic is less than or equal to zero, PROC FREQ computes the left-sided *p*-value, which is the probability of a smaller value of the statistic occurring under the null hypothesis. A small left-sided *p*-value supports the alternative hypothesis that the true value of the proportion is less than p_0. The one-sided *p*-value P_1 can be expressed as

$$P_1 = \begin{cases} \text{Prob}(Z > z) & \text{if } z > 0 \\ \text{Prob}(Z < z) & \text{if } z \leq 0 \end{cases}$$

where Z has a standard normal distribution. The two-sided p-value is computed as $P_2 = 2 \times P_1$.

If you specify the BINOMIAL option in the EXACT statement, PROC FREQ also computes an exact test of the null hypothesis $H_0: p = p_0$. To compute the exact test, PROC FREQ uses the binomial probability function,

$$\text{Prob}(X = x \mid p_0) = \binom{n}{x} p_0^x (1-p_0)^{(n-x)} \quad \text{for} \quad x = 0, 1, 2, \ldots, n$$

where the variable X has a binomial distribution with parameters n and p_0. To compute the left-sided p-value, $\text{Prob}(X \leq n_1)$, PROC FREQ sums the binomial probabilities over x from zero to n_1. To compute the right-sided p-value, $\text{Prob}(X \geq n_1)$, PROC FREQ sums the binomial probabilities over x from n_1 to n. The exact one-sided p-value is the minimum of the left-sided and right-sided p-values,

$$P_1 = \min(\text{Prob}(X \leq n_1 \mid p_0), \text{Prob}(X \geq n_1 \mid p_0))$$

and the exact two-sided p-value is computed as $P_2 = 2 \times P_1$.

Noninferiority Test If you specify the NONINF binomial-option, PROC FREQ provides a noninferiority test for the binomial proportion. The null hypothesis for the noninferiority test is

$$H_0: p - p_0 \leq -\delta$$

versus the alternative

$$H_a: p - p_0 > -\delta$$

where δ is the noninferiority margin and p_0 is the null proportion. Rejection of the null hypothesis indicates that the binomial proportion is not inferior to the null value. See Chow, Shao, and Wang (2003) for more information.

You can specify the value of δ with the MARGIN= binomial-option, and you can specify p_0 with the P= binomial-option. By default, $\delta = 0.2$ and $p_0 = 0.5$.

PROC FREQ provides an asymptotic Wald test for noninferiority. The test statistic is computed as

$$z = (\hat{p} - p_0^*) / se$$

where p_0^* is the noninferiority limit,

$$p_0^* = p_0 - \delta$$

By default, the standard error is computed from the sample proportion as

$$se = \sqrt{\hat{p}(1-\hat{p})/n}$$

If you specify the VAR=NULL binomial-option, the standard error is based on the noninferiority limit (determined by the null proportion and the margin) as

$$se = \sqrt{p_0^*(1-p_0^*)/n}$$

If you specify the CORRECT binomial-option or the BINOMIALC option, PROC FREQ includes a continuity correction in the asymptotic test statistic z. The continuity correction of $(1/2n)$ is subtracted from the numerator of the test statistic if $(\hat{p} - p_0^*)$ is positive; otherwise, the continuity correction is added to the numerator.

The *p*-value for the noninferiority test is

$$P_z = \text{Prob}(Z > z)$$

where Z has a standard normal distribution.

As part of the noninferiority analysis, PROC FREQ provides asymptotic Wald confidence limits for the binomial proportion. These confidence limits are computed as described in the section "Wald Confidence Limits" on page 134 but use the same standard error (VAR=NULL or VAR=SAMPLE) as the noninferiority test statistic z. The confidence coefficient is $100(1-2\alpha)\%$ (Schuirmann 1999). By default, if you do not specify the ALPHA= option, the noninferiority confidence limits are 90% confidence limits. You can compare the confidence limits to the noninferiority limit, $p_0^* = p_0 - \delta$.

If you specify the BINOMIAL option in the EXACT statement, PROC FREQ provides an exact noninferiority test for the binomial proportion. The exact *p*-value is computed by using the binomial probability function with parameters p_0^* and n,

$$P_x = \sum_{k=n_1}^{k=n} \binom{n}{k} (p_0^*)^k (1 - p_0^*)^{(n-k)}$$

See Chow, Shao, Wang (2003, p. 116) for details. If you request exact binomial statistics, PROC FREQ also includes exact (Clopper-Pearson) confidence limits for the binomial proportion in the equivalence analysis display. See the section "Exact (Clopper-Pearson) Confidence Limits" on page 135 for details.

Superiority Test If you specify the SUP binomial-option, PROC FREQ provides a superiority test for the binomial proportion. The null hypothesis for the superiority test is

$$H_0: p - p_0 \leq \delta$$

versus the alternative

$$H_a: p - p_0 > \delta$$

where δ is the superiority margin and p_0 is the null proportion. Rejection of the null hypothesis indicates that the binomial proportion is superior to the null value. You can specify the value of δ with the MARGIN= binomial-option, and you can specify the value of p_0 with the P= binomial-option. By default, $\delta = 0.2$ and $p_0 = 0.5$.

The superiority analysis is identical to the noninferiority analysis but uses a positive value of the margin δ in the null hypothesis. The superiority limit equals $p_0 + \delta$. The superiority computations follow those in the section "Noninferiority Test" on page 137 but replace $-\delta$ with δ. See Chow, Shao, and Wang (2003) for more information.

Equivalence Test If you specify the EQUIV binomial-option, PROC FREQ provides an equivalence test for the binomial proportion. The null hypothesis for the equivalence test is

$$H_0: p - p_0 \leq \delta_L \quad \text{or} \quad p - p_0 \geq \delta_U$$

versus the alternative

$$H_a: \delta_L < p - p_0 < \delta_U$$

where δ_L is the lower margin, δ_U is the upper margin, and p_0 is the null proportion. Rejection of the null hypothesis indicates that the binomial proportion is equivalent to the null value. See Chow, Shao, and Wang (2003) for more information.

You can specify the value of the margins δ_L and δ_U with the MARGIN= binomial-option. If you do not specify MARGIN=, PROC FREQ uses lower and upper margins of -0.2 and 0.2 by default. If you specify a single margin value δ, PROC FREQ uses lower and upper margins of $-\delta$ and δ. You can specify the null proportion p_0 with the P= binomial-option. By default, $p_0 = 0.5$.

PROC FREQ computes two one-sided tests (TOST) for equivalence analysis (Schuirmann 1987). The TOST approach includes a right-sided test for the lower margin and a left-sided test for the upper margin. The overall p-value is taken to be the larger of the two p-values from the lower and upper tests.

For the lower margin, the asymptotic Wald test statistic is computed as

$$z_L = (\hat{p} - p_L^*) / se$$

where the lower equivalence limit is

$$p_L^* = p_0 + \delta_L$$

By default, the standard error is computed from the sample proportion as

$$se = \sqrt{\hat{p}(1 - \hat{p})/n}$$

If you specify the VAR=NULL binomial-option, the standard error is based on the lower equivalence limit (determined by the null proportion and the lower margin) as

$$se = \sqrt{p_L^*(1 - p_L^*)/n}$$

If you specify the CORRECT binomial-option or the BINOMIALC option, PROC FREQ includes a continuity correction in the asymptotic test statistic z_L. The continuity correction of $(1/2n)$ is subtracted from the numerator of the test statistic $(\hat{p} - p_L^*)$ if the numerator is positive; otherwise, the continuity correction is added to the numerator.

The p-value for the lower margin test is

$$P_{z,L} = \text{Prob}(Z > z_L)$$

The asymptotic test for the upper margin is computed similarly. The Wald test statistic is

$$z_U = (\hat{p} - p_U^*) / se$$

where the upper equivalence limit is

$$p_U^* = p_0 + \delta_U$$

By default, the standard error is computed from the sample proportion. If you specify the VAR=NULL binomial-option, the standard error is based on the upper equivalence limit as

$$se = \sqrt{p_U^*(1 - p_U^*)/n}$$

If you specify the CORRECT binomial-option or the BINOMIALC option, PROC FREQ includes a continuity correction of $(1/2n)$ in the asymptotic test statistic z_U.

The *p*-value for the upper margin test is

$$P_{z,U} = \text{Prob}(Z < z_U)$$

Based on the two one-sided tests (TOST), the overall *p*-value for the test of equivalence equals the larger *p*-value from the lower and upper margin tests, which can be expressed as

$$P_z = \max(P_{z,L}, P_{z,U})$$

As part of the equivalence analysis, PROC FREQ provides asymptotic Wald confidence limits for the binomial proportion. These confidence limits are computed as described in the section "Wald Confidence Limits" on page 134, but use the same standard error (VAR=NULL or VAR=SAMPLE) as the equivalence test statistics and have a confidence coefficient of $100(1 - 2\alpha)\%$ (Schuirmann 1999). By default, if you do not specify the ALPHA= option, the equivalence confidence limits are 90% limits. If you specify VAR=NULL, separate standard errors are computed for the lower and upper margin tests, each based on the null proportion and the corresponding (lower or upper) margin. The confidence limits are computed by using the maximum of these two standard errors. You can compare the confidence limits to the equivalence limits, $(p_0 + \delta_L, p_0 + \delta_U)$.

If you specify the BINOMIAL option in the EXACT statement, PROC FREQ also provides an exact equivalence test by using two one-sided exact tests (TOST). The procedure computes lower and upper margin exact tests by using the binomial probability function as described in the section "Noninferiority Test" on page 137. The overall exact *p*-value for the equivalence test is taken to be the larger *p*-value from the lower and upper margin exact tests. If you request exact statistics, PROC FREQ also includes exact (Clopper-Pearson) confidence limits in the equivalence analysis display. The confidence coefficient is $100(1 - 2\alpha)\%$ (Schuirmann 1999). See the section "Exact (Clopper-Pearson) Confidence Limits" on page 135 for details.

Risks and Risk Differences

The RISKDIFF option in the TABLES statement provides estimates of risks (or binomial proportions) and risk differences for 2×2 tables. This analysis might be appropriate when comparing the proportion of some characteristic for two groups, where row 1 and row 2 correspond to the two groups, and the columns correspond to two possible characteristics or outcomes. For example, the row variable might be a treatment or dose, and the column variable might be the response. See Collett (1991), Fleiss, Levin, and Paik (2003), and Stokes, Davis, and Koch (2000) for more information.

Let the frequencies of the 2×2 table be represented as follows.

	Column 1	Column 2	Total
Row 1	n_{11}	n_{12}	$n_{1\cdot}$
Row 2	n_{21}	n_{22}	$n_{2\cdot}$
Total	$n_{\cdot 1}$	$n_{\cdot 2}$	n

For column 1 and column 2, PROC FREQ provides estimates of the row 1 risk (or proportion), the row 2 risk, the overall risk and the risk difference. The risk difference is defined as the row 1 risk minus the row 2 risk. The risks are binomial proportions of their rows (row 1, row 2, or overall), and the computation of their standard errors and confidence limits follow the binomial proportion computations, which are described in the section "Binomial Proportion" on page 133.

The column 1 risk for row 1 is the proportion of row 1 observations classified in column 1,

$$p_1 = n_{11} / n_{1\cdot}$$

This estimates the conditional probability of the column 1 response, given the first level of the row variable.

The column 1 risk for row 2 is the proportion of row 2 observations classified in column 1,

$$p_2 = n_{21} / n_{2\cdot}$$

and the overall column 1 risk is the proportion of all observations classified in column 1,

$$p = n_{\cdot 1} / n$$

The column 1 risk difference compares the risks for the two rows, and it is computed as the column 1 risk for row 1 minus the column 1 risk for row 2,

$$d = p_1 - p_2$$

The risks and risk difference are defined similarly for column 2.

The standard error of the column 1 risk for row i is computed as

$$se(p_i) = \sqrt{p_i \, (1 - p_i) \, / \, n_{1\cdot}}$$

The standard error of the overall column 1 risk is computed as

$$se(p) = \sqrt{p \, (1 - p) \, / \, n}$$

If the two rows represent independent binomial samples, the standard error for the column 1 risk difference is computed as

$$se(d) = \sqrt{var(p_1) + var(p_2)}$$

The standard errors are computed in a similar manner for the column 2 risks and risk difference.

Confidence Limits

By default, the RISKDIFF option provides standard Wald asymptotic confidence limits for the risks (row 1, row 2, and overall) and the risk difference. The risks are equivalent to binomial proportions of their corresponding rows, and the computations follow the methods in the section "Wald Confidence Limits" on page 134.

The standard Wald asymptotic confidence limits are based on the normal approximation to the binomial distribution. PROC FREQ computes the Wald confidence limits for the risks and risk differences as

$$est \pm (z_{\alpha/2} \times \text{se}(est))$$

where est is the estimate, $z_{\alpha/2}$ is the $100(1 - \alpha/2)$th percentile of the standard normal distribution, and se(est) is the standard error of the estimate. The confidence level α is determined from the value of the ALPHA= option, which, by default, equals 0.05 and produces 95% confidence limits.

If you specify the CORRECT *riskdiff-option* or the RISKDIFFC option, PROC FREQ includes continuity corrections in the Wald asymptotic confidence limits for the risks and risk differences. The purpose of a continuity correction is to adjust for the difference between the normal approximation and the binomial distribution, which is discrete. See Fleiss, Levin, and Paik (2003) for more information. With the continuity correction, the asymptotic confidence limits are computed as

$$est \pm (z_{\alpha/2} \times \text{se}(est) + cc)$$

where cc is the continuity correction. For the row 1 risk, $cc = (1/2n_{1\cdot})$; for the row 2 risk, $cc = (1/2n_{2\cdot})$; for the overall risk, $cc = (1/2n)$; and for the risk difference, $cc = ((1/n_{1\cdot} + 1/n_{2\cdot})/2)$. The column 1 and column 2 risks use the same continuity corrections.

PROC FREQ also computes exact (Clopper-Pearson) confidence limits for the column 1, column 2, and overall risks. These confidence limits are constructed by inverting the equal-tailed test based on the binomial distribution. PROC FREQ uses the F distribution to compute the Clopper-Pearson confidence limits. See the section "Exact (Clopper-Pearson) Confidence Limits" on page 135 for details.

PROC FREQ does not provide exact confidence limits for the risk difference by default. If you specify the RISKDIFF option in the EXACT statement, PROC FREQ provides exact unconditional confidence limits for the risk difference, which are described in the section "Exact Unconditional Confidence Limits for the Risk Difference" on page 147. Note that the conditional exact approach, which is the basis for other exact tests provided by PROC FREQ such as Fisher's exact test, does not apply to the risk difference due to nuisance parameters. See Agresti (1992) for more information.

Tests

You can specify *riskdiff-options* to request tests of the risk (or proportion) difference. You can request tests of equality, noninferiority, superiority, and equivalence for the risk difference. The test of equality is a standard Wald asymptotic test, available with or without a continuity correction. For noninferiority, superiority, and equivalence tests of the risk difference, the following test methods are provided: Wald (with and without continuity correction), Hauck-Anderson, Farrington-Manning, and Newcombe score (with and without continuity correction). You can specify the test method with the METHOD= riskdiff-option. By default, PROC FREQ uses METHOD=WALD.

Equality Test If you specify the EQUAL riskdiff-option, PROC FREQ computes a test of equality, or a test of the null hypothesis that the risk difference equals zero. For the column 1 (or 2) risk difference, this test can be expressed as $H_0: d = 0$ versus the alternative $H_a: d \neq 0$, where $d = p_1 - p_2$ denotes the column 1 (or 2) risk difference. PROC FREQ computes a standard Wald asymptotic test, and the test statistic is

$$z = \hat{d}/\text{se}(\hat{d})$$

By default, the standard error is computed from the sample proportions as

$$\text{se}(\hat{d}) = \sqrt{\hat{p}_1(1-\hat{p}_1)/n_{1\cdot} + \hat{p}_2(1-\hat{p}_2)/n_{2\cdot}}$$

If you specify the VAR=NULL riskdiff-option, the standard error is based on the null hypothesis that the row 1 and row 2 risks are equal,

$$\text{se}(\hat{d}) = \sqrt{\hat{p}(1-\hat{p}) \times (1/n_{1\cdot} + 1/n_{2\cdot})}$$

where $\hat{p} = n_{\cdot 1}/n$ estimates the overall column 1 risk.

If you specify the CORRECT riskdiff-option or the RISKDIFFC option, PROC FREQ includes a continuity correction in the asymptotic test statistic. If $\hat{d} > 0$, the continuity correction is subtracted from $\hat{d}$ in the numerator of the test statistic; otherwise, the continuity correction is added to the numerator. The value of the continuity correction is $(1/n_{1\cdot} + 1/n_{2\cdot})/2$.

PROC FREQ computes one-sided and two-sided p-values for this test. When the test statistic z is greater than 0, PROC FREQ displays the right-sided p-value, which is the probability of a larger value occurring under the null hypothesis. The one-sided p-value can be expressed as

$$P_1 = \begin{cases} \text{Prob}(Z > z) & \text{if } z > 0 \\ \text{Prob}(Z < z) & \text{if } z \leq 0 \end{cases}$$

where Z has a standard normal distribution. The two-sided p-value is computed as $P_2 = 2 \times P_1$.

Noninferiority Test If you specify the NONINF riskdiff-option, PROC FREQ provides a noninferiority test for the risk difference, or the difference between two proportions. The null hypothesis for the noninferiority test is

$$H_0: p_1 - p_2 \leq -\delta$$

versus the alternative

$$H_a: p_1 - p_2 > -\delta$$

where δ is the noninferiority margin. Rejection of the null hypothesis indicates that the row 1 risk is not inferior to the row 2 risk. See Chow, Shao, and Wang (2003) for more information.

You can specify the value of δ with the MARGIN= riskdiff-option. By default, $\delta = 0.2$. You can specify the test method with the METHOD= riskdiff-option. The following methods are available for the risk difference noninferiority analysis: Wald (with and without continuity correction), Hauck-Anderson, Farrington-Manning, and Newcombe score (with and without continuity correction). The Wald, Hauck-Anderson, and Farrington-Manning methods provide tests and corresponding test-based confidence limits; the Newcombe score method provides only confidence limits. If you do not specify METHOD=, PROC FREQ uses the Wald test by default.

The confidence coefficient for the test-based confidence limits is $100(1-2\alpha)\%$ (Schuirmann 1999). By default, if you do not specify the ALPHA= option, these are 90% confidence limits. You can compare the confidence limits to the noninferiority limit, $-\delta$.

The following sections describe the noninferiority analysis methods for the risk difference.

Wald Test

If you specify the METHOD=WALD riskdiff-option, PROC FREQ provides an asymptotic Wald test of noninferiority for the risk difference. This is also the default method. The Wald test statistic is computed as

$$z = (\hat{d} + \delta) / \text{se}(\hat{d})$$

where $(\hat{d} = \hat{p}_1 - \hat{p}_2)$ estimates the risk difference and δ is the noninferiority margin.

By default, the standard error for the Wald test is computed from the sample proportions as

$$\text{se}(\hat{d}) = \sqrt{\hat{p}_1(1-\hat{p}_1)/n_{1\cdot} + \hat{p}_2(1-\hat{p}_2)/n_{2\cdot}}$$

If you specify the VAR=NULL riskdiff-option, the standard error is based on the null hypothesis that the risk difference equals $-\delta$ (Dunnett and Gent 1977). The standard error is computed as

$$\text{se}(\hat{d}) = \sqrt{\tilde{p}(1-\tilde{p})/n_{2\cdot} + (\tilde{p}-\delta)(1-\tilde{p}+\delta)/n_{1\cdot}}$$

where

$$\tilde{p} = (n_{11} + n_{21} + \delta n_{1\cdot})/n$$

If you specify the CORRECT riskdiff-option or the RISKDIFFC option, a continuity correction is included in the test statistic. The continuity correction is subtracted from the numerator of the test statistic if the numerator is greater than zero; otherwise, the continuity correction is added to the numerator. The value of the continuity correction is $(1/n_{1\cdot} + 1/n_{2\cdot})/2$.

The p-value for the Wald noninferiority test is $P_z = \text{Prob}(Z > z)$, where Z has a standard normal distribution.

Hauck-Anderson Test

If you specify the METHOD=HA riskdiff-option, PROC FREQ provides the Hauck-Anderson test for noninferiority. The Hauck-Anderson test statistic is computed as

$$z = (\hat{d} + \delta \pm cc) / s(\hat{d})$$

where $\hat{d} = \hat{p}_1 - \hat{p}_2$ and the standard error is computed from the sample proportions as

$$\text{se}(\hat{d}) = \sqrt{\hat{p}_1(1-\hat{p}_1)/(n_{1\cdot}-1) + \hat{p}_2(1-\hat{p}_2)/(n_{2\cdot}-1)}$$

The Hauck-Anderson continuity correction cc is computed as

$$cc = 1 / \left(2 \min(n_{1\cdot}, n_{2\cdot})\right)$$

The p-value for the Hauck-Anderson noninferiority test is $P_z = \text{Prob}(Z > z)$, where Z has a standard normal distribution. See Hauck and Anderson (1986) and Schuirmann (1999) for more information.

Farrington-Manning Test

If you specify the METHOD=FM riskdiff-option, PROC FREQ provides the Farrington-Manning test of noninferiority for the risk difference. The Farrington-Manning test statistic is computed as

$$z = (\hat{d} + \delta) / \text{se}(\hat{d})$$

where $\hat{d} = \hat{p}_1 - \hat{p}_2$ and

$$\text{se}(\hat{d}) = \sqrt{\tilde{p}_1(1 - \tilde{p}_1)/n_{1\cdot} + \tilde{p}_2(1 - \tilde{p}_2)/n_{2\cdot}}.$$

where $\tilde{p}_1$ and $\tilde{p}_2$ are the maximum likelihood estimators of p_1 and p_2 under the null hypothesis that the risk difference equals $-\delta$. The p-value for the Farrington-Manning noninferiority test is then $P_z = \text{Prob}(Z > z)$, where Z has a standard normal distribution.

From Farrington and Manning (1990), the solution to the maximum likelihood equation is

$$\tilde{p}_1 = 2u \cos(w) - b/3a \quad \text{and} \quad \tilde{p}_2 = \tilde{p}_1 + \delta$$

where

$$
\begin{aligned}
w &= (\pi + \cos^{-1}(v/u^3))/3 \\
v &= b^3/(3a)^3 - bc/6a^2 + d/2a \\
u &= \text{sign}(v)\sqrt{b^2/(3a)^2 - c/3a} \\
a &= 1 + \theta \\
b &= -(1 + \theta + \hat{p}_1 + \theta \hat{p}_2 - \delta(\theta + 2)) \\
c &= \delta^2 - \delta(2\hat{p}_1 + \theta + 1) + \hat{p}_1 + \theta \hat{p}_2 \\
d &= \hat{p}_1 \delta(1 - \delta) \\
\theta &= n_{2\cdot}/n_{1\cdot}
\end{aligned}
$$

Newcombe Score Confidence Limits

If you specify the METHOD=SCORE riskdiff-option, PROC FREQ provides the Newcombe hybrid score (or Wilson) confidence limits for the risk difference. The confidence coefficient for the confidence limits is $100(1 - 2\alpha)\%$ (Schuirmann 1999). By default, if you do not specify the ALPHA= option, these are 90% confidence limits. You can compare the confidence limits to the noninferiority limit, $-\delta$.

The Newcombe score confidence limits for the risk difference are constructed from the Wilson score confidence limits for each of the two individual proportions. The confidence limits for the individual proportions are used in the standard error terms of the Wald confidence limits for the proportion difference. See Newcombe (1998) and Barker et al. (2001) for more information.

Wilson score confidence limits for p_1 and p_2 are the roots of

$$|p_i - \hat{p}_i| = z_\alpha \sqrt{p_i(1 - p_i)/n_{i\cdot}}.$$

for $i = 1, 2$. The confidence limits are computed as

$$\left(\hat{p}_i + z_\alpha^2/2n_{i\cdot}\right) \pm \left(z_\alpha \sqrt{\left(\hat{p}_i(1 - \hat{p}_i) + z_\alpha^2\right)/4n_{i\cdot}} \, / \, \left(1 + z_\alpha^2/n_{i\cdot}\right)\right)$$

See the section "Wilson (Score) Confidence Limits" on page 135 for details.

Denote the lower and upper Wilson score confidence limits for p_1 as L_1 and U_1, and denote the lower and upper confidence limits for p_2 as L_2 and U_2. The Newcombe score confidence limits for the proportion difference ($d = p_1 - p_2$) are computed as

$$d_L = (\hat{p}_1 - \hat{p}_2) - z_\alpha \sqrt{L_1(1-L_1)/n_1. + U_2(1-U_2)/n_2.}$$

$$d_U = (\hat{p}_1 - \hat{p}_2) + z_\alpha \sqrt{U_1(1-U_1)/n_1. + L_2(1-L_2)/n_2.}$$

If you specify the CORRECT riskdiff-option, PROC FREQ provides continuity-corrected Newcombe score confidence limits. By including a continuity correction of $1/2n_i.$, the Wilson score confidence limits for the individual proportions are the roots of

$$|p_i - \hat{p}_i| - 1/2n_i. = z_\alpha \sqrt{p_i(1-p_i)/n_i.}$$

These confidence limits for the individual proportions are then used in the standard error terms of the Wald confidence limits for the proportion difference to compute d_L and d_U.

Superiority Test If you specify the SUP riskdiff-option, PROC FREQ provides a superiority test for the risk difference. The null hypothesis is

$$H_0:: p_1 - p_2 \leq \delta$$

versus the alternative

$$H_a: p_1 - p_2 > \delta$$

where δ is the superiority margin. Rejection of the null hypothesis indicates that the row 1 proportion is superior to the row 2 proportion. You can specify the value of δ with the MARGIN= riskdiff-option. By default, $\delta = 0.2$.

The superiority analysis is identical to the noninferiority analysis but uses a positive value of the margin δ in the null hypothesis. The superiority computations follow those in the section "Noninferiority Test" on page 143 by replacing $-\delta$ by δ. See Chow, Shao, and Wang (2003) for more information.

Equivalence Tests If you specify the EQUIV riskdiff-option, PROC FREQ provides an equivalence test for the risk difference, or the difference between two proportions. The null hypothesis for the equivalence test is

$$H_0: p_1 - p_2 \leq -\delta_L \quad \text{or} \quad p_1 - p_2 \geq \delta_U$$

versus the alternative

$$H_a: \delta_L < p_1 - p_2 < \delta_U$$

where δ_L is the lower margin and δ_U is the upper margin. Rejection of the null hypothesis indicates that the two binomial proportions are equivalent. See Chow, Shao, and Wang (2003) for more information.

You can specify the value of the margins δ_L and δ_U with the MARGIN= riskdiff-option. If you do not specify MARGIN=, PROC FREQ uses lower and upper margins of –0.2 and 0.2 by default. If you specify a single margin value δ, PROC FREQ uses lower and upper margins of $-\delta$ and δ. You can specify the test method with the METHOD= riskdiff-option. The following methods are available for the risk difference equivalence analysis: Wald (with and without continuity correction), Hauck-Anderson, Farrington-Manning, and Newcombe's score (with and without continuity correction). The Wald, Hauck-Anderson, and Farrington-Manning methods provide tests and corresponding test-based confidence limits; the Newcombe score method provides only confidence limits. If you do not specify METHOD=, PROC FREQ uses the Wald test by default.

PROC FREQ computes two one-sided tests (TOST) for equivalence analysis (Schuirmann 1987). The TOST approach includes a right-sided test for the lower margin δ_L and a left-sided test for the upper margin δ_U. The overall p-value is taken to be the larger of the two p-values from the lower and upper tests.

The section "Noninferiority Test" on page 143 gives details about the Wald, Hauck-Anderson, Farrington-Manning and Newcombe score methods for the risk difference. The lower margin equivalence test statistic takes the same form as the noninferiority test statistic but uses the lower margin value δ_L in place of $-\delta$. The upper margin equivalence test statistic take the same form as the noninferiority test statistic but uses the upper margin value δ_U in place of $-\delta$.

The test-based confidence limits for the risk difference are computed according to the equivalence test method that you select. If you specify METHOD=WALD with VAR=NULL, or METHOD=FM, separate standard errors are computed for the lower and upper margin tests. In this case, the test-based confidence limits are computed by using the maximum of these two standard errors. The confidence limits have a confidence coefficient of $100(1 - 2\alpha)\%$ (Schuirmann 1999). By default, if you do not specify the ALPHA= option, these are 90% confidence limits. You can compare the confidence limits to the equivalence limits, (δ_L, δ_U).

Exact Unconditional Confidence Limits for the Risk Difference

If you specify the RISKDIFF option in the EXACT statement, PROC FREQ provides exact unconditional confidence limits for the risk difference. Unconditional computations differ from the exact conditional approach that PROC FREQ uses for other exact statistics such as Fisher's exact test. (See the section "Exact Statistics" on page 167 for more information.) Exact conditional inference does not apply to the risk difference due to nuisance parameters. See Agresti (1992) for details. The unconditional approach eliminates nuisance parameters by maximizing the p-value over all possible values of the nuisance parameters (Santner and Snell 1980).

Denote the proportion difference by $d = p_1 - p_2$. For a 2×2 table with row totals n_1 and n_2, the joint probability function can be expressed in terms of the table cell frequencies and the parameters d and p_2,

$$f(n_{11}, n_{21}; n_1, n_2, d, p_2) = \binom{n_1}{n_{11}}(d + p_2)^{n_{11}}(1 - d - p_2)^{n_1 - n_{11}} \times \binom{n_2}{n_{21}} p_2^{n_{21}}(1 - p_2)^{n_2 - n_{21}}$$

When constructing confidence limits for the proportion difference, the parameter of interest is d and p_2 is a nuisance parameter.

Denote the observed value of the proportion difference by $d_0 = \hat{p}_1 - \hat{p}_2$. The $100(1 - \alpha/2)\%$ confidence limits for d are computed as

$$\begin{aligned} d_L &= \sup \, (d_* : P_U(d_*) > \alpha/2) \\ d_U &= \inf \, (d_* : P_L(d_*) > \alpha/2) \end{aligned}$$

where

$$P_U(d_*) = \sup_{p_2} \Big(\sum_{A, D(a) \geq d_0} f(n_{11}, n_{21}; n_1, n_2, d_*, p_2) \Big)$$

$$P_L(d_*) = \sup_{p_2} \Big(\sum_{A, D(a) \leq d_0} f(n_{11}, n_{21}; n_1, n_2, d_*, p_2) \Big)$$

The set A includes all 2×2 tables with row sums equal to n_1 and n_2, and $D(a)$ denotes the value of the proportion difference $(p_1 - p_2)$ for table a in A. To compute $P_U(d_*)$, the sum includes probabilities of those tables for which $(D(a) \geq d_0)$, where d_0 is the observed value of the proportion difference. For a fixed value of d_*, $P_U(d_*)$ is taken to be the maximum sum over all possible values of p_2. See Santner and Snell (1980) and Agresti and Min (2001) for details.

This method of eliminating the nuisance parameter is considered to be a conservative approach (Agresti and Min 2001). Additionally, the confidence limits are conservative for small samples because this is a discrete problem; the confidence coefficient is not exactly $(1 - \alpha)$ but is at least $(1 - \alpha)$. See Agresti (1992) for more information.

Odds Ratio and Relative Risks for 2 x 2 Tables

Odds Ratio (Case-Control Studies)

The odds ratio is a useful measure of association for a variety of study designs. For a retrospective design called a *case-control study*, the odds ratio can be used to estimate the relative risk when the probability of positive response is small (Agresti 2002). In a case-control study, two independent samples are identified based on a binary (yes-no) response variable, and the conditional distribution of a binary explanatory variable is examined, within fixed levels of the response variable. See Stokes, Davis, and Koch (2000) and Agresti (2007).

The odds of a positive response (column 1) in row 1 is n_{11}/n_{12}. Similarly, the odds of a positive response in row 2 is n_{21}/n_{22}. The odds ratio is formed as the ratio of the row 1 odds to the row 2 odds. The odds ratio for a 2×2 table is defined as

$$OR = \frac{n_{11}/n_{12}}{n_{21}/n_{22}} = \frac{n_{11} \, n_{22}}{n_{12} \, n_{21}}$$

The odds ratio can be any nonnegative number. When the row and column variables are independent, the true value of the odds ratio equals 1. An odds ratio greater than 1 indicates that the odds of a positive response are higher in row 1 than in row 2. Values less than 1 indicate the odds of positive response are higher in row 2. The strength of association increases with the deviation from 1.

The transformation $G = (OR - 1)/(OR + 1)$ transforms the odds ratio to the range $(-1, 1)$ with $G = 0$ when $OR = 1$; $G = -1$ when $OR = 0$; and G approaches 1 as OR approaches infinity. G is the gamma statistic, which PROC FREQ computes when you specify the MEASURES option.

The asymptotic $100(1 - \alpha)\%$ confidence limits for the odds ratio are

$$\left(OR \times \exp(-z\sqrt{v}), \ OR \times \exp(z\sqrt{v}) \right)$$

where

$$v = \text{var}(\ln OR) = \frac{1}{n_{11}} + \frac{1}{n_{12}} + \frac{1}{n_{21}} + \frac{1}{n_{22}}$$

and z is the $100(1 - \alpha/2)$th percentile of the standard normal distribution. If any of the four cell frequencies are zero, the estimates are not computed.

When you specify the OR option in the EXACT statement, PROC FREQ computes exact confidence limits for the odds ratio. Because this is a discrete problem, the confidence coefficient for the exact confidence interval is not exactly $(1 - \alpha)$ but is at least $(1 - \alpha)$. Thus, these confidence limits are conservative. See Agresti (1992) for more information.

PROC FREQ computes exact confidence limits for the odds ratio by using an algorithm based on Thomas (1971). See also Gart (1971). The following two equations are solved iteratively to determine the lower and upper confidence limits, ϕ_1 and ϕ_2:

$$\sum_{i=n_{11}}^{n_{\cdot 1}} \binom{n_{1\cdot}}{i}\binom{n_{2\cdot}}{n_{\cdot 1} - i} \phi_1^i \ / \ \sum_{i=0}^{n_{\cdot 1}} \binom{n_{1\cdot}}{i}\binom{n_{2\cdot}}{n_{\cdot 1} - i} \phi_1^i \ = \ \alpha/2$$

$$\sum_{i=0}^{n_{11}} \binom{n_{1\cdot}}{i}\binom{n_{2\cdot}}{n_{\cdot 1} - i} \phi_2^i \ / \ \sum_{i=0}^{n_{\cdot 1}} \binom{n_{1\cdot}}{i}\binom{n_{2\cdot}}{n_{\cdot 1} - i} \phi_2^i \ = \ \alpha/2$$

When the odds ratio equals zero, which occurs when either $n_{11} = 0$ or $n_{22} = 0$, PROC FREQ sets the lower exact confidence limit to zero and determines the upper limit with level α. Similarly, when the odds ratio equals infinity, which occurs when either $n_{12} = 0$ or $n_{21} = 0$, PROC FREQ sets the upper exact confidence limit to infinity and determines the lower limit with level α.

Relative Risks (Cohort Studies)

These measures of relative risk are useful in *cohort* (prospective) study designs, where two samples are identified based on the presence or absence of an explanatory factor. The two samples are observed in future time for the binary (yes-no) response variable under study. Relative risk measures are also useful in cross-sectional studies, where two variables are observed simultaneously. See Stokes, Davis, and Koch (2000) and Agresti (2007) for more information.

The column 1 relative risk is the ratio of the column 1 risk for row 1 to row 2. The column 1 risk for row 1 is the proportion of the row 1 observations classified in column 1,

$$p_{1|1} = n_{11} \ / \ n_{1\cdot}$$

Similarly, the column 1 risk for row 2 is

$$p_{1|2} = n_{21} \ / \ n_{2\cdot}$$

The column 1 relative risk is then computed as

$$RR_1 = p_{1|1} / p_{1|2}$$

A relative risk greater than 1 indicates that the probability of positive response is greater in row 1 than in row 2. Similarly, a relative risk less than 1 indicates that the probability of positive response is less in row 1 than in row 2. The strength of association increases with the deviation from 1.

Asymptotic $100(1 - \alpha)\%$ confidence limits for the column 1 relative risk are computed as

$$\left(RR_1 \times \exp(-z\sqrt{v}), \quad RR_1 \times \exp(z\sqrt{v}) \right)$$

where

$$v = \text{var}(\ln RR_1) = \left((1 - p_{1|1})/n_{11}\right) + \left((1 - p_{1|2})/n_{21}\right)$$

and z is the $100(1 - \alpha/2)$th percentile of the standard normal distribution. If either n_{11} or n_{21} is zero, the estimates are not computed.

PROC FREQ computes the column 2 relative risks in the same way.

Cochran-Armitage Test for Trend

The TREND option in the TABLES statement provides the Cochran-Armitage test for trend, which tests for trend in binomial proportions across levels of a single factor or covariate. This test is appropriate for a two-way table where one variable has two levels and the other variable is ordinal. The two-level variable represents the response, and the other variable represents an explanatory variable with ordered levels. When the two-way has two columns and R rows, PROC FREQ tests for trend across the R levels of the row variable, and the binomial proportion is computed as the proportion of observations in the first column. When the table has two rows and C columns, PROC FREQ tests for trend across the C levels of the column variable, and the binomial proportion is computed as the proportion of observations in the first row.

The trend test is based on the regression coefficient for the weighted linear regression of the binomial proportions on the scores of the explanatory variable levels. See Margolin (1988) and Agresti (2002) for details. If the table has two columns and R rows, the trend test statistic is computed as

$$T = \sum_{i=1}^{R} n_{i1}(R_i - \bar{R}) / \sqrt{p_{\cdot 1}(1 - p_{\cdot 1}) s^2}$$

where R_i is the score of row i, $\bar{R}$ is the average row score, and

$$s^2 = \sum_{i=1}^{R} n_{i\cdot}(R_i - \bar{R})^2$$

The SCORES= option in the TABLES statement determines the type of row scores used in computing the trend test (and other score-based statistics). The default is SCORES=TABLE. See the section "Scores" on page 118 for details. For character variables, the table scores for the row variable are the row numbers (for example, 1 for the first row, 2 for the second row, and so on). For numeric variables, the table score for each row is the numeric value of the row level. When you perform the trend test, the explanatory variable might be numeric (for example, dose of a test substance), and the variable values might be appropriate scores. If the explanatory variable has ordinal levels that are not numeric, you can assign meaningful scores to the variable levels. Sometimes equidistant scores, such as the table scores for a character variable, might be appropriate. For more information on choosing scores for the trend test, see Margolin (1988).

The null hypothesis for the Cochran-Armitage test is no trend, which means that the binomial proportion $p_{i1} = n_{i1}/n_{i\cdot}$ is the same for all levels of the explanatory variable. Under the null hypothesis, the trend statistic has an asymptotic standard normal distribution.

PROC FREQ computes one-sided and two-sided p-values for the trend test. When the test statistic is greater than its null hypothesis expected value of zero, PROC FREQ displays the right-sided p-value, which is the probability of a larger value of the statistic occurring under the null hypothesis. A small right-sided p-value supports the alternative hypothesis of increasing trend in proportions from row 1 to row R. When the test statistic is less than or equal to zero, PROC FREQ displays the left-sided p-value. A small left-sided p-value supports the alternative of decreasing trend.

The one-sided p-value for the trend test is computed as

$$P_1 = \begin{cases} \text{Prob}(Z > T) & \text{if } T > 0 \\ \text{Prob}(Z < T) & \text{if } T \leq 0 \end{cases}$$

where Z has a standard normal distribution. The two-sided p-value is computed as

$$P_2 = \text{Prob}(|Z| > |T|)$$

PROC FREQ also provides exact p-values for the Cochran-Armitage trend test. You can request the exact test by specifying the TREND option in the EXACT statement. See the section "Exact Statistics" on page 167 for more information.

Jonckheere-Terpstra Test

The JT option in the TABLES statement provides the Jonckheere-Terpstra test, which is a nonparametric test for ordered differences among classes. It tests the null hypothesis that the distribution of the response variable does not differ among classes. It is designed to detect alternatives of ordered class differences, which can be expressed as $\tau_1 \leq \tau_2 \leq \cdots \leq \tau_R$ (or $\tau_1 \geq \tau_2 \geq \cdots \geq \tau_R$), with at least one of the inequalities being strict, where τ_i denotes the effect of class i. For such ordered alternatives, the Jonckheere-Terpstra test can be preferable to tests of more general class difference alternatives, such as the Kruskal–Wallis test (produced by the WILCOXON option in the NPAR1WAY procedure). See Pirie (1983) and Hollander and Wolfe (1999) for more information about the Jonckheere-Terpstra test.

The Jonckheere-Terpstra test is appropriate for a two-way table in which an ordinal column variable represents the response. The row variable, which can be nominal or ordinal, represents the classification variable. The levels of the row variable should be ordered according to the ordering you want the test to detect. The order of variable levels is determined by the ORDER= option in the PROC FREQ statement. The default is ORDER=INTERNAL, which orders by unformatted values. If you specify ORDER=DATA, PROC FREQ orders values according to their order in the input data set. For more information about how to order variable levels, see the ORDER= option.

The Jonckheere-Terpstra test statistic is computed by first forming $R(R-1)/2$ Mann-Whitney counts $M_{i,i'}$, where $i < i'$, for pairs of rows in the contingency table,

$$M_{i,i'} = \{ \text{ number of times } X_{i,j} < X_{i',j'}, \quad j = 1,\ldots,n_{i\cdot}; \quad j' = 1,\ldots,n_{i'\cdot} \}$$
$$+ \tfrac{1}{2} \{ \text{ number of times } X_{i,j} = X_{i',j'}, \quad j = 1,\ldots,n_{i\cdot}; \quad j' = 1,\ldots,n_{i'\cdot} \}$$

where $X_{i,j}$ is response j in row i. The Jonckheere-Terpstra test statistic is computed as

$$J = \sum_{1 \le i < i' \le R} \sum M_{i,i'}$$

This test rejects the null hypothesis of no difference among classes for large values of J. Asymptotic p-values for the Jonckheere-Terpstra test are obtained by using the normal approximation for the distribution of the standardized test statistic. The standardized test statistic is computed as

$$J^* = (J - E_0(J)) / \sqrt{\text{var}_0(J)}$$

where $E_0(J)$ and $\text{var}_0(J)$ are the expected value and variance of the test statistic under the null hypothesis,

$$E_0(J) = \left(n^2 - \sum_i n_{i\cdot}^2 \right) / 4$$

$$\text{var}_0(J) = A/72 + B/(36n(n-1)(n-2)) + C/(8n(n-1))$$

where

$$A = n(n-1)(2n+5) - \sum_i n_{i\cdot}(n_{i\cdot}-1)(2n_{i\cdot}+5) - \sum_j n_{\cdot j}(n_{\cdot j}-1)(2n_{\cdot j}+5)$$

$$B = \left(\sum_i n_{i\cdot}(n_{i\cdot}-1)(n_{i\cdot}-2) \right) \left(\sum_j n_{\cdot j}(n_{\cdot j}-1)(n_{\cdot j}-2) \right)$$

$$C = \left(\sum_i n_{i\cdot}(n_{i\cdot}-1) \right) \left(\sum_j n_{\cdot j}(n_{\cdot j}-1) \right)$$

PROC FREQ computes one-sided and two-sided p-values for the Jonckheere-Terpstra test. When the standardized test statistic is greater than its null hypothesis expected value of zero, PROC FREQ displays the right-sided p-value, which is the probability of a larger value of the statistic occurring under the null hypothesis. A small right-sided p-value supports the alternative hypothesis of increasing order from row 1 to row R. When the standardized test statistic is less than or equal to zero, PROC FREQ displays the left-sided p-value. A small left-sided p-value supports the alternative of decreasing order from row 1 to row R.

The one-sided p-value for the Jonckheere-Terpstra test, P_1, is computed as

$$P_1 = \begin{cases} \text{Prob}(Z > J^*) & \text{if } J^* > 0 \\ \text{Prob}(Z < J^*) & \text{if } J^* \leq 0 \end{cases}$$

where Z has a standard normal distribution. The two-sided p-value, P_2, is computed as

$$P_2 = \text{Prob}(|Z| > |J^*|)$$

PROC FREQ also provides exact p-values for the Jonckheere-Terpstra test. You can request the exact test by specifying the JT option in the EXACT statement. See the section "Exact Statistics" on page 167 for more information.

Tests and Measures of Agreement

When you specify the AGREE option in the TABLES statement, PROC FREQ computes tests and measures of agreement for square tables (that is, for tables where the number of rows equals the number of columns). For two-way tables, these tests and measures include McNemar's test for 2×2 tables, Bowker's test of symmetry, the simple kappa coefficient, and the weighted kappa coefficient. For multiple strata (n-way tables, where $n > 2$), PROC FREQ also computes the overall simple kappa coefficient and the overall weighted kappa coefficient, as well as tests for equal kappas (simple and weighted) among strata. Cochran's Q is computed for multiway tables when each variable has two levels, that is, for $h \times 2 \times 2$ tables.

PROC FREQ computes the kappa coefficients (simple and weighted), their asymptotic standard errors, and their confidence limits when you specify the AGREE option in the TABLES statement. If you also specify the KAPPA option in the TEST statement, then PROC FREQ computes the asymptotic test of the hypothesis that simple kappa equals zero. Similarly, if you specify the WTKAP option in the TEST statement, PROC FREQ computes the asymptotic test for weighted kappa.

In addition to the asymptotic tests described in this section, PROC FREQ provides exact p-values for McNemar's test, the simple kappa coefficient test, and the weighted kappa coefficient test. You can request these exact tests by specifying the corresponding options in the EXACT statement. See the section "Exact Statistics" on page 167 for more information.

The following sections provide the formulas that PROC FREQ uses to compute the AGREE statistics. For information about the use and interpretation of these statistics, see Agresti (2002), Agresti (2007), Fleiss, Levin, and Paik (2003), and the other references cited for each statistic.

McNemar's Test

PROC FREQ computes McNemar's test for 2×2 tables when you specify the AGREE option. McNemar's test is appropriate when you are analyzing data from matched pairs of subjects with a dichotomous (yes-no) response. It tests the null hypothesis of marginal homogeneity, or $p_1. = p_{.1}$. McNemar's test is computed as

$$Q_M = (n_{12} - n_{21})^2 / (n_{12} + n_{21})$$

Under the null hypothesis, Q_M has an asymptotic chi-square distribution with one degree of freedom. See McNemar (1947), as well as the general references cited in the preceding section. In addition to the asymptotic test, PROC FREQ also computes the exact p-value for McNemar's test when you specify the MCNEM option in the EXACT statement.

Bowker's Test of Symmetry

For Bowker's test of symmetry, the null hypothesis is that the cell proportions are symmetric, or that $p_{ij} = p_{ji}$ for all pairs of table cells. For 2×2 tables, Bowker's test is identical to McNemar's test, and so PROC FREQ provides Bowker's test for square tables larger than 2×2.

Bowker's test of symmetry is computed as

$$Q_B = \sum\sum_{i<j} (n_{ij} - n_{ji})^2 / (n_{ij} + n_{ji})$$

For large samples, Q_B has an asymptotic chi-square distribution with $R(R-1)/2$ degrees of freedom under the null hypothesis of symmetry. See Bowker (1948) for details.

Simple Kappa Coefficient

The simple kappa coefficient, introduced by Cohen (1960), is a measure of interrater agreement. PROC FREQ computes the simple kappa coefficient as

$$\hat{\kappa} = (P_o - P_e) / (1 - P_e)$$

where $P_o = \sum_i p_{ii}$ and $P_e = \sum_i p_{i.} p_{.i}$. If the two response variables are viewed as two independent ratings of the n subjects, the kappa coefficient equals +1 when there is complete agreement of the raters. When the observed agreement exceeds chance agreement, kappa is positive, with its magnitude reflecting the strength of agreement. Although this is unusual in practice, kappa is negative when the observed agreement is less than chance agreement. The minimum value of kappa is between -1 and 0, depending on the marginal proportions.

The asymptotic variance of the simple kappa coefficient is computed as

$$\text{var}(\hat{\kappa}) = (A + B - C) / (1 - P_e)^2 n$$

where

$$A = \sum_i p_{ii} \left(1 - (p_{i.} + p_{.i})(1 - \hat{\kappa})\right)^2$$

$$B = (1 - \hat{\kappa})^2 \sum\sum_{i \neq j} p_{ij} (p_{.i} + p_{j.})^2$$

$$C = (\hat{\kappa} - P_e(1 - \hat{\kappa}))^2$$

See Fleiss, Cohen, and Everitt (1969) for details.

PROC FREQ computes confidence limits for the simple kappa coefficient as

$$\hat{\kappa} \pm \left(z_{\alpha/2} \times \sqrt{\text{var}(\hat{\kappa})} \right)$$

where $z_{\alpha/2}$ is the $100(1 - \alpha/2)$th percentile of the standard normal distribution. The value of α is determined by the value of the ALPHA= option, which, by default, equals 0.05 and produces 95% confidence limits.

To compute an asymptotic test for the kappa coefficient, PROC FREQ uses the standardized test statistic $\hat{\kappa}^*$, which has an asymptotic standard normal distribution under the null hypothesis that kappa equals zero. The standardized test statistic is computed as

$$\hat{\kappa}^* = \hat{\kappa} \,/\, \sqrt{\text{var}_0(\hat{\kappa})}$$

where $\text{var}_0(\hat{\kappa})$ is the variance of the kappa coefficient under the null hypothesis,

$$\text{var}_0(\hat{\kappa}) = \left(P_e + P_e^2 - \sum_i p_{i\cdot} p_{\cdot i}(p_{i\cdot} + p_{\cdot i}) \right) \,/\, (1 - P_e)^2 \, n$$

See Fleiss, Levin, and Paik (2003) for details.

PROC FREQ also provides an exact test for the simple kappa coefficient. You can request the exact test by specifying the KAPPA or AGREE option in the EXACT statement. See the section "Exact Statistics" on page 167 for more information.

Weighted Kappa Coefficient

The weighted kappa coefficient is a generalization of the simple kappa coefficient that uses weights to quantify the relative difference between categories. For 2×2 tables, the weighted kappa coefficient equals the simple kappa coefficient. PROC FREQ displays the weighted kappa coefficient only for tables larger than 2×2. PROC FREQ computes the kappa weights from the column scores, by using either Cicchetti-Allison weights or Fleiss-Cohen weights, both of which are described in the following section. The weights w_{ij} are constructed so that $0 \leq w_{ij} < 1$ for all $i \neq j$, $w_{ii} = 1$ for all i, and $w_{ij} = w_{ji}$. The weighted kappa coefficient is computed as

$$\hat{\kappa}_w = \left(P_{o(w)} - P_{e(w)} \right) \,/\, \left(1 - P_{e(w)} \right)$$

where

$$P_{o(w)} = \sum_i \sum_j w_{ij} p_{ij}$$

$$P_{e(w)} = \sum_i \sum_j w_{ij} p_{i\cdot} p_{\cdot j}$$

The asymptotic variance of the weighted kappa coefficient is

$$\mathrm{var}(\hat{\kappa}_w) = \left(\sum_i \sum_j p_{ij} \left(w_{ij} - (\overline{w}_{i\cdot} + \overline{w}_{\cdot j})(1 - \hat{\kappa}_w) \right)^2 - \left(\hat{\kappa}_w - P_{e(w)}(1 - \hat{\kappa}_w) \right)^2 \right) / (1 - P_{e(w)})^2 \, n$$

where

$$\overline{w}_{i\cdot} = \sum_j p_{\cdot j} w_{ij}$$

$$\overline{w}_{\cdot j} = \sum_i p_{i\cdot} w_{ij}$$

See Fleiss, Cohen, and Everitt (1969) for details.

PROC FREQ computes confidence limits for the weighted kappa coefficient as

$$\hat{\kappa}_w \pm \left(z_{\alpha/2} \times \sqrt{\mathrm{var}(\hat{\kappa}_w)} \right)$$

where $z_{\alpha/2}$ is the $100(1 - \alpha/2)$th percentile of the standard normal distribution. The value of α is determined by the value of the ALPHA= option, which, by default, equals 0.05 and produces 95% confidence limits.

To compute an asymptotic test for the weighted kappa coefficient, PROC FREQ uses the standardized test statistic $\hat{\kappa}_w^*$, which has an asymptotic standard normal distribution under the null hypothesis that weighted kappa equals zero. The standardized test statistic is computed as

$$\hat{\kappa}_w^* = \hat{\kappa}_w / \sqrt{\mathrm{var}_0(\hat{\kappa}_w)}$$

where $\mathrm{var}_0(\hat{\kappa}_w)$ is the variance of the weighted kappa coefficient under the null hypothesis,

$$\mathrm{var}_0(\hat{\kappa}_w) = \left(\sum_i \sum_j p_{i\cdot} p_{\cdot j} \left(w_{ij} - (\overline{w}_{i\cdot} + \overline{w}_{\cdot j}) \right)^2 - P_{e(w)}^2 \right) / (1 - P_{e(w)})^2 \, n$$

See Fleiss, Levin, and Paik (2003) for details.

PROC FREQ also provides an exact test for the weighted kappa coefficient. You can request the exact test by specifying the WTKAPPA or AGREE option in the EXACT statement. See the section "Exact Statistics" on page 167 for more information.

Weights PROC FREQ computes kappa coefficient weights by using the column scores and one of the two available weight types. The column scores are determined by the SCORES= option in the TABLES statement. The two available types of kappa weights are Cicchetti-Allison and Fleiss-Cohen weights. By default, PROC FREQ uses Cicchetti-Allison weights. If you specify (WT=FC) with the AGREE option, then PROC FREQ uses Fleiss-Cohen weights to compute the weighted kappa coefficient.

PROC FREQ computes Cicchetti-Allison kappa coefficient weights as

$$w_{ij} = 1 - \frac{|C_i - C_j|}{C_C - C_1}$$

where C_i is the score for column i and C is the number of categories or columns. See Cicchetti and Allison (1971) for details.

The SCORES= option in the TABLES statement determines the type of column scores used to compute the kappa weights (and other score-based statistics). The default is SCORES=TABLE. See the section "Scores" on page 118 for details. For numeric variables, table scores are the values of the variable levels. You can assign numeric values to the levels in a way that reflects their level of similarity. For example, suppose you have four levels and order them according to similarity. If you assign them values of 0, 2, 4, and 10, the Cicchetti-Allison kappa weights take the following values: $w_{12} = 0.8$, $w_{13} = 0.6$, $w_{14} = 0$, $w_{23} = 0.8$, $w_{24} = 0.2$, and $w_{34} = 0.4$. Note that when there are only two categories (that is, $C = 2$), the weighted kappa coefficient is identical to the simple kappa coefficient.

If you specify (WT=FC) with the AGREE option in the TABLES statement, PROC FREQ computes Fleiss-Cohen kappa coefficient weights as

$$w_{ij} = 1 - \frac{(C_i - C_j)^2}{(C_C - C_1)^2}$$

See Fleiss and Cohen (1973) for details.

For the preceding example, the Fleiss-Cohen kappa weights are: $w_{12} = 0.96$, $w_{13} = 0.84$, $w_{14} = 0$, $w_{23} = 0.96$, $w_{24} = 0.36$, and $w_{34} = 0.64$.

Overall Kappa Coefficient

When there are multiple strata, PROC FREQ combines the stratum-level estimates of kappa into an overall estimate of the supposed common value of kappa. Assume there are q strata, indexed by $h = 1, 2, \ldots, q$, and let $\text{var}(\hat{\kappa}_h)$ denote the variance of $\hat{\kappa}_h$. The estimate of the overall kappa coefficient is computed as

$$\hat{\kappa}_T = \sum_{h=1}^{q} \frac{\hat{\kappa}_h}{\text{var}(\hat{\kappa}_h)} \Big/ \sum_{h=1}^{q} \frac{1}{\text{var}(\hat{\kappa}_h)}$$

See Fleiss, Levin, and Paik (2003) for details.

PROC FREQ computes an estimate of the overall weighted kappa in the same way.

Tests for Equal Kappa Coefficients

When there are multiple strata, the following chi-square statistic tests whether the stratum-level values of kappa are equal:

$$Q_K = \sum_{h=1}^{q} (\hat{\kappa}_h - \hat{\kappa}_T)^2 \big/ \text{var}(\hat{\kappa}_h)$$

Under the null hypothesis of equal kappas for the q strata, Q_K has an asymptotic chi-square distribution with $q - 1$ degrees of freedom. See Fleiss, Levin, and Paik (2003) for more information. PROC FREQ computes a test for equal weighted kappa coefficients in the same way.

Cochran's Q Test

Cochran's Q is computed for multiway tables when each variable has two levels, that is, for $2 \times 2 \cdots \times 2$ tables. Cochran's Q statistic is used to test the homogeneity of the one-dimensional margins. Let m denote the number of variables and N denote the total number of subjects. Cochran's Q statistic is computed as

$$Q_C = m(m-1) \left(\sum_{j=1}^{m} T_j^2 - T^2 \right) / \left(mT - \sum_{k=1}^{N} S_k^2 \right)$$

where T_j is the number of positive responses for variable j, T is the total number of positive responses over all variables, and S_k is the number of positive responses for subject k. Under the null hypothesis, Cochran's Q has an asymptotic chi-square distribution with $m - 1$ degrees of freedom. See Cochran (1950) for details. When there are only two binary response variables ($m = 2$), Cochran's Q simplifies to McNemar's test. When there are more than two response categories, you can test for marginal homogeneity by using the repeated measures capabilities of the CATMOD procedure.

Tables with Zero Rows and Columns

The AGREE statistics are defined only for square tables, where the number of rows equals the number of columns. If the table is not square, PROC FREQ does not compute AGREE statistics. In the kappa statistic framework, where two independent raters assign ratings to each of n subjects, suppose one of the raters does not use all possible r rating levels. If the corresponding table has r rows but only $r - 1$ columns, then the table is not square and PROC FREQ does not compute AGREE statistics. To create a square table in this situation, use the ZEROS option in the WEIGHT statement, which requests that PROC FREQ include observations with zero weights in the analysis. Include zero-weight observations in the input data set to represent any rating levels that are not used by a rater, so that the input data set has at least one observation for each possible rater and rating combination. The analysis then includes all rating levels, even when all levels are not actually assigned by both raters. The resulting table (of rater 1 by rater 2) is a square table, and AGREE statistics can be computed.

For more information, see the description of the ZEROS option. By default, PROC FREQ does not process observations that have zero weights, because these observations do not contribute to the total frequency count, and because any resulting zero-weight row or column causes many of the tests and measures of association to be undefined. However, kappa statistics are defined for tables with a zero-weight row or column, and the ZEROS option makes it possible to input zero-weight observations and construct the tables needed to compute kappas.

Cochran-Mantel-Haenszel Statistics

The CMH option in the TABLES statement gives a stratified statistical analysis of the relationship between the row and column variables after controlling for the strata variables in a multiway table. For example, for the table request A*B*C*D, the CMH option provides an analysis of the relationship between C and D, after controlling for A and B. The stratified analysis provides a way to adjust for the possible confounding effects of A and B without being forced to estimate parameters for them.

The CMH analysis produces Cochran-Mantel-Haenszel statistics, which include the correlation statistic, the ANOVA (row mean scores) statistic, and the general association statistic. For 2×2 tables, the CMH option also provides Mantel-Haenszel and logit estimates of the common odds ratio and the common relative risks, as well as the Breslow-Day test for homogeneity of the odds ratios.

Exact statistics are also available for stratified 2×2 tables. If you specify the EQOR option in the EXACT statement, PROC FREQ provides Zelen's exact test for equal odds ratios. If you specify the COMOR option in the EXACT statement, PROC FREQ provides exact confidence limits for the common odds ratio and an exact test that the common odds ratio equals one.

Let the number of strata be denoted by q, indexing the strata by $h = 1, 2, \ldots, q$. Each stratum contains a contingency table with X representing the row variable and Y representing the column variable. For table h, denote the cell frequency in row i and column j by n_{hij}, with corresponding row and column marginal totals denoted by $n_{hi\cdot}$ and $n_{h\cdot j}$, and the overall stratum total by n_h.

Because the formulas for the Cochran-Mantel-Haenszel statistics are more easily defined in terms of matrices, the following notation is used. Vectors are presumed to be column vectors unless they are transposed $(')$.

$$\mathbf{n}'_{hi} = (n_{hi1}, n_{hi2}, \ldots, n_{hiC}) \quad (1 \times C)$$

$$\mathbf{n}'_h = (\mathbf{n}'_{h1}, \mathbf{n}'_{h2}, \ldots, \mathbf{n}'_{hR}) \quad (1 \times RC)$$

$$p_{hi\cdot} = n_{hi\cdot} / n_h \quad (1 \times 1)$$

$$p_{h\cdot j} = n_{h\cdot j} / n_h \quad (1 \times 1)$$

$$\mathbf{P}'_{h*\cdot} = (p_{h1\cdot}, p_{h2\cdot}, \ldots, p_{hR\cdot}) \quad (1 \times R)$$

$$\mathbf{P}'_{h\cdot *} = (p_{h\cdot 1}, p_{h\cdot 2}, \ldots, p_{h\cdot C}) \quad (1 \times C)$$

Assume that the strata are independent and that the marginal totals of each stratum are fixed. The null hypothesis, H_0, is that there is no association between X and Y in any of the strata. The corresponding model is the multiple hypergeometric; this implies that, under H_0, the expected value and covariance matrix of the frequencies are, respectively,

$$\mathbf{m}_h = \mathbf{E}[\mathbf{n}_h \mid H_0] = n_h (\mathbf{P}_{h\cdot *} \otimes \mathbf{P}_{h*\cdot})$$

$$\text{var}[\mathbf{n}_h \mid H_0] = c \left((\mathbf{D}_{\mathbf{P}_{h\cdot *}} - \mathbf{P}_{h\cdot *}\mathbf{P}'_{h\cdot *}) \otimes (\mathbf{D}_{\mathbf{P}_{h*\cdot}} - \mathbf{P}_{h*\cdot}\mathbf{P}'_{h*\cdot}) \right)$$

where

$$c = n_h^2 / (n_h - 1)$$

and where $\otimes$ denotes Kronecker product multiplication and $\mathbf{D_a}$ is a diagonal matrix with the elements of $\mathbf{a}$ on the main diagonal.

The generalized CMH statistic (Landis, Heyman, and Koch 1978) is defined as

$$Q_{CMH} = \mathbf{G}' \mathbf{V_G}^{-1} \mathbf{G}$$

where

$$G = \sum_h \mathbf{B}_h (\mathbf{n}_h - \mathbf{m}_h)$$

$$\mathbf{V}_G = \sum_h \mathbf{B}_h \left(Var(\mathbf{n}_h \mid H_0) \right) \mathbf{B}_h'$$

and where

$$\mathbf{B}_h = \mathbf{C}_h \otimes \mathbf{R}_h$$

is a matrix of fixed constants based on column scores $\mathbf{C}_h$ and row scores $\mathbf{R}_h$. When the null hypothesis is true, the CMH statistic has an asymptotic chi-square distribution with degrees of freedom equal to the rank of $\mathbf{B}_h$. If $\mathbf{V}_G$ is found to be singular, PROC FREQ prints a message and sets the value of the CMH statistic to missing.

PROC FREQ computes three CMH statistics by using this formula for the generalized CMH statistic, with different row and column score definitions for each statistic. The CMH statistics that PROC FREQ computes are the correlation statistic, the ANOVA (row mean scores) statistic, and the general association statistic. These statistics test the null hypothesis of no association against different alternative hypotheses. The following sections describe the computation of these CMH statistics.

CAUTION: The CMH statistics have low power for detecting an association in which the patterns of association for some of the strata are in the opposite direction of the patterns displayed by other strata. Thus, a nonsignificant CMH statistic suggests either that there is no association or that no pattern of association has enough strength or consistency to dominate any other pattern.

Correlation Statistic

The correlation statistic, popularized by Mantel and Haenszel (1959) and Mantel (1963), has one degree of freedom and is known as the Mantel-Haenszel statistic.

The alternative hypothesis for the correlation statistic is that there is a linear association between X and Y in at least one stratum. If either X or Y does not lie on an ordinal (or interval) scale, then this statistic is not meaningful.

To compute the correlation statistic, PROC FREQ uses the formula for the generalized CMH statistic with the row and column scores determined by the SCORES= option in the TABLES statement. See the section "Scores" on page 118 for more information about the available score types. The matrix of row scores $\mathbf{R}_h$ has dimension $1 \times R$, and the matrix of column scores $\mathbf{C}_h$ has dimension $1 \times C$.

When there is only one stratum, this CMH statistic reduces to $(n-1)r^2$, where r is the Pearson correlation coefficient between X and Y. When nonparametric (RANK or RIDIT) scores are specified, the statistic reduces to $(n-1)r_s^2$, where r_s is the Spearman rank correlation coefficient between X and Y. When there is more than one stratum, this CMH statistic becomes a stratum-adjusted correlation statistic.

ANOVA (Row Mean Scores) Statistic

The ANOVA statistic can be used only when the column variable Y lies on an ordinal (or interval) scale so that the mean score of Y is meaningful. For the ANOVA statistic, the mean score is computed for each row of the table, and the alternative hypothesis is that, for at least one stratum, the mean scores of the R rows are unequal. In other words, the statistic is sensitive to location differences among the R distributions of Y.

The matrix of column scores $\mathbf{C}_h$ has dimension $1 \times C$, and the column scores are determined by the SCORES= option.

The matrix of row scores $\mathbf{R}_h$ has dimension $(R-1) \times R$ and is created internally by PROC FREQ as

$$\mathbf{R}_h = [\mathbf{I}_{R-1}, -\mathbf{J}_{R-1}]$$

where $\mathbf{I}_{R-1}$ is an identity matrix of rank $R-1$ and $\mathbf{J}_{R-1}$ is an $(R-1) \times 1$ vector of ones. This matrix has the effect of forming $R-1$ independent contrasts of the R mean scores.

When there is only one stratum, this CMH statistic is essentially an analysis of variance (ANOVA) statistic in the sense that it is a function of the variance ratio F statistic that would be obtained from a one-way ANOVA on the dependent variable Y. If nonparametric scores are specified in this case, then the ANOVA statistic is a Kruskal-Wallis test.

If there is more than one stratum, then this CMH statistic corresponds to a stratum-adjusted ANOVA or Kruskal-Wallis test. In the special case where there is one subject per row and one subject per column in the contingency table of each stratum, this CMH statistic is identical to Friedman's chi-square. See Example 3.9 for an illustration.

General Association Statistic

The alternative hypothesis for the general association statistic is that, for at least one stratum, there is some kind of association between X and Y. This statistic is always interpretable because it does not require an ordinal scale for either X or Y.

For the general association statistic, the matrix $\mathbf{R}_h$ is the same as the one used for the ANOVA statistic. The matrix $\mathbf{C}_h$ is defined similarly as

$$\mathbf{C}_h = [\mathbf{I}_{C-1}, -\mathbf{J}_{C-1}]$$

PROC FREQ generates both score matrices internally. When there is only one stratum, then the general association CMH statistic reduces to $Q_P(n-1)/n$, where Q_P is the Pearson chi-square statistic. When there is more than one stratum, then the CMH statistic becomes a stratum-adjusted Pearson chi-square statistic. Note that a similar adjustment can be made by summing the Pearson chi-squares across the strata. However, the latter statistic requires a large sample size in each stratum to support the resulting chi-square distribution with $q(R-1)(C-1)$ degrees of freedom. The CMH statistic requires only a large overall sample size because it has only $(R-1)(C-1)$ degrees of freedom.

See Cochran (1954); Mantel and Haenszel (1959); Mantel (1963); Birch (1965); and Landis, Heyman, and Koch (1978).

Adjusted Odds Ratio and Relative Risk Estimates

The CMH option provides adjusted odds ratio and relative risk estimates for stratified 2 × 2 tables. For each of these measures, PROC FREQ computes a Mantel-Haenszel estimate and a logit estimate. These estimates apply to n-way table requests in the TABLES statement, when the row and column variables both have two levels.

For example, for the table request A*B*C*D, if the row and column variables C and D both have two levels, PROC FREQ provides odds ratio and relative risk estimates, adjusting for the confounding variables A and B.

The choice of an appropriate measure depends on the study design. For case-control (retrospective) studies, the odds ratio is appropriate. For cohort (prospective) or cross-sectional studies, the relative risk is appropriate. See the section "Odds Ratio and Relative Risks for 2 x 2 Tables" on page 148 for more information on these measures.

Throughout this section, z denotes the $100(1-\alpha/2)$th percentile of the standard normal distribution.

Odds Ratio, Case-Control Studies PROC FREQ provides Mantel-Haenszel and logit estimates for the common odds ratio for stratified 2 × 2 tables.

The Mantel-Haenszel estimate of the common odds ratio is computed as

$$OR_{MH} = \left(\sum_h n_{h11} n_{h22}/n_h \right) / \left(\sum_h n_{h12} n_{h21}/n_h \right)$$

It is always computed unless the denominator is zero. See Mantel and Haenszel (1959) and Agresti (2002) for details.

To compute confidence limits for the common odds ratio, PROC FREQ uses the Greenland and Robins (1985) variance estimate for $\ln(OR_{MH})$. The $100(1-\alpha/2)$ confidence limits for the common odds ratio are

$$\left(OR_{MH} \times \exp(-z\hat{\sigma}), \ OR_{MH} \times \exp(z\hat{\sigma}) \right)$$

where

$$\begin{aligned}
\hat{\sigma}^2 &= \widehat{\text{var}}(\ln(OR_{MH})) \\
&= \frac{\sum_h (n_{h11} + n_{h22})(n_{h11} n_{h22})/n_h^2}{2\left(\sum_h n_{h11} n_{h22}/n_h\right)^2} \\
&\quad + \frac{\sum_h [(n_{h11} + n_{h22})(n_{h12} n_{h21}) + (n_{h12} + n_{h21})(n_{h11} n_{h22})]/n_h^2}{2\left(\sum_h n_{h11} n_{h22}/n_h\right)\left(\sum_h n_{h12} n_{h21}/n_h\right)} \\
&\quad + \frac{\sum_h (n_{h12} + n_{h21})(n_{h12} n_{h21})/n_h^2}{2\left(\sum_h n_{h12} n_{h21}/n_h\right)^2}
\end{aligned}$$

Note that the Mantel-Haenszel odds ratio estimator is less sensitive to small n_h than the logit estimator.

The adjusted logit estimate of the common odds ratio (Woolf 1955) is computed as

$$OR_L = \exp\left(\sum_h w_h \ln(OR_h) / \sum_h w_h\right)$$

and the corresponding $100(1-\alpha)\%$ confidence limits are

$$\left(OR_L \times \exp\left(-z/\sqrt{\sum_h w_h}\right),\ OR_L \times \exp\left(z/\sqrt{\sum_h w_h}\right)\right)$$

where OR_h is the odds ratio for stratum h, and

$$w_h = 1/\text{var}(\ln(OR_h))$$

If any table cell frequency in a stratum h is zero, PROC FREQ adds 0.5 to each cell of the stratum before computing OR_h and w_h (Haldane 1955) for the logit estimate. The procedure prints a warning when this occurs.

Relative Risks, Cohort Studies PROC FREQ provides Mantel-Haenszel and logit estimates of the common relative risks for stratified 2×2 tables.

The Mantel-Haenszel estimate of the common relative risk for column 1 is computed as

$$RR_{MH} = \left(\sum_h n_{h11}\, n_{h2\cdot} / n_h\right) / \left(\sum_h n_{h21}\, n_{h1\cdot} / n_h\right)$$

It is always computed unless the denominator is zero. See Mantel and Haenszel (1959) and Agresti (2002) for more information.

To compute confidence limits for the common relative risk, PROC FREQ uses the Greenland and Robins (1985) variance estimate for $\log(RR_{MH})$. The $100(1-\alpha/2)$ confidence limits for the common relative risk are

$$\left(RR_{MH} \times \exp(-z\hat{\sigma}),\ RR_{MH} \times \exp(z\hat{\sigma})\right)$$

where

$$\hat{\sigma}^2 = \widehat{\text{var}}(\ln(RR_{MH})) = \frac{\sum_h (n_{h1\cdot}\, n_{h2\cdot}\, n_{h\cdot 1} - n_{h11}\, n_{h21}\, n_h)/n_h^2}{\left(\sum_h n_{h11}\, n_{h2\cdot}/n_h\right)\left(\sum_h n_{h21}\, n_{h1\cdot}/n_h\right)}$$

The adjusted logit estimate of the common relative risk for column 1 is computed as

$$RR_L = \exp\left(\sum_h w_h \ln(RR_h) / \sum_h w_h\right)$$

and the corresponding $100(1-\alpha)\%$ confidence limits are

$$\left(RR_L \times \exp\left(-z/\sqrt{\sum_h w_h}\right),\ RR_L \times \exp\left(z/\sqrt{\sum_h w_h}\right)\right)$$

where RR_h is the column 1 relative risk estimate for stratum h and

$$w_h = 1 / \text{var}(\ln(RR_h))$$

If n_{h11} or n_{h21} is zero, then PROC FREQ adds 0.5 to each cell of the stratum before computing RR_h and w_h for the logit estimate. The procedure prints a warning when this occurs. See Kleinbaum, Kupper, and Morgenstern (1982, Sections 17.4 and 17.5) for details.

Breslow-Day Test for Homogeneity of the Odds Ratios

When you specify the CMH option, PROC FREQ computes the Breslow-Day test for stratified 2×2 tables. It tests the null hypothesis that the odds ratios for the q strata are equal. When the null hypothesis is true, the statistic has approximately a chi-square distribution with $q - 1$ degrees of freedom. See Breslow and Day (1980) and Agresti (2007) for more information.

The Breslow-Day statistic is computed as

$$Q_{BD} = \sum_h (n_{h11} - \text{E}(n_{h11} \mid OR_{MH}))^2 / \text{var}(n_{h11} \mid OR_{MH})$$

where E and var denote expected value and variance, respectively. The summation does not include any table with a zero row or column. If OR_{MH} equals zero or if it is undefined, then PROC FREQ does not compute the statistic and prints a warning message.

For the Breslow-Day test to be valid, the sample size should be relatively large in each stratum, and at least 80% of the expected cell counts should be greater than 5. Note that this is a stricter sample size requirement than the requirement for the Cochran-Mantel-Haenszel test for $q \times 2 \times 2$ tables, in that each stratum sample size (not just the overall sample size) must be relatively large. Even when the Breslow-Day test is valid, it might not be very powerful against certain alternatives, as discussed in Breslow and Day (1980).

If you specify the BDT option, PROC FREQ computes the Breslow-Day test with Tarone's adjustment, which subtracts an adjustment factor from Q_{BD} to make the resulting statistic asymptotically chi-square. The Breslow-Day-Tarone statistic is computed as

$$Q_{BDT} = Q_{BD} - \left(\sum_h (n_{h11} - \text{E}(n_{h11} \mid OR_{MH})) \right)^2 / \sum_h \text{var}(n_{h11} \mid OR_{MH})$$

See Tarone (1985), Jones et al. (1989), and Breslow (1996) for more information.

Zelen's Exact Test for Equal Odds Ratios

If you specify the EQOR option in the EXACT statement, PROC FREQ computes Zelen's exact test for equal odds ratios for stratified 2×2 tables. Zelen's test is an exact counterpart to the Breslow-Day asymptotic test for equal odds ratios. The reference set for Zelen's test includes all possible $q \times 2 \times 2$ tables with the same row, column, and stratum totals as the observed multiway table and with the same sum of cell (1, 1) frequencies as the observed table. The test statistic is the probability of the observed $q \times 2 \times 2$ table conditional on the fixed margins, which is a product of hypergeometric probabilities.

The *p*-value for Zelen's test is the sum of all table probabilities that are less than or equal to the observed table probability, where the sum is computed over all tables in the reference set determined by the fixed margins and the observed sum of cell (1, 1) frequencies. This test is similar to Fisher's exact test for two-way tables. See Zelen (1971), Hirji (2006), and Agresti (1992) for more information. PROC FREQ computes Zelen's exact test by using the polynomial multiplication algorithm of Hirji et al. (1996).

Exact Confidence Limits for the Common Odds Ratio

If you specify the COMOR option in the EXACT statement, PROC FREQ computes exact confidence limits for the common odds ratio for stratified 2×2 tables. This computation assumes that the odds ratio is constant over all the 2×2 tables. Exact confidence limits are constructed from the distribution of $S = \sum_h n_{h11}$, conditional on the marginal totals of the 2×2 tables.

Because this is a discrete problem, the confidence coefficient for these exact confidence limits is not exactly $(1 - \alpha)$ but is at least $(1 - \alpha)$. Thus, these confidence limits are conservative. See Agresti (1992) for more information.

PROC FREQ computes exact confidence limits for the common odds ratio by using an algorithm based on Vollset, Hirji, and Elashoff (1991). See also Mehta, Patel, and Gray (1985).

Conditional on the marginal totals of 2×2 table h, let the random variable S_h denote the frequency of table cell (1, 1). Given the row totals $n_{h1 \cdot}$ and $n_{h2 \cdot}$ and column totals $n_{h \cdot 1}$ and $n_{h \cdot 2}$, the lower and upper bounds for S_h are l_h and u_h,

$$l_h = \max(0, n_{h1 \cdot} - n_{h \cdot 2})$$
$$u_h = \min(n_{h1 \cdot}, n_{h \cdot 1})$$

Let C_{s_h} denote the hypergeometric coefficient,

$$C_{s_h} = \binom{n_{h \cdot 1}}{s_h} \binom{n_{h \cdot 2}}{n_{h1 \cdot} - s_h}$$

and let ϕ denote the common odds ratio. Then the conditional distribution of S_h is

$$P(S_h = s_h \mid n_{1 \cdot}, n_{\cdot 1}, n_{\cdot 2}) = C_{s_h} \phi^{s_h} \Big/ \sum_{x = l_h}^{x = u_h} C_x \phi^x$$

Summing over all the 2×2 tables, $S = \sum_h S_h$, and the lower and upper bounds of S are l and u,

$$l = \sum_h l_h \quad \text{and} \quad u = \sum_h u_h$$

The conditional distribution of the sum S is

$$P(S = s \mid n_{h1 \cdot}, n_{h \cdot 1}, n_{h \cdot 2}; h = 1, \ldots, q) = C_s \phi^s \Big/ \sum_{x = l}^{x = u} C_x \phi^x$$

where

$$C_s = \sum_{s_1+\ldots+s_q = s} \left(\prod_h C_{s_h} \right)$$

Let s_0 denote the observed sum of cell (1,1) frequencies over the q tables. The following two equations are solved iteratively for lower and upper confidence limits for the common odds ratio, ϕ_1 and ϕ_2:

$$\sum_{x=s_0}^{x=u} C_x \phi_1^x \Big/ \sum_{x=l}^{x=u} C_x \phi_1^x = \alpha/2$$

$$\sum_{x=l}^{x=s_0} C_x \phi_2^x \Big/ \sum_{x=l}^{x=u} C_x \phi_2^x = \alpha/2$$

When the observed sum s_0 equals the lower bound l, PROC FREQ sets the lower confidence limit to zero and determines the upper limit with level α. Similarly, when the observed sum s_0 equals the upper bound u, PROC FREQ sets the upper confidence limit to infinity and determines the lower limit with level α.

When you specify the COMOR option in the EXACT statement, PROC FREQ also computes the exact test that the common odds ratio equals one. Setting $\phi = 1$, the conditional distribution of the sum S under the null hypothesis becomes

$$P_0(S = s \mid n_{h1\cdot},\ n_{h\cdot 1},\ n_{h\cdot 2};\ h = 1, \ldots, q) = C_s \Big/ \sum_{x=l}^{x=u} C_x$$

The point probability for this exact test is the probability of the observed sum s_0 under the null hypothesis, conditional on the marginals of the stratified 2×2 tables, and is denoted by $P_0(s_0)$. The expected value of S under the null hypothesis is

$$E_0(S) = \sum_{x=l}^{x=u} x\, C_x \Big/ \sum_{x=l}^{x=u} C_x$$

The one-sided exact p-value is computed from the conditional distribution as $P_0(S >= s_0)$ or $P_0(S \leq s_0)$, depending on whether the observed sum s_0 is greater or less than $E_0(S)$,

$$P_1 = P_0(S >= s_0) = \sum_{x=s_0}^{x=u} C_x \Big/ \sum_{x=l}^{x=u} C_x \quad \text{if } s_0 > E_0(S)$$

$$P_1 = P_0(S <= s_0) = \sum_{x=l}^{x=s_0} C_x \Big/ \sum_{x=l}^{x=u} C_x \quad \text{if } s_0 \leq E_0(S)$$

PROC FREQ computes two-sided *p*-values for this test according to three different definitions. A two-sided *p*-value is computed as twice the one-sided *p*-value, setting the result equal to one if it exceeds one,

$$P_2^a = 2 \times P_1$$

Additionally, a two-sided *p*-value is computed as the sum of all probabilities less than or equal to the point probability of the observed sum s_0, summing over all possible values of s, $l \leq s \leq u$,

$$P_2^b = \sum_{l \leq s \leq u : P_0(s) \leq P_0(s_0)} P_0(s)$$

Also, a two-sided *p*-value is computed as the sum of the one-sided *p*-value and the corresponding area in the opposite tail of the distribution, equidistant from the expected value,

$$P_2^c = P_0 \left(|S - E_0(S)| \geq |s_0 - E_0(S)| \right)$$

Exact Statistics

Exact statistics can be useful in situations where the asymptotic assumptions are not met, and so the asymptotic *p*-values are not close approximations for the true *p*-values. Standard asymptotic methods involve the assumption that the test statistic follows a particular distribution when the sample size is sufficiently large. When the sample size is not large, asymptotic results might not be valid, with the asymptotic *p*-values differing perhaps substantially from the exact *p*-values. Asymptotic results might also be unreliable when the distribution of the data is sparse, skewed, or heavily tied. See Agresti (2007) and Bishop, Fienberg, and Holland (1975) for more information. Exact computations are based on the statistical theory of exact conditional inference for contingency tables, reviewed by Agresti (1992).

In addition to computation of exact *p*-values, PROC FREQ provides the option of estimating exact *p*-values by Monte Carlo simulation. This can be useful for problems that are so large that exact computations require a great amount of time and memory, but for which asymptotic approximations might not be sufficient.

Exact statistics are available for many PROC FREQ tests. For one-way tables, PROC FREQ provides exact *p*-values for the binomial proportion tests and the chi-square goodness-of-fit test. Exact confidence limits are available for the binomial proportion. For two-way tables, PROC FREQ provides exact *p*-values for the following tests: Pearson chi-square test, likelihood-ratio chi-square test, Mantel-Haenszel chi-square test, Fisher's exact test, Jonckheere-Terpstra test, and Cochran-Armitage test for trend. PROC FREQ also computes exact *p*-values for tests of the following statistics: Pearson correlation coefficient, Spearman correlation coefficient, simple kappa coefficient, and weighted kappa coefficient. For 2 × 2 tables, PROC FREQ provides exact confidence limits for the odds ratio, exact unconditional confidence limits for the proportion difference, and McNemar's exact test. For stratified 2 × 2 tables, PROC FREQ provides Zelen's exact test for equal odds ratios, exact confidence limits for the common odds ratio, and an exact test for the common odds ratio.

The following sections summarize the exact computational algorithms, define the exact *p*-values that PROC FREQ computes, discuss the computational resource requirements, and describe the Monte Carlo estimation option.

Computational Algorithms

PROC FREQ computes exact p-values for general $R \times C$ tables by using the network algorithm developed by Mehta and Patel (1983). This algorithm provides a substantial advantage over direct enumeration, which can be very time-consuming and feasible only for small problems. See Agresti (1992) for a review of algorithms for computation of exact p-values, and see Mehta, Patel, and Tsiatis (1984) and Mehta, Patel, and Senchaudhuri (1991) for information about the performance of the network algorithm.

The reference set for a given contingency table is the set of all contingency tables with the observed marginal row and column sums. Corresponding to this reference set, the network algorithm forms a directed acyclic network consisting of nodes in a number of stages. A path through the network corresponds to a distinct table in the reference set. The distances between nodes are defined so that the total distance of a path through the network is the corresponding value of the test statistic. At each node, the algorithm computes the shortest and longest path distances for all the paths that pass through that node. For statistics that can be expressed as a linear combination of cell frequencies multiplied by increasing row and column scores, PROC FREQ computes shortest and longest path distances by using the algorithm of Agresti, Mehta, and Patel (1990). For statistics of other forms, PROC FREQ computes an upper bound for the longest path and a lower bound for the shortest path by following the approach of Valz and Thompson (1994).

The longest and shortest path distances or bounds for a node are compared to the value of the test statistic to determine whether all paths through the node contribute to the p-value, none of the paths through the node contribute to the p-value, or neither of these situations occurs. If all paths through the node contribute, the p-value is incremented accordingly, and these paths are eliminated from further analysis. If no paths contribute, these paths are eliminated from the analysis. Otherwise, the algorithm continues, still processing this node and the associated paths. The algorithm finishes when all nodes have been accounted for.

In applying the network algorithm, PROC FREQ uses full numerical precision to represent all statistics, row and column scores, and other quantities involved in the computations. Although it is possible to use rounding to improve the speed and memory requirements of the algorithm, PROC FREQ does not do this because it can result in reduced accuracy of the p-values.

For one-way tables, PROC FREQ computes the exact chi-square goodness-of-fit test by the method of Radlow and Alf (1975). PROC FREQ generates all possible one-way tables with the observed total sample size and number of categories. For each possible table, PROC FREQ compares its chi-square value with the value for the observed table. If the table's chi-square value is greater than or equal to the observed chi-square, PROC FREQ increments the exact p-value by the probability of that table, which is calculated under the null hypothesis by using the multinomial frequency distribution. By default, the null hypothesis states that all categories have equal proportions. If you specify null hypothesis proportions or frequencies by using the TESTP= or TESTF= option in the TABLES statement, then PROC FREQ calculates the exact chi-square test based on that null hypothesis.

Other exact computations are described in sections about the individual statistics. See the section "Binomial Proportion" on page 133 for details about how PROC FREQ computes exact confidence limits and tests for the binomial proportion. See the section "Odds Ratio and Relative Risks for 2 x 2 Tables" on page 148 for information about computation of exact confidence limits for the odds ratio for 2×2 tables. Also, see the sections "Exact Unconditional Confidence Limits for the Risk

Difference" on page 147, "Exact Confidence Limits for the Common Odds Ratio" on page 165, and "Zelen's Exact Test for Equal Odds Ratios" on page 164.

Definition of p-Values

For several tests in PROC FREQ, the test statistic is nonnegative, and large values of the test statistic indicate a departure from the null hypothesis. Such nondirectional tests include the Pearson chi-square, the likelihood-ratio chi-square, the Mantel-Haenszel chi-square, Fisher's exact test for tables larger than 2 × 2, McNemar's test, and the one-way chi-square goodness-of-fit test. The exact p-value for a nondirectional test is the sum of probabilities for those tables having a test statistic greater than or equal to the value of the observed test statistic.

There are other tests where it might be appropriate to test against either a one-sided or a two-sided alternative hypothesis. For example, when you test the null hypothesis that the true parameter value equals 0 ($T = 0$), the alternative of interest might be one-sided ($T \leq 0$, or $T \geq 0$) or two-sided ($T \neq 0$). Such tests include the Pearson correlation coefficient, Spearman correlation coefficient, Jonckheere-Terpstra test, Cochran-Armitage test for trend, simple kappa coefficient, and weighted kappa coefficient. For these tests, PROC FREQ displays the right-sided p-value when the observed value of the test statistic is greater than its expected value. The right-sided p-value is the sum of probabilities for those tables for which the test statistic is greater than or equal to the observed test statistic. Otherwise, when the observed test statistic is less than or equal to the expected value, PROC FREQ displays the left-sided p-value. The left-sided p-value is the sum of probabilities for those tables for which the test statistic is less than or equal to the one observed. The one-sided p-value P_1 can be expressed as

$$P_1 = \begin{cases} \text{Prob(Test Statistic} \geq t) & \text{if } t > E_0(T) \\ \text{Prob(Test Statistic} \leq t) & \text{if } t \leq E_0(T) \end{cases}$$

where t is the observed value of the test statistic and $E_0(T)$ is the expected value of the test statistic under the null hypothesis. PROC FREQ computes the two-sided p-value as the sum of the one-sided p-value and the corresponding area in the opposite tail of the distribution of the statistic, equidistant from the expected value. The two-sided p-value P_2 can be expressed as

$$P_2 = \text{Prob}(\,|\text{Test Statistic} - E_0(T)| \geq |t - E_0(T)|\,)$$

If you specify the POINT option in the EXACT statement, PROC FREQ also displays exact point probabilities for the test statistics. The exact point probability is the exact probability that the test statistic equals the observed value.

Computational Resources

PROC FREQ uses relatively fast and efficient algorithms for exact computations. These recently developed algorithms, together with improvements in computer power, now make it feasible to perform exact computations for data sets where previously only asymptotic methods could be applied. Nevertheless, there are still large problems that might require a prohibitive amount of time and memory for exact computations, depending on the speed and memory available on your computer. For large problems, consider whether exact methods are really needed or whether asymptotic methods might give results quite close to the exact results, while requiring much less computer time and memory. When asymptotic methods might not be sufficient for such large problems, consider using Monte Carlo estimation of exact *p*-values, as described in the section "Monte Carlo Estimation" on page 170.

A formula does not exist that can predict in advance how much time and memory are needed to compute an exact *p*-value for a certain problem. The time and memory required depend on several factors, including which test is being performed, the total sample size, the number of rows and columns, and the specific arrangement of the observations into table cells. Generally, larger problems (in terms of total sample size, number of rows, and number of columns) tend to require more time and memory. Additionally, for a fixed total sample size, time and memory requirements tend to increase as the number of rows and columns increases, because this corresponds to an increase in the number of tables in the reference set. Also for a fixed sample size, time and memory requirements increase as the marginal row and column totals become more homogeneous. See Agresti, Mehta, and Patel (1990) and Gail and Mantel (1977) for more information.

At any time while PROC FREQ is computing exact *p*-values, you can terminate the computations by pressing the system interrupt key sequence (see the *SAS Companion* for your system) and choosing to stop computations. After you terminate exact computations, PROC FREQ completes all other remaining tasks. The procedure produces the requested output and reports missing values for any exact *p*-values that were not computed by the time of termination.

You can also use the MAXTIME= option in the EXACT statement to limit the amount of time PROC FREQ uses for exact computations. You specify a MAXTIME= value that is the maximum amount of clock time (in seconds) that PROC FREQ can use to compute an exact *p*-value. If PROC FREQ does not finish computing an exact *p*-value within that time, it terminates the computation and completes all other remaining tasks.

Monte Carlo Estimation

If you specify the option MC in the EXACT statement, PROC FREQ computes Monte Carlo estimates of the exact *p*-values instead of directly computing the exact *p*-values. Monte Carlo estimation can be useful for large problems that require a great amount of time and memory for exact computations but for which asymptotic approximations might not be sufficient. To describe the precision of each Monte Carlo estimate, PROC FREQ provides the asymptotic standard error and $100(1 - \alpha)\%$ confidence limits. The confidence level α is determined by the ALPHA= option in the EXACT statement, which, by default, equals 0.01 and produces 99% confidence limits. The N=*n* option in the EXACT statement specifies the number of samples that PROC FREQ uses for Monte Carlo estimation; the default is 10000 samples. You can specify a larger value for *n* to improve the precision of the Monte Carlo estimates. Because larger values of *n* generate more samples, the computation time increases. Alternatively, you can specify a smaller value of *n* to reduce the computation time.

To compute a Monte Carlo estimate of an exact *p*-value, PROC FREQ generates a random sample of tables with the same total sample size, row totals, and column totals as the observed table. PROC FREQ uses the algorithm of Agresti, Wackerly, and Boyett (1979), which generates tables in proportion to their hypergeometric probabilities conditional on the marginal frequencies. For each sample table, PROC FREQ computes the value of the test statistic and compares it to the value for the observed table. When estimating a right-sided *p*-value, PROC FREQ counts all sample tables for which the test statistic is greater than or equal to the observed test statistic. Then the *p*-value estimate equals the number of these tables divided by the total number of tables sampled.

$$\hat{P}_{MC} = M / N$$
$$M = \text{number of samples with (Test Statistic} \geq t)$$
$$N = \text{total number of samples}$$
$$t = \text{observed Test Statistic}$$

PROC FREQ computes left-sided and two-sided p-value estimates in a similar manner. For left-sided p-values, PROC FREQ evaluates whether the test statistic for each sampled table is less than or equal to the observed test statistic. For two-sided p-values, PROC FREQ examines the sample test statistics according to the expression for P_2 given in the section "Definition of p-Values" on page 169.

The variable M is a binomially distributed variable with N trials and success probability p. It follows that the asymptotic standard error of the Monte Carlo estimate is

$$\text{se}(\hat{P}_{MC}) = \sqrt{\hat{P}_{MC}(1 - \hat{P}_{MC}) / (N - 1)}$$

PROC FREQ constructs asymptotic confidence limits for the p-values according to

$$\hat{P}_{MC} \pm \left(z_{\alpha/2} \times \text{se}(\hat{P}_{MC}) \right)$$

where $z_{\alpha/2}$ is the $100(1 - \alpha/2)$th percentile of the standard normal distribution and the confidence level α is determined by the ALPHA= option in the EXACT statement.

When the Monte Carlo estimate $\hat{P}_{MC}$ equals 0, PROC FREQ computes the confidence limits for the p-value as

$$(0, \ 1 - \alpha^{(1/N)})$$

When the Monte Carlo estimate $\hat{P}_{MC}$ equals 1, PROC FREQ computes the confidence limits as

$$(\alpha^{(1/N)}, \ 1)$$

Computational Resources

For each variable in a table request, PROC FREQ stores all of the levels in memory. If all variables are numeric and not formatted, this requires about 84 bytes for each variable level. When there are character variables or formatted numeric variables, the memory that is required depends on the formatted variable lengths, with longer formatted lengths requiring more memory. The number of levels for each variable is limited only by the largest integer that your operating environment can store.

For any single crosstabulation table requested, PROC FREQ builds the entire table in memory, regardless of whether the table has zero cell counts. Thus, if the numeric variables A, B, and C each have 10 levels, PROC FREQ requires 2520 bytes to store the variable levels for the table request A*B*C, as follows:

```
3 variables * 10 levels/variable * 84 bytes/level
```

In addition, PROC FREQ requires 8000 bytes to store the table cell frequencies

```
1000 cells * 8 bytes/cell
```

even though there might be only 10 observations.

When the variables have many levels or when there are many multiway tables, your computer might not have enough memory to construct the tables. If PROC FREQ runs out of memory while constructing tables, it stops collecting levels for the variable with the most levels and returns the memory that is used by that variable. The procedure then builds the tables that do not contain the disabled variables.

If there is not enough memory for your table request and if increasing the available memory is impractical, you can reduce the number of multiway tables or variable levels. If you are not using the CMH or AGREE option in the TABLES statement to compute statistics across strata, reduce the number of multiway tables by using PROC SORT to sort the data set by one or more of the variables or by using the DATA step to create an index for the variables. Then remove the sorted or indexed variables from the TABLES statement and include a BY statement that uses these variables. You can also reduce memory requirements by using a FORMAT statement in the PROC FREQ step to reduce the number of levels. Additionally, reducing the formatted variable lengths reduces the amount of memory that is needed to store the variable levels. For more information about using formats, see the section "Grouping with Formats" on page 113.

Output Data Sets

PROC FREQ produces two types of output data sets that you can use with other statistical and reporting procedures. You can request these data sets as follows:

- Specify the OUT= option in a TABLES statement. This creates an output data set that contains frequency or crosstabulation table counts and percentages

- Specify an OUTPUT statement. This creates an output data set that contains statistics.

PROC FREQ does not display the output data sets. Use PROC PRINT, PROC REPORT, or any other SAS reporting tool to display an output data set.

In addition to these two output data sets, you can create a SAS data set from any piece of PROC FREQ output by using the Output Delivery System. See the section "ODS Table Names" on page 183 for more information.

Contents of the TABLES Statement Output Data Set

The OUT= option in the TABLES statement creates an output data set that contains one observation for each combination of variable values (or table cell) in the last table request. By default, each observation contains the frequency and percentage for the table cell. When the input data set contains missing values, the output data set also contains an observation with the frequency of missing values. The output data set includes the following variables:

- BY variables
- table request variables, such as A, B, C, and D in the table request A*B*C*D
- COUNT, which contains the table cell frequency
- PERCENT, which contains the table cell percentage

If you specify the OUTEXPECT option in the TABLES statement for a two-way or multiway table, the output data set also includes expected frequencies. If you specify the OUTPCT option for a two-way or multiway table, the output data set also includes row, column, and table percentages. The additional variables are as follows:

- EXPECTED, which contains the expected frequency
- PCT_TABL, which contains the percentage of two-way table frequency, for n-way tables where $n > 2$
- PCT_ROW, which contains the percentage of row frequency
- PCT_COL, which contains the percentage of column frequency

If you specify the OUTCUM option in the TABLES statement for a one-way table, the output data set also includes cumulative frequencies and cumulative percentages. The additional variables are as follows:

- CUM_FREQ, which contains the cumulative frequency
- CUM_PCT, which contains the cumulative percentage

The OUTCUM option has no effect for two-way or multiway tables.

The following PROC FREQ statements create an output data set of frequencies and percentages:

```
proc freq;
   tables A A*B / out=D;
run;
```

The output data set D contains frequencies and percentages for the table of A by B, which is the last table request listed in the TABLES statement. If A has two levels (1 and 2), B has three levels (1,2, and 3), and no table cell count is zero or missing, then the output data set D includes six observations, one for each combination of A and B levels. The first observation corresponds to A=1 and B=1; the second observation corresponds to A=1 and B=2; and so on. The data set includes the variables COUNT and PERCENT. The value of COUNT is the number of observations with the given combination of A and B levels. The value of PERCENT is the percentage of the total number of observations with that A and B combination.

When PROC FREQ combines different variable values into the same formatted level, the output data set contains the smallest internal value for the formatted level. For example, suppose a variable X has the values 1.1., 1.4, 1.7, 2.1, and 2.3. When you submit the statement

```
format X 1.;
```

in a PROC FREQ step, the formatted levels listed in the frequency table for X are 1 and 2. If you create an output data set with the frequency counts, the internal values of the levels of X are 1.1 and 1.7. To report the internal values of X when you display the output data set, use a format of 3.1 for X.

Contents of the OUTPUT Statement Output Data Set

The OUTPUT statement creates a SAS data set that contains the statistics that PROC FREQ computes for the last table request. You specify which statistics to store in the output data set. There is an observation with the specified statistics for each stratum or two-way table. If PROC FREQ computes summary statistics for a stratified table, the output data set also contains a summary observation with those statistics.

The OUTPUT data set can include the following variables.

- BY variables
- variables that identify the stratum, such as A and B in the table request A*B*C*D
- variables that contain the specified statistics

The output data set also includes variables with the *p*-values and degrees of freedom, asymptotic standard error (ASE), or confidence limits when PROC FREQ computes these values for a specified statistic.

The variable names for the specified statistics in the output data set are the names of the options enclosed in underscores. PROC FREQ forms variable names for the corresponding *p*-values, degrees of freedom, or confidence limits by combining the name of the option with the appropriate prefix from the following list:

DF_	degrees of freedom
E_	asymptotic standard error (ASE)
L_	lower confidence limit
U_	upper confidence limit
E0_	ASE under the null hypothesis
Z_	standardized value

P_	*p*-value
P2_	two-sided *p*-value
PL_	left-sided *p*-value
PR_	right-sided *p*-value
XP_	exact *p*-value
XP2_	exact two-sided *p*-value
XPL_	exact left-sided *p*-value
XPR_	exact right-sided *p*-value
XPT_	exact point probability
XL_	exact lower confidence limit
XU_	exact upper confidence limit

For example, variable names created for the Pearson chi-square, its degrees of freedom, and its *p*-values are _PCHI_, DF_PCHI, and P_PCHI, respectively.

If the length of the prefix plus the statistic option exceeds eight characters, PROC FREQ truncates the option so that the name of the new variable is eight characters long.

Displayed Output

Number of Variable Levels Table

If you specify the NLEVELS option in the PROC FREQ statement, PROC FREQ displays the "Number of Variable Levels" table. This table provides the number of levels for all variables named in the TABLES statements. PROC FREQ determines the variable levels from the formatted variable values. See "Grouping with Formats" on page 113 for details. The "Number of Variable Levels" table contains the following information:

- Variable name

- Levels, which is the total number of levels of the variable

- Number of Nonmissing Levels, if there are missing levels for any of the variables

- Number of Missing Levels, if there are missing levels for any of the variables

One-Way Frequency Tables

PROC FREQ displays one-way frequency tables for all one-way table requests in the TABLES statements, unless you specify the NOPRINT option in the PROC statement or the NOPRINT option in the TABLES statement. For a one-way table showing the frequency distribution of a single variable, PROC FREQ displays the name of the variable and its values. For each variable value or level, PROC FREQ displays the following information:

- Frequency count, which is the number of observations in the level

- Test Frequency count, if you specify the CHISQ and TESTF= options to request a chi-square goodness-of-fit test for specified frequencies

- Percent, which is the percentage of the total number of observations. (The NOPERCENT option suppresses this information.)

- Test Percent, if you specify the CHISQ and TESTP= options to request a chi-square goodness-of-fit test for specified percents. (The NOPERCENT option suppresses this information.)

- Cumulative Frequency count, which is the sum of the frequency counts for that level and all other levels listed above it in the table. The last cumulative frequency is the total number of nonmissing observations. (The NOCUM option suppresses this information.)

- Cumulative Percent, which is the percentage of the total number of observations in that level and in all other levels listed above it in the table. (The NOCUM or the NOPERCENT option suppresses this information.)

The one-way table also displays the Frequency Missing, which is the number of observations with missing values.

Statistics for One-Way Frequency Tables

For one-way tables, two statistical options are available in the TABLES statement. The CHISQ option provides a chi-square goodness-of-fit test, and the BINOMIAL option provides binomial proportion statistics and tests. PROC FREQ displays the following information, unless you specify the NOPRINT option in the PROC statement:

- If you specify the CHISQ option for a one-way table, PROC FREQ provides a chi-square goodness-of-fit test, displaying the Chi-Square statistic, the degrees of freedom (DF), and the probability value (Pr > ChiSq). If you specify the CHISQ option in the EXACT statement, PROC FREQ also displays the exact probability value for this test. If you specify the POINT option with the CHISQ option in the EXACT statement, PROC FREQ displays the exact point probability for the test statistic.

- If you specify the BINOMIAL option for a one-way table, PROC FREQ displays the estimate of the binomial Proportion, which is the proportion of observations in the first class listed in the one-way table. PROC FREQ also displays the asymptotic standard error (ASE) and the asymptotic (Wald) and exact (Clopper-Pearson) confidence limits by default. For the binomial proportion test, PROC FREQ displays the asymptotic standard error under the null hypothesis (ASE Under H0), the standardized test statistic (Z), and the one-sided and two-sided probability values.

    If you specify the BINOMIAL option in the EXACT statement, PROC FREQ also displays the exact one-sided and two-sided probability values for this test. If you specify the POINT option with the BINOMIAL option in the EXACT statement, PROC FREQ displays the exact point probability for the test.

- If you request additional binomial confidence limits by specifying *binomial-options*, PROC FREQ provides a table that displays the lower and upper confidence limits for each type that you request. In addition to the Wald and exact (Clopper-Pearson) confidence limits, you can request Agresti-Coull, Jeffreys, and Wilson (score) confidence limits for the binomial proportion.

- If you request a binomial noninferiority or superiority test by specifying the NONINF or SUP binomial-option, PROC FREQ displays the following information: the binomial Proportion, the test ASE (under H0 or Sample), the test statistic Z, the probability value, the noninferiority or superiority limit, and the test confidence limits. If you specify the BINOMIAL option in the EXACT statement, PROC FREQ also provides the exact probability value for the test, and exact test confidence limits.

- If you request a binomial equivalence test by specifying the EQUIV binomial-option, PROC FREQ displays the binomial Proportion and the test ASE (under H0 or Sample). PROC FREQ displays two one-sided tests (TOST) for equivalence, which include test statistics (Z) and probability values for the Lower and Upper tests, together with the Overall probability value. PROC FREQ also displays the equivalence limits and the test-based confidence limits. If you specify the BINOMIAL option in the EXACT statement, PROC FREQ provides exact probability values for the TOST and exact test-based confidence limits.

Multiway Tables

PROC FREQ displays all multiway table requests in the TABLES statements, unless you specify the NOPRINT option in the PROC statement or the NOPRINT option in the TABLES statement.

For two-way to multiway crosstabulation tables, the values of the last variable in the table request form the table columns. The values of the next-to-last variable form the rows. Each level (or combination of levels) of the other variables forms one stratum.

There are three ways to display multiway tables in PROC FREQ. By default, PROC FREQ displays multiway tables as separate two-way crosstabulation tables for each stratum of the multiway table. Also by default, PROC FREQ displays these two-way crosstabulation tables in table cell format. Alternatively, if you specify the CROSSLIST option, PROC FREQ displays the two-way crosstabulation tables in ODS column format. If you specify the LIST option, PROC FREQ displays multiway tables in list format, which presents the entire multiway crosstabulation in a single table.

Crosstabulation Tables

By default, PROC FREQ displays two-way crosstabulation tables in table cell format. The row variable values are listed down the side of the table, the column variable values are listed across the top of the table, and each row and column variable level combination forms a table cell.

Each cell of a crosstabulation table can contain the following information:

- Frequency, which is the number of observations in the table cell. (The NOFREQ option suppresses this information.)

- Expected frequency under the hypothesis of independence, if you specify the EXPECTED option

- Deviation of the cell frequency from the expected value, if you specify the DEVIATION option

- Cell Chi-Square, which is the cell's contribution to the total chi-square statistic, if you specify the CELLCHI2 option

- Tot Pct, which is the cell's percentage of the total multiway table frequency, for *n*-way tables when $n > 2$, if you specify the TOTPCT option

- Percent, which is the cell's percentage of the total (two-way table) frequency. (The NOPERCENT option suppresses this information.)

- Row Pct, or the row percentage, which is the cell's percentage of the total frequency for its row. (The NOROW option suppresses this information.)

- Col Pct, or column percentage, which is the cell's percentage of the total frequency for its column. (The NOCOL option suppresses this information.)

- Cumulative Col%, or cumulative column percentage, if you specify the CUMCOL option

The table also displays the Frequency Missing, which is the number of observations with missing values.

CROSSLIST Tables

If you specify the CROSSLIST option, PROC FREQ displays two-way crosstabulation tables in ODS column format. The CROSSLIST column format is different from the default crosstabulation table cell format, but the CROSSLIST table provides the same information (frequencies, percentages, and other statistics) as the default crosstabulation table.

In the CROSSLIST table format, the rows of the display correspond to the crosstabulation table cells, and the columns of the display correspond to descriptive statistics such as frequencies and percentages. Each table cell is identified by the values of its TABLES row and column variable levels, with all column variable levels listed within each row variable level. The CROSSLIST table also provides row totals, column totals, and overall table totals.

For a crosstabulation table in CROSSLIST format, PROC FREQ displays the following information:

- the row variable name and values

- the column variable name and values

- Frequency, which is the number of observations in the table cell. (The NOFREQ option suppresses this information.)

- Expected cell frequency under the hypothesis of independence, if you specify the EXPECTED option

- Deviation of the cell frequency from the expected value, if you specify the DEVIATION option

- Cell Chi-Square, which is the cell's contribution to the total chi-square statistic, if you specify the CELLCHI2 option

- Total Percent, which is the cell's percentage of the total multiway table frequency, for n-way tables when $n > 2$, if you specify the TOTPCT option

- Percent, which is the cell's percentage of the total (two-way table) frequency. (The NOPERCENT option suppresses this information.)

- Row Percent, which is the cell's percentage of the total frequency for its row. (The NOROW option suppresses this information.)

- Column Percent, the cell's percentage of the total frequency for its column. (The NOCOL option suppresses this information.)

The table also displays the Frequency Missing, which is the number of observations with missing values.

LIST Tables

If you specify the LIST option in the TABLES statement, PROC FREQ displays multiway tables in a list format rather than as crosstabulation tables. The LIST option displays the entire multiway table in one table, instead of displaying a separate two-way table for each stratum. The LIST option is not available when you also request statistical options. Unlike the default crosstabulation output, the LIST output does not display row percentages, column percentages, and optional information such as expected frequencies and cell chi-squares.

For a multiway table in list format, PROC FREQ displays the following information:

- the variable names and values

- Frequency, which is the number of observations in the level (with the indicated variable values)

- Percent, which is the level's percentage of the total number of observations. (The NOPERCENT option suppresses this information.)

- Cumulative Frequency, which is the accumulated frequency of the level and all other levels listed above it in the table. The last cumulative frequency in the table is the total number of nonmissing observations. (The NOCUM option suppresses this information.)

- Cumulative Percent, which is the accumulated percentage of the level and all other levels listed above it in the table. (The NOCUM or the NOPERCENT option suppresses this information.)

The table also displays the Frequency Missing, which is the number of observations with missing values.

Statistics for Multiway Tables

PROC FREQ computes statistical tests and measures for crosstabulation tables, depending on which statements and options you specify. You can suppress the display of all these results by specifying the NOPRINT option in the PROC statement. With any of the following information, PROC FREQ also displays the Sample Size and the Frequency Missing.

- If you specify the SCOROUT option, PROC FREQ displays the Row Scores and Column Scores that it uses for statistical computations. The Row Scores table displays the row variable values and the Score corresponding to each value. The Column Scores table displays the column variable values and the corresponding Scores. PROC FREQ also identifies the score type used to compute the row and column scores. You can specify the score type with the SCORES= option in the TABLES statement.

- If you specify the CHISQ option, PROC FREQ displays the following statistics for each two-way table: Pearson Chi-Square, Likelihood-Ratio Chi-Square, Continuity-Adjusted Chi-Square (for 2×2 tables), Mantel-Haenszel Chi-Square, the Phi Coefficient, the Contingency Coefficient, and Cramer's V. For each test statistic, PROC FREQ also displays the degrees of freedom (DF) and the probability value (Prob).

- If you specify the CHISQ option for 2×2 tables, PROC FREQ also displays Fisher's exact test. The test output includes the cell (1,1) frequency (F), the exact left-sided and right-sided probability values, the table probability (P), and the exact two-sided probability value.

- If you specify the FISHER option in the TABLES statement (or, equivalently, the FISHER option in the EXACT statement), PROC FREQ displays Fisher's exact test for tables larger than 2×2. The test output includes the table probability (P) and the probability value. In addition, PROC FREQ displays the CHISQ output listed earlier, even if you do not also specify the CHISQ option.

- If you specify the PCHI, LRCHI, or MHCHI option in the EXACT statement, PROC FREQ also displays the corresponding exact test: Pearson Chi-Square, Likelihood-Ratio Chi-Square, or Mantel-Haenszel Chi-Square, respectively. The test output includes the test statistic, the degrees of freedom (DF), and the asymptotic and exact probability values. If you also specify the POINT option in the EXACT statement, PROC FREQ displays the point probability for each exact test requested. If you specify the CHISQ option in the EXACT statement, PROC FREQ displays exact probability values for all three of these chi-square tests.

- If you specify the MEASURES option, PROC FREQ displays the following statistics and their asymptotic standard errors (ASE) for each two-way table: Gamma, Kendall's Tau-b, Stuart's Tau-c, Somers' $D(C|R)$, Somers' $D(R|C)$, Pearson Correlation, Spearman Correlation, Lambda Asymmetric $(C|R)$, Lambda Asymmetric $(R|C)$, Lambda Symmetric, Uncertainty Coefficient $(C|R)$, Uncertainty Coefficient $(R|C)$, and Uncertainty Coefficient Symmetric. If you specify the CL option, PROC FREQ also displays confidence limits for these measures.

- If you specify the PLCORR option, PROC FREQ displays the tetrachoric correlation for 2×2 tables or the polychoric correlation for larger tables. In addition, PROC FREQ displays the MEASURES output listed earlier, even if you do not also specify the MEASURES option.

- If you specify the option GAMMA, KENTB, STUTC, SMDCR, SMDRC, PCORR, or SCORR in the TEST statement, PROC FREQ displays asymptotic tests for Gamma, Kendall's Tau-b, Stuart's Tau-c, Somers' $D(C|R)$, Somers' $D(R|C)$, the Pearson Correlation, or the Spearman Correlation, respectively. If you specify the MEASURES option in the TEST statement, PROC FREQ displays all these asymptotic tests. The test output includes the statistic, its asymptotic standard error (ASE), Confidence Limits, the ASE under the null hypothesis H0, the standardized test statistic (Z), and the one-sided and two-sided probability values.

- If you specify the PCORR or SCORR option in the EXACT statement, PROC FREQ displays asymptotic and exact tests for the Pearson Correlation or the Spearman Correlation, respectively. The test output includes the correlation, its asymptotic standard error (ASE), Confidence Limits, the ASE under the null hypothesis H0, the standardized test statistic (Z), and the asymptotic and exact one-sided and two-sided probability values. If you also specify the POINT option in the EXACT statement, PROC FREQ displays the point probability for each exact test requested.

- If you specify the RISKDIFF option for 2×2 tables, PROC FREQ displays the Column 1 and Column 2 Risk Estimates. For each column, PROC FREQ displays the Row 1 Risk, Row 2 Risk, Total Risk, and Risk Difference, together with their asymptotic standard errors (ASE) and Asymptotic Confidence Limits. PROC FREQ also displays Exact Confidence Limits for the Row 1 Risk, Row 2 Risk, and Total Risk. If you specify the RISKDIFF option in the EXACT statement, PROC FREQ provides unconditional Exact Confidence Limits for the Risk Difference.

- If you request a noninferiority or superiority test for the proportion difference by specifying the NONINF or SUP riskdiff-option, and if you specify METHOD=HA (Hauck-Anderson), METHOD=FM (Farrington-Manning), or METHOD=WALD (Wald), PROC FREQ displays the following information: the Proportion Difference, the test ASE (H0, Sample, Sample H-A, or FM, depending on the method you specify), the test statistic Z, the probability value, the Noninferiority or Superiority Limit, and the test-based Confidence Limits. If you specify METHOD=SCORE (Newcombe score), PROC FREQ displays the Proportion Difference, the Noninferiority or Superiority Limit, and the score Confidence Limits.

- If you request an equivalence test for the proportion difference by specifying the EQUIV riskdiff-option, and if you specify METHOD=HA (Hauck-Anderson), METHOD=FM (Farrington-Manning), or METHOD=WALD (Wald), PROC FREQ displays the following information: the Proportion Difference and the test ASE (H0, Sample, Sample H-A, or FM, depending on the method you specify). PROC FREQ displays a two one-sided test (TOST) for equivalence, which includes test statistics (Z) and probability values for the Lower and Upper tests, together with the Overall probability value. PROC FREQ also displays the Equivalence Limits and the test-based Confidence Limits. If you specify METHOD=SCORE (Newcombe score), PROC FREQ displays the Proportion Difference, the Equivalence Limits, and the score Confidence Limits.

- If you request an equality test for the proportion difference by specifying the EQUAL riskdiff-option, PROC FREQ displays the following information: the Proportion Difference and the test ASE (H0 or Sample), the test statistic Z, the One-Sided probability value (Pr > Z or Pr < Z), and the Two-Sided probability value, Pr > |Z|.

- If you specify the MEASURES option or the RELRISK option for 2×2 tables, PROC FREQ displays Estimates of the Relative Risk for Case-Control and Cohort studies, together with their Confidence Limits. These measures are also known as the Odds Ratio and the Column 1 and 2 Relative Risks. If you specify the OR option in the EXACT statement, PROC FREQ also displays Exact Confidence Limits for the Odds Ratio.

- If you specify the TREND option, PROC FREQ displays the Cochran-Armitage Trend Test for tables that are $2 \times C$ or $R \times 2$. For this test, PROC FREQ gives the Statistic (Z) and the one-sided and two-sided probability values. If you specify the TREND option in the EXACT statement, PROC FREQ also displays the exact one-sided and two-sided probability values for this test. If you specify the POINT option with the TREND option in the EXACT statement, PROC FREQ displays the exact point probability for the test statistic.

- If you specify the JT option, PROC FREQ displays the Jonckheere-Terpstra Test, showing the Statistic (JT), the standardized test statistic (Z), and the one-sided and two-sided probability values. If you specify the JT option in the EXACT statement, PROC FREQ also displays the exact one-sided and two-sided probability values for this test. If you specify the POINT option with the JT option in the EXACT statement, PROC FREQ displays the exact point probability for the test statistic.

- If you specify the AGREE option and the PRINTKWT option, PROC FREQ displays the Kappa Coefficient Weights for square tables greater than 2×2.

- If you specify the AGREE option, for two-way tables PROC FREQ displays McNemar's Test and the Simple Kappa Coefficient for 2×2 tables. For square tables larger than 2×2, PROC FREQ displays Bowker's Test of Symmetry, the Simple Kappa Coefficient, and the Weighted Kappa Coefficient. For McNemar's Test and Bowker's Test of Symmetry, PROC FREQ displays the Statistic (S), the degrees of freedom (DF), and the probability value (Pr > S). If you specify the MCNEM option in the EXACT statement, PROC FREQ also displays the exact probability value for McNemar's test. If you specify the POINT option with the MCNEM option in the EXACT statement, PROC FREQ displays the exact point probability for the test statistic. For the simple and weighted kappa coefficients, PROC FREQ displays the kappa values, asymptotic standard errors (ASE), and Confidence Limits.

- If you specify the KAPPA or WTKAP option in the TEST statement, PROC FREQ displays asymptotic tests for the simple kappa coefficient or the weighted kappa coefficient, respectively. If you specify the AGREE option in the TEST statement, PROC FREQ displays both these asymptotic tests. The test output includes the kappa coefficient, its asymptotic standard error (ASE), Confidence Limits, the ASE under the null hypothesis H0, the standardized test statistic (Z), and the one-sided and two-sided probability values.

- If you specify the KAPPA or WTKAP option in the EXACT statement, PROC FREQ displays asymptotic and exact tests for the simple kappa coefficient or the weighted kappa coefficient, respectively. The test output includes the kappa coefficient, its asymptotic standard error (ASE), Confidence Limits, the ASE under the null hypothesis H0, the standardized test statistic (Z), and the asymptotic and exact one-sided and two-sided probability values. If you specify the POINT option in the EXACT statement, PROC FREQ displays the point probability for each exact test requested.

- If you specify the MC option in the EXACT statement, PROC FREQ displays Monte Carlo estimates for all exact *p*-values requested by *statistic-options* in the EXACT statement. The

Monte Carlo output includes the *p*-value Estimate, its Confidence Limits, the Number of Samples used to compute the Monte Carlo estimate, and the Initial Seed for random number generation.

- If you specify the AGREE option, for multiple strata PROC FREQ displays Overall Simple and Weighted Kappa Coefficients, with their asymptotic standard errors (ASE) and Confidence Limits. PROC FREQ also displays Tests for Equal Kappa Coefficients, giving the Chi-Squares, degrees of freedom (DF), and probability values (Pr > ChiSq) for the Simple Kappa and Weighted Kappa. For multiple strata of 2×2 tables, PROC FREQ displays Cochran's Q, giving the Statistic (Q), the degrees of freedom (DF), and the probability value (Pr > Q).

- If you specify the CMH option, PROC FREQ displays Cochran-Mantel-Haenszel Statistics for the following three alternative hypotheses: Nonzero Correlation, Row Mean Scores Differ (ANOVA Statistic), and General Association. For each of these statistics, PROC FREQ gives the degrees of freedom (DF) and the probability value (Prob). For 2×2 tables, PROC FREQ also displays Estimates of the Common Relative Risk for Case-Control and Cohort studies, together with their confidence limits. These include both Mantel-Haenszel and Logit stratum-adjusted estimates of the common Odds Ratio, Column 1 Relative Risk, and Column 2 Relative Risk. Also for 2×2 tables, PROC FREQ displays the Breslow-Day Test for Homogeneity of the Odds Ratios. For this test, PROC FREQ gives the Chi-Square, the degrees of freedom (DF), and the probability value (Pr > ChiSq).

- If you specify the CMH option in the TABLES statement and also specify the COMOR option in the EXACT statement, PROC FREQ displays exact confidence limits for the Common Odds Ratio for multiple strata of 2×2 tables. PROC FREQ also displays the Exact Test of H0: Common Odds Ratio = 1. The test output includes the Cell (1,1) Sum (S), Mean of S Under H0, One-sided Pr <= S, and Point Pr = S. PROC FREQ also provides exact two-sided probability values for the test, computed according to the following three methods: 2 * One-sided, Sum of probabilities <= Point probability, and Pr >= |S - Mean|.

- If you specify the CMH option in the TABLES statement and also specify the EQOR option in the EXACT statement, PROC FREQ computes Zelen's exact test for equal odds ratios for $h \times 2 \times 2$ tables. PROC FREQ displays Zelen's test along with the asymptotic Breslow-Day test produced by the CMH option. PROC FREQ displays the test statistic, Zelen's Exact Test (P), and the probability value, Exact Pr <= P.

ODS Table Names

PROC FREQ assigns a name to each table it creates. You can use these names to reference the table when you use the Output Delivery System (ODS) to select tables and create output data sets. For more information about ODS, see Chapter 20, "Using the Output Delivery System" (*SAS/STAT User's Guide*).

Table 3.14 lists the ODS table names together with their descriptions and the options required to produce the tables. Note that the ALL option in the TABLES statement invokes the CHISQ, MEASURES, and CMH options.

Table 3.14 ODS Tables Produced by PROC FREQ

ODS Table Name	Description	Statement	Option
BinomialCLs	Binomial confidence limits	TABLES	BINOMIAL(AC | J | W)
BinomialEquiv	Binomial equivalence analysis	TABLES	BINOMIAL(EQUIV)
BinomialEquivLimits	Binomial equivalence limits	TABLES	BINOMIAL(EQUIV)
BinomialEquivTest	Binomial equivalence test	TABLES	BINOMIAL(EQUIV)
BinomialNoninf	Binomial noninferiority test	TABLES	BINOMIAL(NONINF)
BinomialProp	Binomial proportion	TABLES	BINOMIAL (one-way table
BinomialPropTest	Binomial proportion test	TABLES	BINOMIAL (one-way table
BinomialSup	Binomial superiority test	TABLES	BINOMIAL(SUP)
BreslowDayTest	Breslow-Day test	TABLES	CMH ($h \times 2 \times 2$ tables)
CMH	Cochran-Mantel-Haenszel statistics	TABLES	CMH
ChiSq	Chi-square tests	TABLES	CHISQ
CochransQ	Cochran's Q	TABLES	AGREE ($h \times 2 \times 2$ tables)
ColScores	Column scores	TABLES	SCOROUT
CommonOddsRatioCL	Exact confidence limits for the common odds ratio	EXACT	COMOR ($h \times 2 \times 2$ tables)
CommonOddsRatioTest	Common odds ratio exact test	EXACT	COMOR ($h \times 2 \times 2$ tables)
CommonRelRisks	Common relative risks	TABLES	CMH ($h \times 2 \times 2$ tables)
CrossList	Crosstabulation table in column format	TABLES	CROSSLIST (n-way table request, $n > 1$
CrossTabFreqs	Crosstabulation table	TABLES	(n-way table request, $n > 1$
EqualKappaTest	Test for equal simple kappas	TABLES	AGREE ($h \times 2 \times 2$ tables)
EqualKappaTests	Tests for equal kappas	TABLES	AGREE ($h \times r \times r$ tables, $r > 2$)
EqualOddsRatios	Tests for equal odds ratios	EXACT	EQOR ($h \times 2 \times 2$ tables)
FishersExact	Fisher's exact test	EXACT or TABLES or TABLES	FISHER FISHER or EXACT CHISQ (2×2 tables)
FishersExactMC	Monte Carlo estimates for Fisher's exact test	EXACT	FISHER / MC
Gamma	Gamma	TEST	GAMMA
GammaTest	Gamma test	TEST	GAMMA
JTTest	Jonckheere-Terpstra test	TABLES	JT
JTTestMC	Monte Carlo estimates for Jonckheere-Terpstra exact test	EXACT	JT / MC
KappaStatistics	Kappa statistics	TABLES	AGREE ($r \times r$ tables, $r > 2$, no TEST or EXACT)
KappaWeights	Kappa weights	TABLES	AGREE and PRINTKWT
List	List format multiway table	TABLES	LIST
LRChiSq	Likelihood-ratio chi-square exact test	EXACT	LRCHI
LRChiSqMC	Monte Carlo exact test for likelihood-ratio chi-square	EXACT	LRCHI / MC

Table 3.14 continued

ODS Table Name	Description	Statement	Option
McNemarsTest	McNemar's test	TABLES	AGREE (2 × 2 tables)
Measures	Measures of association	TABLES	MEASURES
MHChiSq	Mantel-Haenszel chi-square exact test	EXACT	MHCHI
MHChiSqMC	Monte Carlo exact test for Mantel-Haenszel chi-square	EXACT	MHCHI / MC
NLevels	Number of variable levels	PROC	NLEVELS
OddsRatioCL	Exact confidence limits for the odds ratio	EXACT	OR (2 × 2 tables)
OneWayChiSq	One-way chi-square test	TABLES	CHISQ (one-way tables)
OneWayChiSqMC	Monte Carlo exact test for one-way chi-square	EXACT	CHISQ / MC (one-way tables)
OneWayFreqs	One-way frequencies	PROC or TABLES	(with no TABLES stmt) (one-way table request)
OverallKappa	Overall simple kappa	TABLES	AGREE (h × 2 × 2 tables)
OverallKappas	Overall kappa coefficients	TABLES	AGREE ($h \times r \times r$ tables, $r > 2$)
PdiffEquiv	Equivalence analysis for the proportion difference	TABLES	RISKDIFF(EQUIV) (2 × 2 tables)
PdiffEquivLimits	Equivalence limits for the proportion difference	TABLES	RISKDIFF(EQUIV) (2 × 2 tables)
PdiffEquivTest	Equivalence test for the proportion difference	TABLES	RISKDIFF(EQUIV) (2 × 2 tables)
PdiffNoninf	Noninferiority test for the proportion difference	TABLES	RISKDIFF(NONINF) (2 × 2 tables)
PdiffSup	Superiority test for the proportion difference	TABLES	RISKDIFF(SUP) (2 × 2 tables)
PdiffTest	Proportion difference test	TABLES	RISKDIFF(EQUAL) (2 × 2 tables)
PearsonChiSq	Pearson chi-square exact test	EXACT	PCHI
PearsonChiSqMC	Monte Carlo exact test for Pearson chi-square	EXACT	PCHI / MC
PearsonCorr	Pearson correlation	TEST or EXACT	PCORR PCORR
PearsonCorrMC	Monte Carlo exact test for Pearson correlation	EXACT	PCORR / MC
PearsonCorrTest	Pearson correlation test	TEST or EXACT	PCORR PCORR
RelativeRisks	Relative risk estimates	TABLES	RELRISK or MEASURES (2 × 2 tables)
RiskDiffCol1	Column 1 risk estimates	TABLES	RISKDIFF (2 × 2 tables)
RiskDiffCol2	Column 2 risk estimates	TABLES	RISKDIFF (2 × 2 tables)
RowScores	Row scores	TABLES	SCOROUT
SimpleKappa	Simple kappa coefficient	TEST or EXACT	KAPPA KAPPA

Table 3.14 *continued*

ODS Table Name	Description	Statement	Option
SimpleKappaMC	Monte Carlo exact test for simple kappa	EXACT	KAPPA / MC
SimpleKappaTest	Simple kappa test	TEST or EXACT	KAPPA KAPPA
SomersDCR	Somers' $D(C\|R)$	TEST	SMDCR
SomersDCRTest	Somers' $D(C\|R)$ test	TEST	SMDCR
SomersDRC	Somers' $D(R\|C)$	TEST	SMDRC
SomersDRCTest	Somers' $D(R\|C)$ test	TEST	SMDRC
SpearmanCorr	Spearman correlation	TEST or EXACT	SCORR SCORR
SpearmanCorrMC	Monte Carlo exact test for Spearman correlation	EXACT	SCORR / MC
SpearmanCorrTest	Spearman correlation test	TEST or EXACT	SCORR SCORR
SymmetryTest	Test of symmetry	TABLES	AGREE
TauB	Kendall's tau-b	TEST	KENTB
TauBTest	Kendall's tau-b test	TEST	KENTB
TauC	Stuart's tau-c	TEST	STUTC
TauCTest	Stuart's tau-c test	TEST	STUTC
TrendTest	Cochran-Armitage trend test	TABLES	TREND
TrendTestMC	Monte Carlo exact test for trend	EXACT	TREND / MC
WeightedKappa	Weighted kappa	TEST or EXACT	WTKAP WTKAP
WeightedKappaMC	Monte Carlo exact test for weighted kappa	EXACT	WTKAP / MC
WeightedKappaTest	Weighted kappa test	TEST or EXACT	WTKAP WTKAP

* The ALL option in the TABLES statement invokes CHISQ, MEASURES, and CMH.

ODS Graphics

PROC FREQ assigns a name to each graph it creates with ODS Graphics. You can use these names to reference the graphs. Table 3.15 lists the names of the graphs that PROC FREQ generates, along with the corresponding analysis options.

To request graphics with PROC FREQ, you must first enable ODS Graphics by specifying the ODS GRAPHICS ON statement. See Chapter 21, "Statistical Graphics Using ODS" (*SAS/STAT User's Guide*), for more information. When you have enabled ODS Graphics, you can request specific plots with the PLOTS= option in the TABLES statement. If you do not specify the PLOTS= option but have enabled ODS Graphics, then PROC FREQ produces all plots associated with the analyses you request.

Table 3.15 ODS Graphics Produced by PROC FREQ

ODS Graph Name	Plot Description	TABLES Statement Option
CumFreqPlot	Cumulative frequency plot	One-way table request
DeviationPlot	Deviation plot	CHISQ and a one-way table request
FreqPlot	Frequency plot	Any table request
KappaPlot	Kappa plot	AGREE ($h \times r \times r$ table)
ORPlot	Odds ratio plot	MEASURES or RELRISK ($h \times 2 \times 2$ table)
WtKappaPlot	Weighted kappa plot	AGREE ($h \times r \times r$ table, $r > 2$)

Examples: FREQ Procedure

Example 3.1: Output Data Set of Frequencies

The eye and hair color of children from two different regions of Europe are recorded in the data set Color. Instead of recording one observation per child, the data are recorded as cell counts, where the variable Count contains the number of children exhibiting each of the 15 eye and hair color combinations. The data set does not include missing combinations.

The following DATA step statements create the SAS data set Color:

```
data Color;
   input Region Eyes $ Hair $ Count @@;
      label Eyes  ='Eye Color'
            Hair  ='Hair Color'
            Region='Geographic Region';
      datalines;
1 blue   fair    23  1 blue   red     7  1 blue   medium 24
1 blue   dark    11  1 green  fair   19  1 green  red     7
1 green  medium  18  1 green  dark   14  1 brown  fair   34
1 brown  red      5  1 brown  medium 41  1 brown  dark   40
1 brown  black    3  2 blue   fair   46  2 blue   red    21
2 blue   medium  44  2 blue   dark   40  2 blue   black   6
2 green  fair    50  2 green  red    31  2 green  medium 37
2 green  dark    23  2 brown  fair   56  2 brown  red    42
2 brown  medium  53  2 brown  dark   54  2 brown  black  13
;
```

The following PROC FREQ statements read the Color data set and create an output data set that contains the frequencies, percentages, and expected cell frequencies of the two-way table of Eyes by Hair. The TABLES statement requests three tables: a frequency table for Eyes, a frequency table for Hair, and a crosstabulation table for Eyes by Hair. The OUT= option creates the FreqCount data set, which contains the crosstabulation table frequencies. The OUTEXPECT option outputs the expected table cell frequencies to FreqCount, and the SPARSE option includes zero cell frequencies in the output data set. The WEIGHT statement specifies that the variable Count contains the observation weights. These statements create Output 3.1.1 through Output 3.1.3.

```
proc freq data=Color;
   tables Eyes Hair Eyes*Hair / out=FreqCount outexpect sparse;
   weight Count;
   title 'Eye and Hair Color of European Children';
run;

proc print data=FreqCount noobs;
   title2 'Output Data Set from PROC FREQ';
run;
```

Output 3.1.1 displays the two frequency tables produced by PROC FREQ: one showing the distribution of eye color, and one showing the distribution of hair color. By default, PROC FREQ lists the variables values in alphabetical order. The 'Eyes*Hair' specification produces a crosstabulation table, shown in Output 3.1.2, with eye color defining the table rows and hair color defining the table columns. A zero cell frequency for green eyes and black hair indicates that this eye and hair color combination does not occur in the data.

The output data set FreqCount (Output 3.1.3) contains frequency counts and percentages for the last table requested in the TABLES statement, Eyes by Hair. Because the SPARSE option is specified, the data set includes the observation with a zero frequency. The variable Expected contains the expected frequencies, as requested by the OUTEXPECT option.

Output 3.1.1 Frequency Tables

```
             Eye and Hair Color of European Children

                       The FREQ Procedure

                           Eye Color

                                       Cumulative    Cumulative
     Eyes      Frequency     Percent    Frequency       Percent
     ---------------------------------------------------------------
     blue           222       29.13          222         29.13
     brown          341       44.75          563         73.88
     green          199       26.12          762        100.00


                           Hair Color

                                       Cumulative    Cumulative
     Hair      Frequency     Percent    Frequency       Percent
     ---------------------------------------------------------------
     black           22        2.89           22          2.89
     dark           182       23.88          204         26.77
     fair           228       29.92          432         56.69
     medium         217       28.48          649         85.17
     red            113       14.83          762        100.00
```

Output 3.1.2 Crosstabulation Table

```
                       Table of Eyes by Hair

      Eyes(Eye Color)       Hair(Hair Color)

      Frequency|
      Percent  |
      Row Pct  |
      Col Pct  |black   |dark    |fair    |medium  |red     |  Total
      ---------+--------+--------+--------+--------+--------+
      blue     |     6  |    51  |    69  |    68  |    28  |    222
               |  0.79  |  6.69  |  9.06  |  8.92  |  3.67  |  29.13
               |  2.70  | 22.97  | 31.08  | 30.63  | 12.61  |
               | 27.27  | 28.02  | 30.26  | 31.34  | 24.78  |
      ---------+--------+--------+--------+--------+--------+
      brown    |    16  |    94  |    90  |    94  |    47  |    341
               |  2.10  | 12.34  | 11.81  | 12.34  |  6.17  |  44.75
               |  4.69  | 27.57  | 26.39  | 27.57  | 13.78  |
               | 72.73  | 51.65  | 39.47  | 43.32  | 41.59  |
      ---------+--------+--------+--------+--------+--------+
      green    |     0  |    37  |    69  |    55  |    38  |    199
               |  0.00  |  4.86  |  9.06  |  7.22  |  4.99  |  26.12
               |  0.00  | 18.59  | 34.67  | 27.64  | 19.10  |
               |  0.00  | 20.33  | 30.26  | 25.35  | 33.63  |
      ---------+--------+--------+--------+--------+--------+
      Total         22      182      228      217      113       762
                  2.89    23.88    29.92    28.48    14.83    100.00
```

Output 3.1.3 Output Data Set of Frequencies

```
              Eye and Hair Color of European Children
                   Output Data Set from PROC FREQ

         Eyes      Hair       COUNT     EXPECTED      PERCENT

         blue      black         6        6.409       0.7874
         blue      dark         51       53.024       6.6929
         blue      fair         69       66.425       9.0551
         blue      medium       68       63.220       8.9239
         blue      red          28       32.921       3.6745
         brown     black        16        9.845       2.0997
         brown     dark         94       81.446      12.3360
         brown     fair         90      102.031      11.8110
         brown     medium       94       97.109      12.3360
         brown     red          47       50.568       6.1680
         green     black         0        5.745       0.0000
         green     dark         37       47.530       4.8556
         green     fair         69       59.543       9.0551
         green     medium       55       56.671       7.2178
         green     red          38       29.510       4.9869
```

Example 3.2: Frequency Dot Plots

This example produces frequency dot plots for the children's eye and hair color data from Example 3.1.

PROC FREQ produces plots by using ODS Graphics to create graphs as part of the procedure output. Frequency plots are available for any frequency or crosstabulation table request. You can display frequency plots as bar charts or dot plots. You can use *plot-options* to specify the orientation (vertical or horizontal), scale, and layout of the plots.

The following PROC FREQ statements request frequency tables and dot plots. The first TABLES statement requests a one-way frequency table of Hair and a crosstabulation table of Eyes by Hair. The PLOTS= option requests frequency plots for the tables, and the TYPE=DOT plot-option specifies dot plots. By default, frequency plots are produced as bar charts.

The second TABLES statement requests a crosstabulation table of Region by Hair and a frequency dot plot for this table. The SCALE=PERCENT plot-option plots percentages instead of frequency counts. SCALE=LOG and SCALE=SQRT plot-options are also available to plot log frequencies and square roots of frequencies, respectively.

The ORDER=FREQ option in the PROC FREQ statement orders the variable levels by frequency. This order applies to the frequency and crosstabulation table displays and also to the corresponding frequency plots.

Before requesting plots, you must enable ODS Graphics with the **ODS GRAPHICS ON** statement.

```
ods graphics on;
proc freq data=Color order=freq;
   tables Hair Eyes*Hair / plots=freqplot(type=dot);
   tables Region*Hair / plots=freqplot(type=dot scale=percent);
   weight Count;
   title 'Eye and Hair Color of European Children';
run;
ods graphics off;
```

Output 3.2.1, Output 3.2.2, and Output 3.2.3 display the dot plots produced by PROC FREQ. By default, the orientation of dot plots is horizontal, which places the variable levels on the *y*-axis. You can specify the ORIENT=VERTICAL plot-option to request a vertical orientation. For two-way plots, you can use the TWOWAY= plot-option to specify the plot layout. The default layout (shown in Output 3.2.2 and Output 3.2.3) is GROUPVERTICAL. Two-way layouts STACKED and GROUPHORIZONTAL are also available.

Output 3.2.1 One-Way Frequency Dot Plot

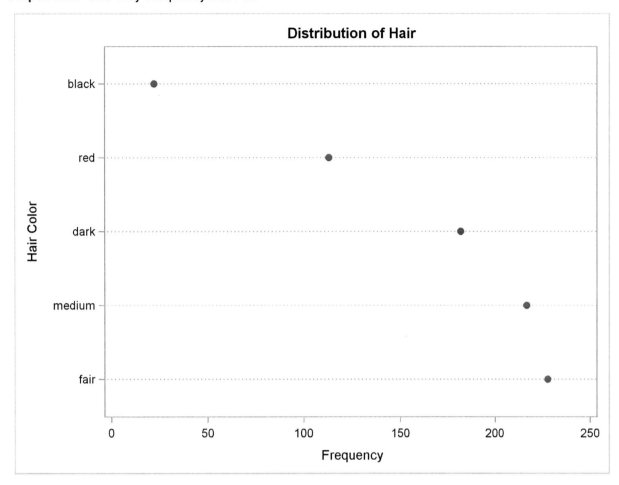

Output 3.2.2 Two-Way Frequency Dot Plot

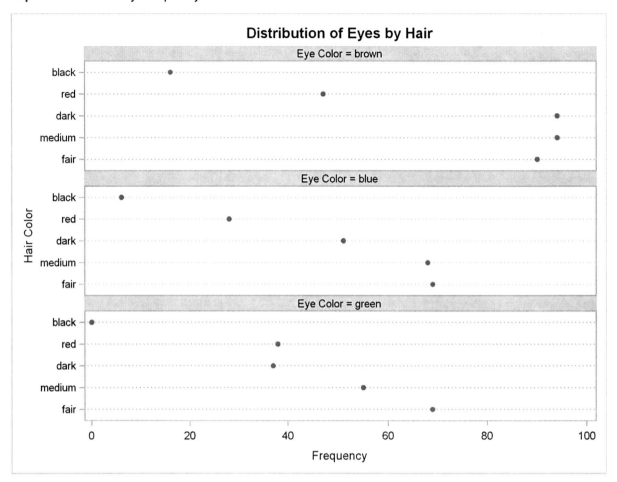

Output 3.2.3 Two-Way Percent Dot Plot

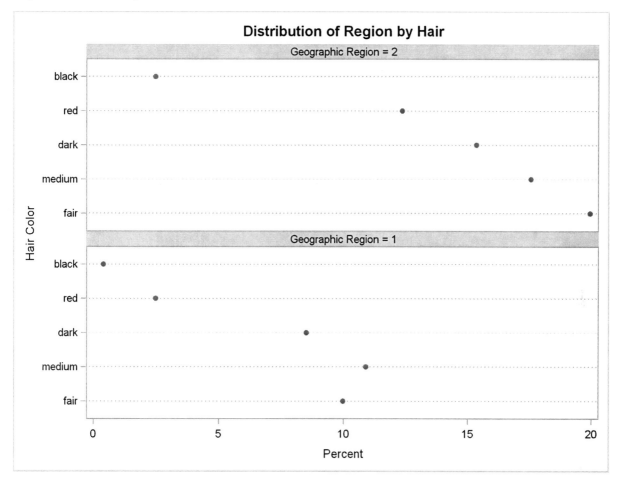

Example 3.3: Chi-Square Goodness-of-Fit Tests

This example examines whether the children's hair color (from Example 3.1) has a specified multinomial distribution for the two geographical regions. The hypothesized distribution of hair color is 30% fair, 12% red, 30% medium, 25% dark, and 3% black.

In order to test the hypothesis for each region, the data are first sorted by Region. Then the FREQ procedure uses a BY statement to produce a separate table for each BY group (Region). The option ORDER=DATA orders the variable values (hair color) in the frequency table by their order in the input data set. The TABLES statement requests a frequency table for hair color, and the option NOCUM suppresses the display of the cumulative frequencies and percentages.

The CHISQ option requests a chi-square goodness-of-fit test for the frequency table of Hair. The TESTP= option specifies the hypothesized (or test) percentages for the chi-square test; the number of percentages listed equals the number of table levels, and the percentages sum to 100%. The TESTP= percentages are listed in the same order as the corresponding variable levels appear in frequency table.

The PLOTS= option requests a deviation plot, which is associated with the CHISQ option and displays the relative deviations from the test frequencies. The TYPE=DOT plot-option requests a dot plot instead of the default type, which is a bar chart. The ONLY plot-option requests that PROC FREQ produce only the deviation plot. By default, PROC FREQ produces all plots associated with the requested analyses. A frequency plot is associated with a one-way table request but is not produced in this example because ONLY is specified with the DEVIATIONPLOT request. Note that ODS Graphics must be enabled before requesting plots. These statements produce Output 3.3.1 through Output 3.3.4.

```
proc sort data=Color;
   by Region;
run;

ods graphics on;
proc freq data=Color order=data;
   tables Hair / nocum chisq testp=(30 12 30 25 3)
                 plots(only)=deviationplot(type=dot);
   weight Count;
   by Region;
   title 'Hair Color of European Children';
run;
ods graphics off;
```

Output 3.3.1 Frequency Table and Chi-Square Test for Region 1

```
                    Hair Color of European Children

--------------------------- Geographic Region=1 ---------------------------

                          The FREQ Procedure

                               Hair Color

                                                     Test
             Hair      Frequency      Percent      Percent
             ---------------------------------------------
             fair            76        30.89        30.00
             red             19         7.72        12.00
             medium          83        33.74        30.00
             dark            65        26.42        25.00
             black            3         1.22         3.00


--------------------------- Geographic Region=1 ---------------------------

                             Chi-Square Test
                        for Specified Proportions
                        -------------------------
                        Chi-Square          7.7602
                        DF                       4
                        Pr > ChiSq          0.1008
```

Output 3.3.1 shows the frequency table and chi-square test for Region 1. The frequency table lists the variable values (hair color) in the order in which they appear in the data set. The "Test Percent" column lists the hypothesized percentages for the chi-square test. Always check that you have ordered the TESTP= percentages to correctly match the order of the variable levels.

Output 3.3.2 shows the deviation plot for Region 1, which displays the relative deviations from the hypothesized values. The relative deviation for a level is the difference between the observed and hypothesized (test) percentage divided by the test percentage. You can suppress the chi-square p-value that is displayed by default in the deviation plot by specifying the NOSTATS plot-option.

Output 3.3.2 Deviation Plot for Region 1

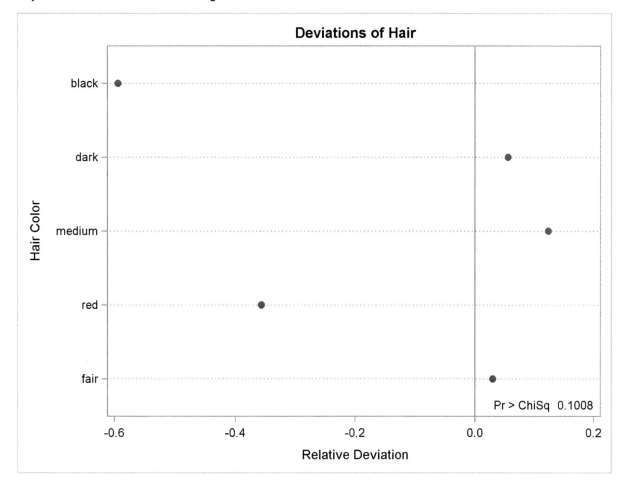

Output 3.3.3 and Output 3.3.4 show the results for Region 2. PROC FREQ computes a chi-square statistic for each region. The chi-square statistic is significant at the 0.05 level for Region 2 ($p=0.0003$) but not for Region 1. This indicates a significant departure from the hypothesized percentages in Region 2.

Output 3.3.3 Frequency Table and Chi-Square Test for Region 2

```
                        Hair Color of European Children

----------------------------- Geographic Region=2 -----------------------------

                              The FREQ Procedure

                                  Hair Color

                                                       Test
            Hair         Frequency       Percent       Percent
            ---------------------------------------------------
            fair              152          29.46        30.00
            red                94          18.22        12.00
            medium            134          25.97        30.00
            dark              117          22.67        25.00
            black              19           3.68         3.00


----------------------------- Geographic Region=2 -----------------------------

                                Chi-Square Test
                            for Specified Proportions
                            -------------------------
                            Chi-Square       21.3824
                            DF                     4
                            Pr > ChiSq        0.0003
```

Output 3.3.4 Deviation Plot for Region 2

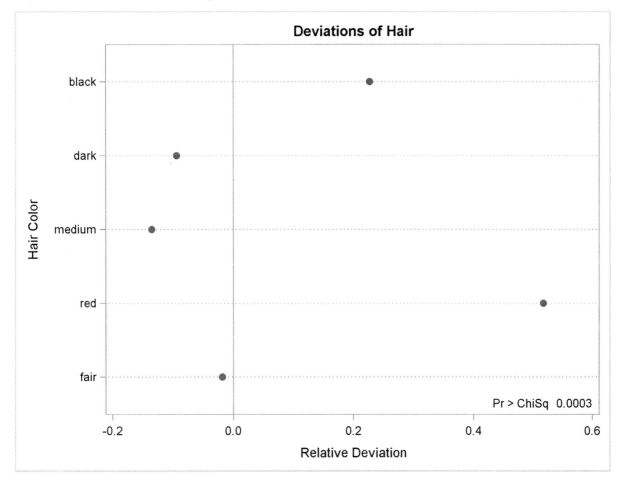

Example 3.4: Binomial Proportions

In this example, PROC FREQ computes binomial proportions, confidence limits, and tests. The example uses the eye and hair color data from Example 3.1. By default, PROC FREQ computes the binomial proportion as the proportion of observations in the first level of the one-way table. You can designate a different level by using the LEVEL= binomial-option.

The following PROC FREQ statements compute the proportion of children with brown eyes (from the data set in Example 3.1) and test the null hypothesis that the population proportion equals 50%. These statements also compute an equivalence for the proportion of children with fair hair.

The first TABLES statement requests a one-way frequency table for the variable Eyes. The BINOMIAL option requests the binomial proportion, confidence limits, and test. PROC FREQ computes the proportion with Eyes = 'brown', which is the first level displayed in the table. The AC, WILSON, and EXACT binomial-options request the following confidence limits types: Agresti-Coull, Wilson (score), and exact (Clopper-Pearson). By default, PROC FREQ provides Wald and exact (Clopper-Pearson) confidence limits for the binomial proportion. The BINOMIAL option also produces an asymptotic Wald test that the proportion equals 0.5. You can specify a different test proportion with the P= binomial-option. The ALPHA=0.1 option specifies that $\alpha = 10\%$, which produces 90% confidence limits.

The second TABLES statement requests a one-way frequency table for the variable Hair. The BINOMIAL option requests the proportion for the first level, Hair = 'fair'. The EQUIV binomial-option requests an equivalence test for the binomial proportion. The P=.28 option specifies 0.28 as the null hypothesis proportion, and the MARGIN=.1 option specifies 0.1 as the equivalence test margin.

```
proc freq data=Color order=freq;
   tables Eyes / binomial(ac wilson exact) alpha=.1;
   tables Hair / binomial(equiv p=.28 margin=.1);
   weight Count;
   title 'Hair and Eye Color of European Children';
run;
```

Output 3.4.1 displays the results for eye color, and Output 3.4.2 displays the results for hair color.

Output 3.4.1 Binomial Proportion for Eye Color

```
              Hair and Eye Color of European Children

                         The FREQ Procedure

                              Eye Color

                                       Cumulative      Cumulative
         Eyes      Frequency    Percent    Frequency       Percent
         ---------------------------------------------------------
         brown         341       44.75          341         44.75
         blue          222       29.13          563         73.88
         green         199       26.12          762        100.00


                         Binomial Proportion
                          for Eyes = brown
                         -------------------
                         Proportion    0.4475
                         ASE           0.0180

         Type                      90% Confidence Limits

         Wilson                        0.4181      0.4773
         Agresti-Coull                 0.4181      0.4773
         Clopper-Pearson (Exact)       0.4174      0.4779

                    Test of H0: Proportion = 0.5

                    ASE under H0              0.0181
                    Z                        -2.8981
                    One-sided Pr <  Z         0.0019
                    Two-sided Pr > |Z|        0.0038
```

The frequency table in Output 3.4.1 displays the values of Eyes in order of descending frequency count. PROC FREQ computes the proportion of children in the first level displayed in the frequency table, Eyes = 'brown'. Output 3.4.1 displays the binomial proportion confidence limits and test. The confidence limits are 90% confidence limits. If you do not specify the ALPHA= option, PROC FREQ computes 95% confidence limits by default. Because the value of Z is less than zero, PROC FREQ displays the a left-sided *p*-value (0.0019). This small *p*-value supports the alternative hypothesis that the true value of the proportion of children with brown eyes is less than 50%.

Output 3.4.2 displays the equivalence test results produced by the second TABLES statement. The null hypothesis proportion is 0.28 and the equivalence margins are −0.1 and 0.1, which yield equivalence limits of 0.18 and 0.38. PROC FREQ provides two one-sided tests (TOST) for equivalence. The small *p*-value indicates rejection of the null hypothesis in favor of the alternative that the proportion is equivalent to the null value.

Output 3.4.2 Binomial Proportion for Hair Color

```
                            Hair Color

                                      Cumulative     Cumulative
     Hair       Frequency    Percent   Frequency      Percent
     ---------------------------------------------------------
     fair          228        29.92       228          29.92
     medium        217        28.48       445          58.40
     dark          182        23.88       627          82.28
     red           113        14.83       740          97.11
     black          22         2.89       762         100.00

                       Equivalence Analysis

           H0: P - p0 <= Lower Margin or >= Upper Margin
           Ha: Lower Margin < P - p0 < Upper Margin

     p0 = 0.28    Lower Margin = -0.1    Upper Margin = 0.1

                    Proportion     ASE (Sample)

                      0.2992          0.0166

                   Two One-Sided Tests (TOST)

           Test                 Z         P-Value

           Lower Margin      7.1865    Pr > Z   <.0001
           Upper Margin     -4.8701    Pr < Z   <.0001
           Overall                              <.0001

           Equivalence Limits     90% Confidence Limits

              0.1800    0.3800       0.2719    0.3265
```

Example 3.5: Analysis of a 2x2 Contingency Table

This example computes chi-square tests and Fisher's exact test to compare the probability of coronary heart disease for two types of diet. It also estimates the relative risks and computes exact confidence limits for the odds ratio.

The data set FatComp contains hypothetical data for a case-control study of high fat diet and the risk of coronary heart disease. The data are recorded as cell counts, where the variable Count contains the frequencies for each exposure and response combination. The data set is sorted in descending order by the variables Exposure and Response, so that the first cell of the 2×2 table contains the frequency of positive exposure and positive response. The FORMAT procedure creates formats to identify the type of exposure and response with character values.

```
proc format;
   value ExpFmt 1='High Cholesterol Diet'
                0='Low Cholesterol Diet';
   value RspFmt 1='Yes'
                0='No';
run;
data FatComp;
   input Exposure Response Count;
   label Response='Heart Disease';
   datalines;
0 0  6
0 1  2
1 0  4
1 1 11
;
proc sort data=FatComp;
   by descending Exposure descending Response;
run;
```

In the following PROC FREQ statements, ORDER=DATA option orders the contingency table values by their order in the input data set. The TABLES statement requests a two-way table of Exposure by Response. The CHISQ option produces several chi-square tests, while the RELRISK option produces relative risk measures. The EXACT statement requests the exact Pearson chi-square test and exact confidence limits for the odds ratio.

```
proc freq data=FatComp order=data;
   format Exposure ExpFmt. Response RspFmt.;
   tables Exposure*Response / chisq relrisk;
   exact pchi or;
   weight Count;
   title 'Case-Control Study of High Fat/Cholesterol Diet';
run;
```

The contingency table in Output 3.5.1 displays the variable values so that the first table cell contains the frequency for the first cell in the data set (the frequency of positive exposure and positive response).

Output 3.5.1 Contingency Table

```
              Case-Control Study of High Fat/Cholesterol Diet

                           The FREQ Procedure

                       Table of Exposure by Response

         Exposure              Response(Heart Disease)

         Frequency        |
         Percent          |
         Row Pct          |
         Col Pct          |Yes     |No      |  Total
         -----------------+--------+--------+
         High Cholesterol |    11  |     4  |     15
              Diet        |  47.83 |  17.39 |  65.22
                          |  73.33 |  26.67 |
                          |  84.62 |  40.00 |
         -----------------+--------+--------+
         Low Cholesterol  |     2  |     6  |      8
              Diet        |   8.70 |  26.09 |  34.78
                          |  25.00 |  75.00 |
                          |  15.38 |  60.00 |
         -----------------+--------+--------+
         Total                 13       10        23
                             56.52    43.48    100.00
```

Output 3.5.2 displays the chi-square statistics. Because the expected counts in some of the table cells are small, PROC FREQ gives a warning that the asymptotic chi-square tests might not be appropriate. In this case, the exact tests are appropriate. The alternative hypothesis for this analysis states that coronary heart disease is more likely to be associated with a high fat diet, so a one-sided test is desired. Fisher's exact right-sided test analyzes whether the probability of heart disease in the high fat group exceeds the probability of heart disease in the low fat group; because this p-value is small, the alternative hypothesis is supported.

The odds ratio, displayed in Output 3.5.3, provides an estimate of the relative risk when an event is rare. This estimate indicates that the odds of heart disease is 8.25 times higher in the high fat diet group; however, the wide confidence limits indicate that this estimate has low precision.

Output 3.5.2 Chi-Square Statistics

```
Statistic                      DF        Value      Prob
------------------------------------------------------------
Chi-Square                      1       4.9597    0.0259
Likelihood Ratio Chi-Square     1       5.0975    0.0240
Continuity Adj. Chi-Square      1       3.1879    0.0742
Mantel-Haenszel Chi-Square      1       4.7441    0.0294
Phi Coefficient                         0.4644
Contingency Coefficient                 0.4212
Cramer's V                              0.4644

WARNING: 50% of the cells have expected counts less than 5.
         (Asymptotic) Chi-Square may not be a valid test.


              Pearson Chi-Square Test
         ----------------------------------
         Chi-Square                   4.9597
         DF                                1
         Asymptotic Pr >  ChiSq       0.0259
         Exact       Pr >= ChiSq      0.0393


                Fisher's Exact Test
         ----------------------------------
         Cell (1,1) Frequency (F)         11
         Left-sided Pr <= F           0.9967
         Right-sided Pr >= F          0.0367

         Table Probability (P)        0.0334
         Two-sided Pr <= P            0.0393
```

Output 3.5.3 Relative Risk

```
           Estimates of the Relative Risk (Row1/Row2)

Type of Study              Value        95% Confidence Limits
-------------------------------------------------------------
Case-Control (Odds Ratio)  8.2500       1.1535        59.0029
Cohort (Col1 Risk)         2.9333       0.8502        10.1204
Cohort (Col2 Risk)         0.3556       0.1403         0.9009


              Odds Ratio (Case-Control Study)
         ----------------------------------------
         Odds Ratio                   8.2500

         Asymptotic Conf Limits
         95% Lower Conf Limit         1.1535
         95% Upper Conf Limit        59.0029

         Exact Conf Limits
         95% Lower Conf Limit         0.8677
         95% Upper Conf Limit       105.5488
```

Example 3.6: Output Data Set of Chi-Square Statistics

This example uses the Color data from Example 3.1 to output the Pearson chi-square and the likelihood-ratio chi-square statistics to a SAS data set. The following PROC FREQ statements create a two-way table of eye color versus hair color.

```
proc freq data=Color order=data;
   tables Eyes*Hair / expected cellchi2 norow nocol chisq;
   output out=ChiSqData n nmiss pchi lrchi;
   weight Count;
   title 'Chi-Square Tests for 3 by 5 Table of Eye and Hair Color';
run;

proc print data=ChiSqData noobs;
   title1 'Chi-Square Statistics for Eye and Hair Color';
   title2 'Output Data Set from the FREQ Procedure';
run;
```

The EXPECTED option displays expected cell frequencies in the crosstabulation table, and the CELLCHI2 option displays the cell contribution to the overall chi-square. The NOROW and NO-COL options suppress the display of row and column percents in the crosstabulation table. The CHISQ option produces chi-square tests.

The OUTPUT statement creates the ChiSqData output data set and specifies the statistics to include. The N option requests the number of nonmissing observations, the NMISS option stores the number of missing observations, and the PCHI and LRCHI options request Pearson and likelihood-ratio chi-square statistics, respectively, together with their degrees of freedom and p-values.

The preceding statements produce Output 3.6.1 and Output 3.6.2. The contingency table in Output 3.6.1 displays eye and hair color in the order in which they appear in the Color data set. The Pearson chi-square statistic in Output 3.6.2 provides evidence of an association between eye and hair color ($p=0.0073$). The cell chi-square values show that most of the association is due to more green-eyed children with fair or red hair and fewer with dark or black hair. The opposite occurs with the brown-eyed children.

Output 3.6.3 displays the output data set created by the OUTPUT statement. It includes one observation that contains the sample size, the number of missing values, and the chi-square statistics and corresponding degrees of freedom and p-values as in Output 3.6.2.

Output 3.6.1 Contingency Table

```
           Chi-Square Tests for 3 by 5 Table of Eye and Hair Color

                             The FREQ Procedure

                             Table of Eyes by Hair

 Eyes(Eye Color)       Hair(Hair Color)

 Frequency      |
 Expected       |
 Cell Chi-Square|
 Percent        |fair    |red     |medium  |dark    |black   |  Total
 ---------------+--------+--------+--------+--------+--------+
 blue           |     69 |     28 |     68 |     51 |      6 |    222
                | 66.425 | 32.921 |  63.22 | 53.024 | 6.4094 |
                | 0.0998 | 0.7357 | 0.3613 | 0.0772 | 0.0262 |
                |   9.06 |   3.67 |   8.92 |   6.69 |   0.79 |  29.13
 ---------------+--------+--------+--------+--------+--------+
 green          |     69 |     38 |     55 |     37 |      0 |    199
                | 59.543 |  29.51 | 56.671 |  47.53 | 5.7454 |
                | 1.5019 | 2.4422 | 0.0492 | 2.3329 | 5.7454 |
                |   9.06 |   4.99 |   7.22 |   4.86 |   0.00 |  26.12
 ---------------+--------+--------+--------+--------+--------+
 brown          |     90 |     47 |     94 |     94 |     16 |    341
                | 102.03 | 50.568 | 97.109 | 81.446 | 9.8451 |
                | 1.4187 | 0.2518 | 0.0995 |  1.935 | 3.8478 |
                |  11.81 |   6.17 |  12.34 |  12.34 |   2.10 |  44.75
 ---------------+--------+--------+--------+--------+--------+
 Total               228      113      217      182       22      762
                   29.92    14.83    28.48    23.88     2.89   100.00
```

Output 3.6.2 Chi-Square Statistics

```
       Statistic                     DF       Value      Prob
       ------------------------------------------------------
       Chi-Square                     8     20.9248    0.0073
       Likelihood Ratio Chi-Square    8     25.9733    0.0011
       Mantel-Haenszel Chi-Square     1      3.7838    0.0518
       Phi Coefficient                       0.1657
       Contingency Coefficient               0.1635
       Cramer's V                            0.1172
```

Output 3.6.3 Output Data Set

```
              Chi-Square Statistics for Eye and Hair Color
                 Output Data Set from the FREQ Procedure

  N    NMISS    _PCHI_    DF_PCHI     P_PCHI      _LRCHI_   DF_LRCHI    P_LRCHI

 762     0     20.9248       8      .007349898   25.9733       8      .001061424
```

Example 3.7: Cochran-Mantel-Haenszel Statistics

The data set Migraine contains hypothetical data for a clinical trial of migraine treatment. Subjects of both genders receive either a new drug therapy or a placebo. Their response to treatment is coded as 'Better' or 'Same'. The data are recorded as cell counts, and the number of subjects for each treatment and response combination is recorded in the variable Count.

```
data Migraine;
   input Gender $ Treatment $ Response $ Count @@;
   datalines;
female Active  Better 16   female Active  Same 11
female Placebo Better  5   female Placebo Same 20
male   Active  Better 12   male   Active  Same 16
male   Placebo Better  7   male   Placebo Same 19
;
```

The following PROC FREQ statements create a multiway table stratified by Gender, where Treatment forms the rows and Response forms the columns. The CMH option produces the Cochran-Mantel-Haenszel statistics. For this stratified 2 × 2 table, estimates of the common relative risk and the Breslow-Day test for homogeneity of the odds ratios are also displayed. The NOPRINT option suppresses the display of the contingency tables. These statements produce Output 3.7.1 through Output 3.7.3.

```
proc freq data=Migraine;
   tables Gender*Treatment*Response / cmh;
   weight Count;
   title 'Clinical Trial for Treatment of Migraine Headaches';
run;
```

Output 3.7.1 Cochran-Mantel-Haenszel Statistics

```
             Clinical Trial for Treatment of Migraine Headaches

                              The FREQ Procedure

                  Summary Statistics for Treatment by Response
                             Controlling for Gender

          Cochran-Mantel-Haenszel Statistics (Based on Table Scores)

          Statistic    Alternative Hypothesis      DF      Value       Prob
          ---------------------------------------------------------------------
              1        Nonzero Correlation          1      8.3052     0.0040
              2        Row Mean Scores Differ       1      8.3052     0.0040
              3        General Association          1      8.3052     0.0040
```

For a stratified 2 × 2 table, the three CMH statistics displayed in Output 3.7.1 test the same hypothesis. The significant p-value (0.004) indicates that the association between treatment and response remains strong after adjusting for gender.

The CMH option also produces a table of relative risks, as shown in Output 3.7.2. Because this is a prospective study, the relative risk estimate assesses the effectiveness of the new drug; the "Cohort (Col1 Risk)" values are the appropriate estimates for the first column (the risk of improvement). The probability of migraine improvement with the new drug is just over two times the probability of improvement with the placebo.

The large *p*-value for the Breslow-Day test (0.2218) in Output 3.7.3 indicates no significant gender difference in the odds ratios.

Output 3.7.2 CMH Option: Relative Risks

```
           Estimates of the Common Relative Risk (Row1/Row2)

Type of Study    Method            Value      95% Confidence Limits
---------------------------------------------------------------------
Case-Control     Mantel-Haenszel   3.3132     1.4456       7.5934
 (Odds Ratio)    Logit             3.2941     1.4182       7.6515

Cohort           Mantel-Haenszel   2.1636     1.2336       3.7948
 (Col1 Risk)     Logit             2.1059     1.1951       3.7108

Cohort           Mantel-Haenszel   0.6420     0.4705       0.8761
 (Col2 Risk)     Logit             0.6613     0.4852       0.9013
```

Output 3.7.3 CMH Option: Breslow-Day Test

```
            Breslow-Day Test for
        Homogeneity of the Odds Ratios
        ------------------------------
        Chi-Square              1.4929
        DF                           1
        Pr > ChiSq              0.2218
```

Example 3.8: Cochran-Armitage Trend Test

The data set Pain contains hypothetical data for a clinical trial of a drug therapy to control pain. The clinical trial investigates whether adverse responses increase with larger drug doses. Subjects receive either a placebo or one of four drug doses. An adverse response is recorded as Adverse='Yes'; otherwise, it is recorded as Adverse='No'. The number of subjects for each drug dose and response combination is contained in the variable Count.

```
data pain;
   input Dose Adverse $ Count @@;
   datalines;
0 No 26   0 Yes  6
1 No 26   1 Yes  7
2 No 23   2 Yes  9
3 No 18   3 Yes 14
4 No  9   4 Yes 23
;
```

The following PROC FREQ statements provide a trend analysis. The TABLES statement requests a table of Adverse by Dose. The MEASURES option produces measures of association, and the CL option produces confidence limits for these measures. The TREND option tests for a trend across the ordinal values of the variable Dose with the Cochran-Armitage test. The EXACT statement produces exact p-values for this test, and the MAXTIME= option terminates the exact computations if they do not complete within 60 seconds. The TEST statement computes an asymptotic test for Somers' $D(R|C)$.

The PLOTS= option requests a frequency plot for the table of Adverse by Dose. By default, PROC FREQ provides a bar chart for the frequency plot. The TWOWAY=STACKED option requests a stacked layout, where the bars correspond to the column variable (Dose) values, and the row variable (Adverse) frequencies are stacked within each bar.

```
ods graphics on;
proc freq data=Pain;
   tables Adverse*Dose / trend measures cl
         plots=freqplot(twoway=stacked);
   test smdrc;
   exact trend / maxtime=60;
   weight Count;
   title 'Clinical Trial for Treatment of Pain';
run;
ods graphics off;
```

Output 3.8.1 through Output 3.8.4 display the results of the analysis. The "Col Pct" values in Output 3.8.1 show the expected increasing trend in the proportion of adverse effects with the increasing dosage (from 18.75% to 71.88%). The corresponding frequency bar chart (Output 3.8.2) also shows this increasing trend.

Output 3.8.1 Contingency Table

```
                 Clinical Trial for Treatment of Pain

                         The FREQ Procedure

                       Table of Adverse by Dose

          Adverse     Dose

          Frequency|
          Percent  |
          Row Pct  |
          Col Pct  |        0|        1|        2|        3|        4|  Total
          ---------+--------+--------+--------+--------+--------+
          No       |     26 |     26 |     23 |     18 |      9 |    102
                   |  16.15 |  16.15 |  14.29 |  11.18 |   5.59 |  63.35
                   |  25.49 |  25.49 |  22.55 |  17.65 |   8.82 |
                   |  81.25 |  78.79 |  71.88 |  56.25 |  28.13 |
          ---------+--------+--------+--------+--------+--------+
          Yes      |      6 |      7 |      9 |     14 |     23 |     59
                   |   3.73 |   4.35 |   5.59 |   8.70 |  14.29 |  36.65
                   |  10.17 |  11.86 |  15.25 |  23.73 |  38.98 |
                   |  18.75 |  21.21 |  28.13 |  43.75 |  71.88 |
          ---------+--------+--------+--------+--------+--------+
          Total          32       33       32       32       32     161
                      19.88    20.50    19.88    19.88    19.88  100.00
```

Output 3.8.2 Stacked Bar Chart of Frequencies

Output 3.8.3 displays the measures of association produced by the MEASURES option. Somers' $D(R|C)$ measures the association treating the row variable (Adverse) as the response and the column variable (Dose) as a predictor. Because the asymptotic 95% confidence limits do not contain zero, this indicates a strong positive association. Similarly, the Pearson and Spearman correlation coefficients show evidence of a strong positive association, as hypothesized.

The Cochran-Armitage test (Output 3.8.4) supports the trend hypothesis. The small left-sided p-values for the Cochran-Armitage test indicate that the probability of the Row 1 level (Adverse='No') decreases as Dose increases or, equivalently, that the probability of the Row 2 level (Adverse='Yes') increases as Dose increases. The two-sided p-value tests against either an increasing or decreasing alternative. This is an appropriate hypothesis when you want to determine whether the drug has progressive effects on the probability of adverse effects but the direction is unknown.

Output 3.8.3 Measures of Association

```
                                                        95%
Statistic                        Value      ASE    Confidence Limits
---------------------------------------------------------------------
Gamma                           0.5313    0.0935    0.3480    0.7146
Kendall's Tau-b                 0.3373    0.0642    0.2114    0.4631
Stuart's Tau-c                  0.4111    0.0798    0.2547    0.5675

Somers' D C|R                   0.4427    0.0837    0.2786    0.6068
Somers' D R|C                   0.2569    0.0499    0.1592    0.3547

Pearson Correlation             0.3776    0.0714    0.2378    0.5175
Spearman Correlation            0.3771    0.0718    0.2363    0.5178

Lambda Asymmetric C|R           0.1250    0.0662    0.0000    0.2547
Lambda Asymmetric R|C           0.2373    0.0837    0.0732    0.4014
Lambda Symmetric                0.1604    0.0621    0.0388    0.2821

Uncertainty Coefficient C|R     0.0515    0.0191    0.0140    0.0890
Uncertainty Coefficient R|C     0.1261    0.0467    0.0346    0.2175
Uncertainty Coefficient Symmetric 0.0731  0.0271    0.0199    0.1262


                       Somers' D R|C
          ---------------------------------------
          Somers' D R|C                   0.2569
          ASE                             0.0499
          95% Lower Conf Limit            0.1592
          95% Upper Conf Limit            0.3547

            Test of H0: Somers' D R|C = 0

          ASE under H0                    0.0499
          Z                               5.1511
          One-sided Pr >  Z              <.0001
          Two-sided Pr > |Z|             <.0001
```

Output 3.8.4 Trend Test

```
             Cochran-Armitage Trend Test
          ---------------------------------
          Statistic (Z)              -4.7918

          Asymptotic Test
          One-sided Pr <  Z           <.0001
          Two-sided Pr > |Z|          <.0001

          Exact Test
          One-sided Pr <=  Z         7.237E-07
          Two-sided Pr >= |Z|        1.324E-06
```

Example 3.9: Friedman's Chi-Square Test

Friedman's test is a nonparametric test for treatment differences in a randomized complete block design. Each block of the design might be a subject or a homogeneous group of subjects. If blocks are groups of subjects, the number of subjects in each block must equal the number of treatments. Treatments are randomly assigned to subjects within each block. If there is one subject per block, then the subjects are repeatedly measured once under each treatment. The order of treatments is randomized for each subject.

In this setting, Friedman's test is identical to the ANOVA (row means scores) CMH statistic when the analysis uses rank scores (SCORES=RANK). The three-way table uses subject (or subject group) as the stratifying variable, treatment as the row variable, and response as the column variable. PROC FREQ handles ties by assigning midranks to tied response values. If there are multiple subjects per treatment in each block, the ANOVA CMH statistic is a generalization of Friedman's test.

The data set Hypnosis contains data from a study investigating whether hypnosis has the same effect on skin potential (measured in millivolts) for four emotions (Lehmann 1975, p. 264). Eight subjects are asked to display fear, joy, sadness, and calmness under hypnosis. The data are recorded as one observation per subject for each emotion.

```
data Hypnosis;
   length Emotion $ 10;
   input Subject Emotion $ SkinResponse @@;
   datalines;
1 fear 23.1  1 joy 22.7  1 sadness 22.5  1 calmness 22.6
2 fear 57.6  2 joy 53.2  2 sadness 53.7  2 calmness 53.1
3 fear 10.5  3 joy  9.7  3 sadness 10.8  3 calmness  8.3
4 fear 23.6  4 joy 19.6  4 sadness 21.1  4 calmness 21.6
5 fear 11.9  5 joy 13.8  5 sadness 13.7  5 calmness 13.3
6 fear 54.6  6 joy 47.1  6 sadness 39.2  6 calmness 37.0
7 fear 21.0  7 joy 13.6  7 sadness 13.7  7 calmness 14.8
8 fear 20.3  8 joy 23.6  8 sadness 16.3  8 calmness 14.8
;
```

In the following PROC FREQ statements, the TABLES statement creates a three-way table stratified by Subject and a two-way table; the variables Emotion and SkinResponse form the rows and columns of each table. The CMH2 option produces the first two Cochran-Mantel-Haenszel statistics, the option SCORES=RANK specifies that rank scores are used to compute these statistics, and the NOPRINT option suppresses the contingency tables. These statements produce Output 3.9.1 and Output 3.9.2.

```
proc freq data=Hypnosis;
   tables Subject*Emotion*SkinResponse /
          cmh2 scores=rank noprint;
run;
```

```
proc freq data=Hypnosis;
   tables Emotion*SkinResponse /
          cmh2 scores=rank noprint;
run;
```

Because the CMH statistics in Output 3.9.1 are based on rank scores, the Row Mean Scores Differ statistic is identical to Friedman's chi-square ($Q = 6.45$). The *p*-value of 0.0917 indicates that differences in skin potential response for different emotions are significant at the 10% level but not at the 5% level.

When you do not stratify by subject, the Row Mean Scores Differ CMH statistic is identical to a Kruskal-Wallis test and is not significant ($p=0.9038$ in Output 3.9.2). Thus, adjusting for subject is critical to reducing the background variation due to subject differences.

Output 3.9.1 CMH Statistics: Stratifying by Subject

```
                   Clinical Trial for Treatment of Pain

                           The FREQ Procedure

              Summary Statistics for Emotion by SkinResponse
                         Controlling for Subject

         Cochran-Mantel-Haenszel Statistics (Based on Rank Scores)

         Statistic    Alternative Hypothesis    DF      Value      Prob
         ---------------------------------------------------------------
             1        Nonzero Correlation        1     0.2400     0.6242
             2        Row Mean Scores Differ     3     6.4500     0.0917
```

Output 3.9.2 CMH Statistics: No Stratification

```
                   Clinical Trial for Treatment of Pain

                           The FREQ Procedure

              Summary Statistics for Emotion by SkinResponse

         Cochran-Mantel-Haenszel Statistics (Based on Rank Scores)

         Statistic    Alternative Hypothesis    DF      Value      Prob
         ---------------------------------------------------------------
             1        Nonzero Correlation        1     0.0001     0.9933
             2        Row Mean Scores Differ     3     0.5678     0.9038
```

Example 3.10: Cochran's Q Test

When a binary response is measured several times or under different conditions, Cochran's Q tests that the marginal probability of a positive response is unchanged across the times or conditions. When there are more than two response categories, you can use the CATMOD procedure to fit a repeated-measures model.

The data set Drugs contains data for a study of three drugs to treat a chronic disease (Agresti 2002). Forty-six subjects receive drugs A, B, and C. The response to each drug is either favorable ('F') or unfavorable ('U').

```
proc format;
   value $ResponseFmt 'F'='Favorable'
                      'U'='Unfavorable';
run;

data drugs;
   input Drug_A $ Drug_B $ Drug_C $ Count @@;
   datalines;
F F F  6    U F F  2
F F U 16    U F U  4
F U F  2    U U F  6
F U U  4    U U U  6
;
```

The following statements create one-way frequency tables of the responses to each drug. The AGREE option produces Cochran's Q and other measures of agreement for the three-way table. These statements produce Output 3.10.1 through Output 3.10.5.

```
proc freq data=Drugs;
   tables Drug_A Drug_B Drug_C / nocum;
   tables Drug_A*Drug_B*Drug_C / agree noprint;
   format Drug_A Drug_B Drug_C $ResponseFmt.;
   weight Count;
   title 'Study of Three Drug Treatments for a Chronic Disease';
run;
```

The one-way frequency tables in Output 3.10.1 provide the marginal response for each drug. For drugs A and B, 61% of the subjects reported a favorable response while 35% of the subjects reported a favorable response to drug C. Output 3.10.2 and Output 3.10.3 display measures of agreement for the 'Favorable' and 'Unfavorable' levels of drug A, respectively. McNemar's test shows a strong discordance between drugs B and C when the response to drug A is favorable.

Output 3.10.1 One-Way Frequency Tables

```
            Study of Three Drug Treatments for a Chronic Disease

                           The FREQ Procedure

            Drug_A             Frequency         Percent
            ------------------------------------------------
            Favorable               28            60.87
            Unfavorable             18            39.13

            Drug_B             Frequency         Percent
            ------------------------------------------------
            Favorable               28            60.87
            Unfavorable             18            39.13

            Drug_C             Frequency         Percent
            ------------------------------------------------
            Favorable               16            34.78
            Unfavorable             30            65.22
```

Output 3.10.2 Measures of Agreement for Drug A Favorable

```
                         McNemar's Test
                    ------------------------
                    Statistic (S)    10.8889
                    DF                     1
                    Pr > S            0.0010

                    Simple Kappa Coefficient
                    ------------------------
                    Kappa                -0.0328
                    ASE                   0.1167
                    95% Lower Conf Limit -0.2615
                    95% Upper Conf Limit  0.1960
```

Output 3.10.3 Measures of Agreement for Drug A Unfavorable

```
                         McNemar's Test
                    ------------------------
                    Statistic (S)     0.4000
                    DF                     1
                    Pr > S            0.5271

                    Simple Kappa Coefficient
                    ------------------------
                    Kappa                -0.1538
                    ASE                   0.2230
                    95% Lower Conf Limit -0.5909
                    95% Upper Conf Limit  0.2832
```

Output 3.10.4 displays the overall kappa coefficient. The small negative value of kappa indicates no agreement between drug B response and drug C response.

Cochran's Q is statistically significant ($p=0.0144$ in Output 3.10.5), which leads to rejection of the hypothesis that the probability of favorable response is the same for the three drugs.

Output 3.10.4 Overall Measures of Agreement

```
            Overall Kappa Coefficient
    -----------------------------------
    Kappa                        -0.0588
    ASE                           0.1034
    95% Lower Conf Limit         -0.2615
    95% Upper Conf Limit          0.1439

             Test for Equal Kappa
                 Coefficients
          --------------------------
          Chi-Square       0.2314
          DF                    1
          Pr > ChiSq       0.6305
```

Output 3.10.5 Cochran's Q Test

```
           Cochran's Q, for Drug_A
              by Drug_B by Drug_C
         ---------------------------
         Statistic (Q)        8.4706
         DF                        2
         Pr > Q               0.0145
```

References

Agresti, A. (1992), "A Survey of Exact Inference for Contingency Tables," *Statistical Science*, 7(1), 131–177.

Agresti, A. (2002), *Categorical Data Analysis*, Second Edition, New York: John Wiley & Sons.

Agresti, A. (2007), *An Introduction to Categorical Data Analysis*, Second Edition, New York: John Wiley & Sons.

Agresti, A. and Coull, B. A. (1998), "Approximate is Better than "Exact" for Interval Estimation of Binomial Proportions," *The American Statistician*, 52, 119–126.

Agresti, A., Mehta, C. R. and Patel, N. R. (1990), "Exact Inference for Contingency Tables with Ordered Categories," *Journal of the American Statistical Association*, 85, 453–458.

Agresti, A. and Min, Y. (2001), "On Small-Sample Confidence Intervals for Parameters in Discrete Distributions," *Biometrics*, 57, 963–971.

Agresti, A., Wackerly, D., and Boyett, J. M. (1979), "Exact Conditional Tests for Cross-Classifications: Approximation of Attained Significance Levels," *Psychometrika*, 44, 75–83.

Barker, L., Rolka, H., Rolka, D., and Brown, C. (2001), "Equivalence Testing for Binomial Random Variables: Which Test to Use?," *The American Statistician*, 55, 279–287.

Berger, J. O. (1985), *Statistical Decision Theory and Bayesian Analysis*, Second Edition, New York: Springer-Verlag.

Birch, M. W. (1965), "The Detection of Partial Association, II: The General Case," *Journal of the Royal Statistical Society, B*, 27, 111–124.

Bishop, Y., Fienberg, S. E., and Holland, P. W. (1975), *Discrete Multivariate Analysis: Theory and Practice*, Cambridge, MA: MIT Press.

Bowker, A. H. (1948), "Bowker's Test for Symmetry," *Journal of the American Statistical Association*, 43, 572–574.

Breslow, N. E. (1996), "Statistics in Epidemiology: The Case-Control Study," *Journal of the American Statistical Association*, 91, 14–26.

Breslow, N. E. and Day, N. E. (1980), *Statistical Methods in Cancer Research, Volume I: The Analysis of Case-Control Studies*, IARC Scientific Publications, No. 32, Lyon, France: International Agency for Research on Cancer.

Breslow, N. E. and Day, N. E. (1987), *Statistical Methods in Cancer Research, Volume II: The Design and Analysis of Cohort Studies*, IARC Scientific Publications, No. 82, Lyon, France: International Agency for Research on Cancer.

Bross, I. D. J. (1958), "How to Use Ridit Analysis," *Biometrics*, 14, 18–38.

Brown, L. D., Cai, T. T., and DasGupta, A. (2001), "Interval Estimation for a Binomial Proportion," *Statistical Science* 16, 101–133.

Brown, M. B. and Benedetti, J. K. (1977), "Sampling Behavior of Tests for Correlation in Two-Way Contingency Tables," *Journal of the American Statistical Association*, 72, 309–315.

Chow, S., Shao, J., and Wang, H. (2003), *Sample Size Calculations in Clinical Research*, Boca Raton, FL: CRC Press.

Cicchetti, D. V. and Allison, T. (1971), "A New Procedure for Assessing Reliability of Scoring EEG Sleep Recordings," *American Journal of EEG Technology*, 11, 101–109.

Clopper, C. J., and Pearson, E. S. (1934), "The Use of Confidence or Fiducial Limits Illustrated in the Case of the Binomial," *Biometrika* 26, 404–413.

Cochran, W. G. (1950), "The Comparison of Percentages in Matched Samples," *Biometrika*, 37, 256–266.

Cochran, W. G. (1954), "Some Methods for Strengthening the Common χ^2 Tests," *Biometrics*, 10, 417–451.

Collett, D. (1991), *Modelling Binary Data*, London: Chapman & Hall.

Cohen, J. (1960), "A Coefficient of Agreement for Nominal Scales," *Educational and Psychological Measurement*, 20, 37–46.

Drasgow, F. (1986), "Polychoric and Polyserial Correlations" in *Encyclopedia of Statistical Sciences*, vol. 7, ed. S. Kotz and N. L. Johnson, New York: John Wiley & Sons, 68–74.

Dunnett, C. W., and Gent, M. (1977), "Significance Testing to Establish Equivalence Between Treatments, with Special Reference to Data in the Form of 2 × 2 Tables," *Biometrics*, 33, 593–602.

Farrington, C. P., and Manning, G. (1990), "Test Statistics and Sample Size Formulae for Comparative Binomial Trials with Null Hypothesis of Non-zero Risk Difference or Non-unity Relative Risk," *Statistics in Medicine*, 9, 1447–1454.

Fienberg, S. E. (1980), *The Analysis of Cross-Classified Data*, Second Edition, Cambridge, MA: MIT Press.

Fleiss, J. L., Levin, B., and Paik, M. C. (2003), *Statistical Methods for Rates and Proportions*, Third Edition, New York: John Wiley & Sons.

Fleiss, J. L. and Cohen, J. (1973), "The Equivalence of Weighted Kappa and the Intraclass Correlation Coefficient as Measures of Reliability," *Educational and Psychological Measurement*, 33, 613–619.

Fleiss, J. L., Cohen, J., and Everitt, B. S. (1969), "Large-Sample Standard Errors of Kappa and Weighted Kappa," *Psychological Bulletin*, 72, 323–327.

Freeman, G. H. and Halton, J. H. (1951), "Note on an Exact Treatment of Contingency, Goodness of Fit and Other Problems of Significance," *Biometrika*, 38, 141–149.

Gail, M. and Mantel, N. (1977), "Counting the Number of $r \times c$ Contingency Tables with Fixed Margins," *Journal of the American Statistical Association*, 72, 859–862.

Gart, J. J. (1971), "The Comparison of Proportions: A Review of Significance Tests, Confidence Intervals and Adjustments for Stratification," *Review of the International Statistical Institute*, 39(2), 148–169.

Goodman, L. A. and Kruskal, W. H. (1979), *Measures of Association for Cross Classification*, New York: Springer-Verlag.

Greenland, S. and Robins, J. M. (1985), "Estimators of the Mantel-Haenszel Variance Consistent in Both Sparse Data and Large-Strata Limiting Models," *Biometrics*, 42, 311–323.

Haldane, J. B. S. (1955), "The Estimation and Significance of the Logarithm of a Ratio of Frequencies," *Annals of Human Genetics*, 20, 309–314.

Hauck, W. W. and Anderson, S. (1986), "A Comparison of Large-Sample Confidence Interval Methods for the Difference of Two Binomial Probabilities," *The American Statistician*, 40, 318–322.

Hirji, K. F. (2006), *Exact Analysis of Discrete Data*, Boca Raton, FL: Chapman & Hall/CRC.

Hirji, K. F., Vollset, S. E., Reis, I. M., and Afifi, A. A. (1996), "Exact Tests for Interaction in Several 2×2 Tables," *Journal of Computational and Graphical Statistics*, 5, 209–224.

Hollander, M. and Wolfe, D. A. (1999), *Nonparametric Statistical Methods*, Second Edition, New York: John Wiley & Sons.

Jones, M. P., O'Gorman, T. W., Lemka, J. H., and Woolson, R. F. (1989), "A Monte Carlo Investigation of Homogeneity Tests of the Odds Ratio Under Various Sample Size Configurations," *Biometrics*, 45, 171–181.

Kendall, M. (1955), *Rank Correlation Methods*, Second Edition, London: Charles Griffin and Co.

Kendall, M. and Stuart, A. (1979), *The Advanced Theory of Statistics*, vol. 2, New York: Macmillan.

Kleinbaum, D. G., Kupper, L. L., and Morgenstern, H. (1982), *Epidemiologic Research: Principles and Quantitative Methods*, Research Methods Series, New York: Van Nostrand Reinhold.

Landis, R. J., Heyman, E. R., and Koch, G. G. (1978), "Average Partial Association in Threeway Contingency Tables: A Review and Discussion of Alternative Tests," *International Statistical Review*, 46, 237–254.

Leemis, L. M. and Trivedi, K. S. (1996), "A Comparison of Approximate Interval Estimators for the Bernoulli Parameter," *The American Statistician*, 50, 63–68.

Lehmann, E. L. (1975), *Nonparametrics: Statistical Methods Based on Ranks*, San Francisco: Holden-Day.

Liebetrau, A. M. (1983), *Measures of Association, Quantitative Application in the Social Sciences*, vol. 32, Beverly Hills: Sage Publications.

Mack, G. A. and Skillings, J. H. (1980), "A Friedman-Type Rank Test for Main Effects in a Two-Factor ANOVA," *Journal of the American Statistical Association*, 75, 947–951.

Mantel, N. (1963), "Chi-square Tests with One Degree of Freedom: Extensions of the Mantel-Haenszel Procedure," *Journal of the American Statistical Association*, 58, 690–700.

Mantel, N. and Haenszel, W. (1959), "Statistical Aspects of the Analysis of Data from Retrospective Studies of Disease," *Journal of the National Cancer Institute*, 22, 719–748.

Margolin, B. H. (1988), "Test for Trend in Proportions," in *Encyclopedia of Statistical Sciences*, vol. 9, ed. S. Kotz and N. L. Johnson, New York: John Wiley & Sons, 334–336.

McNemar, Q. (1947), "Note on the Sampling Error of the Difference Between Correlated Proportions or Percentages," *Psychometrika*, 12, 153–157.

Mehta, C. R. and Patel, N. R. (1983), "A Network Algorithm for Performing Fisher's Exact Test in $r \times c$ Contingency Tables," *Journal of the American Statistical Association*, 78, 427–434.

Mehta, C. R., Patel, N. R., and Gray, R. (1985), "On Computing an Exact Confidence Interval for the Common Odds Ratio in Several 2 × 2 Contingency Tables," *Journal of the American Statistical Association*, 80, 969–973.

Mehta, C. R., Patel, N. R., and Senchaudhuri, P. (1991), "Exact Stratified Linear Rank Tests for Binary Data," *Computing Science and Statistics: Proceedings of the 23rd Symposium on the Interface*, ed. E.M. Keramidas, 200–207.

Mehta, C. R., Patel, N. R., and Tsiatis, A. A. (1984), "Exact Significance Testing to Establish Treatment Equivalence with Ordered Categorical Data," *Biometrics*, 40, 819–825.

Narayanan, A. and Watts, D. (1996), "Exact Methods in the NPAR1WAY Procedure," in *Proceedings of the Twenty-First Annual SAS Users Group International Conference*, Cary, NC: SAS Institute Inc., 1290–1294.

Newcombe, R. G. (1998), "Two-sided Confidence Intervals for the Single Proportion: Comparison of Seven Methods," *Statistics in Medicine*, 17, 857–872.

Newcombe, R. G. (1998), "Interval Estimation for the Difference Between Independent Proportions: Comparison of Eleven Methods," *Statistics in Medicine*, 17, 873–890.

Olsson, U. (1979), "Maximum Likelihood Estimation of the Polychoric Correlation Coefficient," *Psychometrika*, 12, 443–460.

Pirie, W. (1983), "Jonckheere Tests for Ordered Alternatives," in *Encyclopedia of Statistical Sciences*, vol. 4, ed. S. Kotz and N. L. Johnson, New York: John Wiley & Sons, 315–318.

Radlow, R. and Alf, E. F. (1975), "An Alternate Multinomial Assessment of the Accuracy of the Chi-Square Test of Goodness of Fit," *Journal of the American Statistical Association*, 70, 811–813.

Robins, J. M., Breslow, N., and Greenland, S. (1986), "Estimators of the Mantel-Haenszel Variance Consistent in Both Sparse Data and Large-Strata Limiting Models," *Biometrics*, 42, 311–323.

Santner, T. J. and Snell, M. K. (1980), "Small-Sample Confidence Intervals for p1-p2 and p1/p2 in 2 × 2 Contingency Tables," *Journal of the American Statistical Association*, 75, 386–394.

Schuirmann, D. J. (1987), "A Comparison of the Two One-Sided Tests Procedure and the Power Approach for Assessing the Equivalence of Average Bioavailability," *Journal of Pharmacokinetics and Biopharmaceutics*, 15, 657–680.

Schuirmann, D. J. (1999), "Confidence Interval Methods for Bioequivalence Testing with Binomial Endpoints," *Proceedings of the Biopharmaceutical Section, ASA*, 227–232.

Snedecor, G. W. and Cochran, W. G. (1989), *Statistical Methods*, Eighth Edition, Ames: Iowa State University Press.

Somers, R. H. (1962), "A New Asymmetric Measure of Association for Ordinal Variables," *American Sociological Review*, 27, 799–811.

Stokes, M. E., Davis, C. S., and Koch, G. G. (2000), *Categorical Data Analysis Using the SAS System*, Second Edition, Cary, NC: SAS Institute Inc.

Tarone, R. E. (1985), "On Heterogeneity Tests Based on Efficient Scores," *Biometrika*, 72, 1, 91–95.

Theil, H. (1972), *Statistical Decomposition Analysis*, Amsterdam: North-Holland Publishing Company.

Thomas, D. G. (1971), "Algorithm AS-36. Exact Confidence Limits for the Odds Ratio in a 2×2 Table," *Applied Statistics*, 20, 105–110.

Valz, P. D. and Thompson, M. E. (1994), "Exact Inference for Kendall's S and Spearman's Rho with Extensions to Fisher's Exact Test in $r \times c$ Contingency Tables," *Journal of Computational and Graphical Statistics*, 3(4), 459–472.

van Elteren, P. H. (1960), "On the Combination of Independent Two-Sample Tests of Wilcoxon," *Bulletin of the International Statistical Institute*, 37, 351–361.

Vollset, S. E., Hirji, K. F., and Elashoff, R. M. (1991), "Fast Computation of Exact Confidence Limits for the Common Odds Ratio in a Series of 2×2 Tables," *Journal of the American Statistical Association*, 86, 404–409.

Wilson, E. B. (1927), "Probable Inference, the Law of Succession, and Statistical Inference," *Journal of the American Statistical Association*, 22, 209–212.

Woolf, B. (1955), "On Estimating the Relationship Between Blood Group and Disease," *Annals of Human Genetics*, 19, 251–253.

Zelen, M. (1971), "The Analysis of Several 2×2 Contingency Tables," *Biometrika*, 58, 129–137.

Chapter 4
The UNIVARIATE Procedure

Contents

Overview: UNIVARIATE Procedure	**223**
Getting Started: UNIVARIATE Procedure	**224**
Capabilities of PROC UNIVARIATE	224
Summarizing a Data Distribution	225
Exploring a Data Distribution	226
Modeling a Data Distribution	229
Syntax: UNIVARIATE Procedure	**233**
PROC UNIVARIATE Statement	233
BY Statement	240
CDFPLOT Statement	241
CLASS Statement	251
FREQ Statement	253
HISTOGRAM Statement	253
ID Statement	271
INSET Statement	271
OUTPUT Statement	280
PPPLOT Statement	284
PROBPLOT Statement	294
QQPLOT Statement	305
VAR Statement	315
WEIGHT Statement	315
Dictionary of Common Options	317
Details: UNIVARIATE Procedure	**323**
Missing Values	323
Rounding	324
Descriptive Statistics	325
Calculating the Mode	328
Calculating Percentiles	328
Tests for Location	331
Confidence Limits for Parameters of the Normal Distribution	333
Robust Estimators	333
Creating Line Printer Plots	337
Creating High-Resolution Graphics	339
Using the CLASS Statement to Create Comparative Plots	343

Positioning Insets	344
Formulas for Fitted Continuous Distributions	348
Goodness-of-Fit Tests	356
Kernel Density Estimates	360
Construction of Quantile-Quantile and Probability Plots	362
Interpretation of Quantile-Quantile and Probability Plots	363
Distributions for Probability and Q-Q Plots	364
Estimating Shape Parameters Using Q-Q Plots	368
Estimating Location and Scale Parameters Using Q-Q Plots	368
Estimating Percentiles Using Q-Q Plots	369
Input Data Sets	370
OUT= Output Data Set in the OUTPUT Statement	370
OUTHISTOGRAM= Output Data Set	372
OUTKERNEL= Output Data Set	373
OUTTABLE= Output Data Set	374
Tables for Summary Statistics	376
ODS Table Names	376
ODS Tables for Fitted Distributions	377
ODS Graphics (Experimental)	378
Computational Resources	379
Examples: UNIVARIATE Procedure	**380**
Example 4.1: Computing Descriptive Statistics for Multiple Variables	380
Example 4.2: Calculating Modes	382
Example 4.3: Identifying Extreme Observations and Extreme Values	384
Example 4.4: Creating a Frequency Table	386
Example 4.5: Creating Plots for Line Printer Output	387
Example 4.6: Analyzing a Data Set With a FREQ Variable	393
Example 4.7: Saving Summary Statistics in an OUT= Output Data Set	394
Example 4.8: Saving Percentiles in an Output Data Set	396
Example 4.9: Computing Confidence Limits for the Mean, Standard Deviation, and Variance	397
Example 4.10: Computing Confidence Limits for Quantiles and Percentiles	399
Example 4.11: Computing Robust Estimates	401
Example 4.12: Testing for Location	403
Example 4.13: Performing a Sign Test Using Paired Data	404
Example 4.14: Creating a Histogram	405
Example 4.15: Creating a One-Way Comparative Histogram	407
Example 4.16: Creating a Two-Way Comparative Histogram	409
Example 4.17: Adding Insets with Descriptive Statistics	411
Example 4.18: Binning a Histogram	414
Example 4.19: Adding a Normal Curve to a Histogram	417
Example 4.20: Adding Fitted Normal Curves to a Comparative Histogram	419
Example 4.21: Fitting a Beta Curve	421
Example 4.22: Fitting Lognormal, Weibull, and Gamma Curves	423

Example 4.23:	Computing Kernel Density Estimates	428
Example 4.24:	Fitting a Three-Parameter Lognormal Curve	429
Example 4.25:	Annotating a Folded Normal Curve	431
Example 4.26:	Creating Lognormal Probability Plots	437
Example 4.27:	Creating a Histogram to Display Lognormal Fit	442
Example 4.28:	Creating a Normal Quantile Plot	444
Example 4.29:	Adding a Distribution Reference Line	446
Example 4.30:	Interpreting a Normal Quantile Plot	447
Example 4.31:	Estimating Three Parameters from Lognormal Quantile Plots	449
Example 4.32:	Estimating Percentiles from Lognormal Quantile Plots . . .	454
Example 4.33:	Estimating Parameters from Lognormal Quantile Plots . . .	455
Example 4.34:	Comparing Weibull Quantile Plots	456
Example 4.35:	Creating a Cumulative Distribution Plot	459
Example 4.36:	Creating a P-P Plot	461
References	. . . .	**462**

Overview: UNIVARIATE Procedure

The UNIVARIATE procedure provides the following:

- descriptive statistics based on moments (including skewness and kurtosis), quantiles or percentiles (such as the median), frequency tables, and extreme values

- histograms that optionally can be fitted with probability density curves for various distributions and with kernel density estimates

- cumulative distribution function plots (cdf plots). Optionally, these can be superimposed with probability distribution curves for various distributions.

- quantile-quantile plots (Q-Q plots), probability plots, and probability-probability plots (P-P plots). These plots facilitate the comparison of a data distribution with various theoretical distributions.

- goodness-of-fit tests for a variety of distributions including the normal

- the ability to inset summary statistics on plots

- the ability to analyze data sets with a frequency variable

- the ability to create output data sets containing summary statistics, histogram intervals, and parameters of fitted curves

You can use the PROC UNIVARIATE statement, together with the VAR statement, to compute summary statistics. See the section "Getting Started: UNIVARIATE Procedure" on page 224 for introductory examples. In addition, you can use the following statements to request plots:

- the CDFPLOT statement for creating cdf plots
- the HISTOGRAM statement for creating histograms
- the PPPLOT statement for creating P-P plots
- the PROBPLOT statement for creating probability plots
- the QQPLOT statement for creating Q-Q plots
- the CLASS statement together with any of these plot statements for creating comparative plots
- the INSET statement with any of the plot statements for enhancing the plot with an inset table of summary statistics

The UNIVARIATE procedure produces two kinds of graphical output:

- traditional graphics, which are produced by default
- ODS Statistical Graphics output (supported on an experimental basis for SAS 9.2), which is produced when you specify the ODS GRAPHICS statement prior to your procedure statements statements.

See the section "Creating High-Resolution Graphics" on page 339 for more information about producing traditional graphics and ODS Graphics output.

Getting Started: UNIVARIATE Procedure

The following examples demonstrate how you can use the UNIVARIATE procedure to analyze the distributions of variables through the use of descriptive statistical measures and graphical displays, such as histograms.

Capabilities of PROC UNIVARIATE

The UNIVARIATE procedure provides a variety of descriptive measures, graphical displays, and statistical methods, which you can use to summarize, visualize, analyze, and model the statistical distributions of numeric variables. These tools are appropriate for a broad range of tasks and applications:

- Exploring the distributions of the variables in a data set is an important preliminary step in data analysis, data warehousing, and data mining. With the UNIVARIATE procedure you can use tables and graphical displays, such as histograms and nonparametric density estimates, to find key features of distributions, identify outliers and extreme observations, determine the need for data transformations, and compare distributions.

- Modeling the distributions of data and validating distributional assumptions are basic steps in statistical analysis. You can use the UNIVARIATE procedure to fit parametric distributions (beta, exponential, gamma, lognormal, normal, Johnson S_B, Johnson S_U, and Weibull) and to compute probabilities and percentiles from these models. You can assess goodness of fit with hypothesis tests and with graphical displays such as probability plots and quantile-quantile plots. You can also use the UNIVARIATE procedure to validate distributional assumptions for other types of statistical analysis. When standard assumptions are not met, you can use the UNIVARIATE procedure to perform nonparametric tests and compute robust estimates of location and scale.

- Summarizing the distribution of the data is often helpful for creating effective statistical reports and presentations. You can use the UNIVARIATE procedure to create tables of summary measures, such as means and percentiles, together with graphical displays, such as histograms and comparative histograms, which facilitate the interpretation of the report.

The following examples illustrate a few of the tasks that you can carry out with the UNIVARIATE procedure.

Summarizing a Data Distribution

Figure 4.1 shows a table of basic summary measures and a table of extreme observations for the loan-to-value ratios of 5,840 home mortgages. The ratios are saved as values of the variable LoanToValueRatio in a data set named HomeLoans. The following statements request a univariate analysis:

```
ods select BasicMeasures ExtremeObs;
proc univariate data=HomeLoans;
   var LoanToValueRatio;
run;
```

The ODS SELECT statement restricts the default output to the tables for basic statistical measures and extreme observations.

Figure 4.1 Basic Measures and Extreme Observations

```
                  The UNIVARIATE Procedure
         Variable:   LoanToValueRatio   (Loan to Value Ratio)

                     Basic Statistical Measures

         Location                      Variability

     Mean     0.292512     Std Deviation           0.16476
     Median   0.248050     Variance                0.02715
     Mode     0.250000     Range                   1.24780
                           Interquartile Range     0.16419
```

Figure 4.1 *continued*

```
                     Extreme Observations

               -------Lowest------      -----Highest-----

                  Value        Obs         Value       Obs

                0.0651786        1       1.13976      5776
                0.0690157        3       1.14209      5791
                0.0699755       59       1.14286      5801
                0.0702412       84       1.17090      5799
                0.0704787        4       1.31298      5811
```

The tables in Figure 4.1 show, in particular, that the average ratio is 0.2925 and the minimum and maximum ratios are 0.06518 and 1.1398, respectively.

Exploring a Data Distribution

Figure 4.2 shows a histogram of the loan-to-value ratios. The histogram reveals features of the ratio distribution, such as its skewness and the peak at 0.175, which are not evident from the tables in the previous example. The following statements create the histogram:

```
   title 'Home Loan Analysis';
proc univariate data=HomeLoans noprint;
   histogram LoanToValueRatio;
   inset n = 'Number of Homes' / position=ne;
run;
```

By default, PROC UNIVARIATE produces traditional graphics output, and the basic appearance of the histogram is determined by the prevailing ODS style. The NOPRINT option suppresses the display of summary statistics. The INSET statement inserts the total number of analyzed home loans in the upper right (northeast) corner of the plot.

Figure 4.2 Histogram for Loan-to-Value Ratio

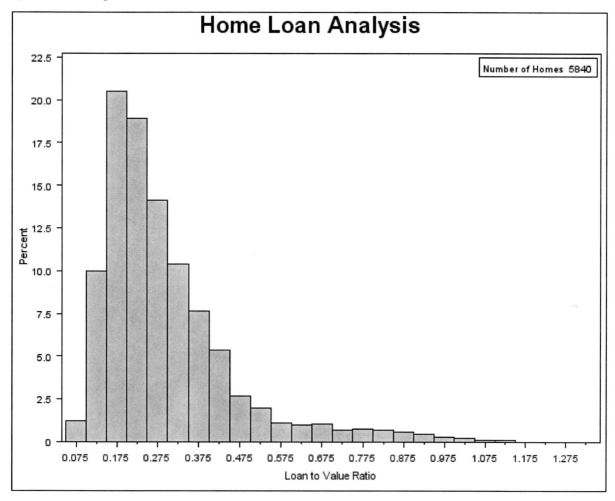

The data set HomeLoans contains a variable named LoanType that classifies the loans into two types: Gold and Platinum. It is useful to compare the distributions of LoanToValueRatio for the two types. The following statements request quantiles for each distribution and a comparative histogram, which are shown in Figure 4.3 and Figure 4.4.

```
title 'Comparison of Loan Types';
options nogstyle;
ods select Quantiles MyHist;
proc univariate data=HomeLoans;
   var LoanToValueRatio;
   class LoanType;
   histogram LoanToValueRatio / kernel(color=red)
                                cfill=ltgray
                                name='MyHist';
   inset n='Number of Homes' median='Median Ratio' (5.3) / position=ne;
   label LoanType = 'Type of Loan';
run;
options gstyle;
```

The ODS SELECT statement restricts the default output to the tables of quantiles and the graph produced by the HISTOGRAM statement, which is identified by the value specified by the NAME= option. The CLASS statement specifies LoanType as a classification variable for the quantile computations and comparative histogram. The KERNEL option adds a smooth nonparametric estimate of the ratio density to each histogram. The INSET statement specifies summary statistics to be displayed directly in the graph.

The NOGSTYLE system option specifies that the ODS style not influence the appearance of the histogram. Instead, the CFILL= option determines the color of the histogram bars and the COLOR= option specifies the color of the kernel density curve.

Figure 4.3 Quantiles for Loan-to-Value Ratio

```
                    Comparison of Loan Types

                      The UNIVARIATE Procedure
          Variable: LoanToValueRatio   (Loan to Value Ratio)
                         LoanType = Gold


                      Quantiles (Definition 5)

                      Quantile         Estimate

                      100% Max         1.0617647
                      99%              0.8974576
                      95%              0.6385908
                      90%              0.4471369
                      75% Q3           0.2985099
                      50% Median       0.2217033
                      25% Q1           0.1734568
                      10%              0.1411130
                      5%               0.1213079
                      1%               0.0942167
                      0% Min           0.0651786

                    Comparison of Loan Types

                      The UNIVARIATE Procedure
          Variable: LoanToValueRatio   (Loan to Value Ratio)
                        LoanType = Platinum


                      Quantiles (Definition 5)

                      Quantile         Estimate

                      100% Max         1.312981
                      99%              1.050000
                      95%              0.691803
                      90%              0.549273
                      75% Q3           0.430160
                      50% Median       0.366168
                      25% Q1           0.314452
                      10%              0.273670
                      5%               0.253124
                      1%               0.231114
                      0% Min           0.215504
```

The output in Figure 4.3 shows that the median ratio for Platinum loans (0.366) is greater than the median ratio for Gold loans (0.222). The comparative histogram in Figure 4.4 enables you to compare the two distributions more easily. It shows that the ratio distributions are similar except for a shift of about 0.14.

Figure 4.4 Comparative Histogram for Loan-to-Value Ratio

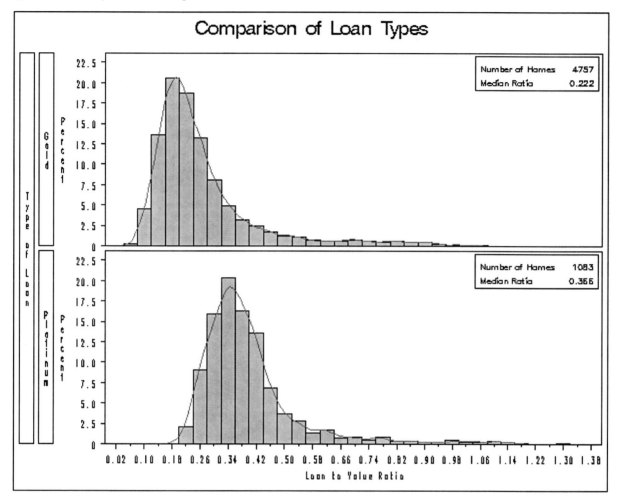

A sample program for this example, *univar1.sas*, is available in the SAS Sample Library for Base SAS software.

Modeling a Data Distribution

In addition to summarizing a data distribution as in the preceding example, you can use PROC UNIVARIATE to statistically model a distribution based on a random sample of data. The following statements create a data set named **Aircraft** that contains the measurements of a position deviation for a sample of 30 aircraft components.

```
data Aircraft;
   input Deviation @@;
   label Deviation = 'Position Deviation';
   datalines;
-.00653  0.00141 -.00702 -.00734 -.00649 -.00601
-.00631 -.00148 -.00731 -.00764 -.00275 -.00497
-.00741 -.00673 -.00573 -.00629 -.00671 -.00246
-.00222 -.00807 -.00621 -.00785 -.00544 -.00511
-.00138 -.00609  0.00038 -.00758 -.00731 -.00455
;
run;
```

An initial question in the analysis is whether the measurement distribution is normal. The following statements request a table of moments, the tests for normality, and a normal probability plot, which are shown in Figure 4.5 and Figure 4.6:

```
title 'Position Deviation Analysis';
ods graphics on;
ods select Moments TestsForNormality ProbPlot;
proc univariate data=Aircraft normaltest;
   var Deviation;
   probplot Deviation / normal (mu=est sigma=est)
                       square;
   label Deviation = 'Position Deviation';
   inset  mean std / format=6.4;
run;
ods graphics off;
```

The ODS GRAPHICS statement causes the procedure to produce ODS Graphics output rather than traditional graphics. (See the section "Alternatives for Producing Graphics" on page 339 for information about traditional graphics and ODS Graphics.) The INSET statement displays the sample mean and standard deviation on the probability plot.

Figure 4.5 Moments and Tests for Normality

```
                    Position Deviation Analysis

                       The UNIVARIATE Procedure
              Variable:  Deviation   (Position Deviation)

                               Moments

        N                          30    Sum Weights                 30
        Mean                -0.0053067   Sum Observations       -0.1592
        Std Deviation       0.00254362   Variance             6.47002E-6
        Skewness             1.2562507   Kurtosis             0.69790426
        Uncorrected SS      0.00103245   Corrected SS         0.00018763
        Coeff Variation     -47.932613   Std Error Mean        0.0004644
```

Figure 4.5 *continued*

```
                     Tests for Normality

       Test                    --Statistic---     -----p Value------

       Shapiro-Wilk            W     0.845364     Pr < W       0.0005
       Kolmogorov-Smirnov      D     0.208921     Pr > D      <0.0100
       Cramer-von Mises        W-Sq  0.329274     Pr > W-Sq   <0.0050
       Anderson-Darling        A-Sq  1.784881     Pr > A-Sq   <0.0050
```

All four goodness-of-fit tests in Figure 4.5 reject the hypothesis that the measurements are normally distributed.

Figure 4.6 shows a normal probability plot for the measurements. A linear pattern of points following the diagonal reference line would indicate that the measurements are normally distributed. Instead, the curved point pattern suggests that a skewed distribution, such as the lognormal, is more appropriate than the normal distribution.

A lognormal distribution for Deviation is fitted in Example 4.26.

A sample program for this example, *univar2.sas*, is available in the SAS Sample Library for Base SAS software.

Figure 4.6 Normal Probability Plot

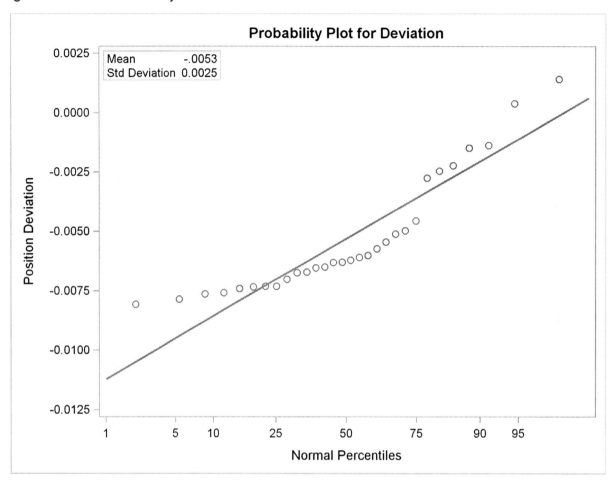

Syntax: UNIVARIATE Procedure

PROC UNIVARIATE <*options*> ;
    BY *variables* ;
    CDFPLOT <*variables*> < / *options*> ;
    CLASS *variable-1* < (*v-options*)> < *variable-2* < (*v-options*)>> </ KEYLEVEL= *value1* | (*value1 value2*)> ;
    FREQ *variable* ;
    HISTOGRAM <*variables*> < / *options*> ;
    ID *variables* ;
    INSET *keyword-list* </ *options*> ;
    OUTPUT < OUT=*SAS-data-set*> < *keyword1*=*names* ... *keywordk*=*names*> < *percentile-options*> ;
    PPPLOT <*variables*> < / *options*> ;
    PROBPLOT <*variables*> < / *options*> ;
    QQPLOT <*variables*> < / *options*> ;
    VAR *variables* ;
    WEIGHT *variable* ;

The PROC UNIVARIATE statement invokes the procedure. The VAR statement specifies the numeric variables to be analyzed, and it is required if the OUTPUT statement is used to save summary statistics in an output data set. If you do not use the VAR statement, all numeric variables in the data set are analyzed. The plot statements CDFPLOT, HISTOGRAM, PPPLOT, PROBPLOT, and QQPLOT create graphical displays, and the INSET statement enhances these displays by adding a table of summary statistics directly on the graph. You can specify one or more of each of the plot statements, the INSET statement, and the OUTPUT statement. If you use a VAR statement, the variables listed in a plot statement must be a subset of the variables listed in the VAR statement.

You can use a CLASS statement to specify one or two variables that group the data into classification levels. The analysis is carried out for each combination of levels. You can use the CLASS statement with plot statements to create comparative displays, in which each *cell* contains a plot for one combination of classification levels.

You can specify a BY statement to obtain separate analyses for each BY group. The FREQ statement specifies a variable whose values provide the frequency for each observation. The WEIGHT statement specifies a variable whose values are used to weight certain statistics. The ID statement specifies one or more variables to identify the extreme observations.

PROC UNIVARIATE Statement

PROC UNIVARIATE <*options*> ;

The PROC UNIVARIATE statement is required to invoke the UNIVARIATE procedure. You can use the PROC UNIVARIATE statement by itself to request a variety of statistics for summarizing the data distribution of each analysis variable:

- sample moments
- basic measures of location and variability
- confidence intervals for the mean, standard deviation, and variance
- tests for location
- tests for normality
- trimmed and Winsorized means
- robust estimates of scale
- quantiles and related confidence intervals
- extreme observations and extreme values
- frequency counts for observations
- missing values

In addition, you can use options in the PROC UNIVARIATE statement to do the following:

- specify the input data set to be analyzed
- specify a graphics catalog for saving traditional graphics output
- specify rounding units for variable values
- specify the definition used to calculate percentiles
- specify the divisor used to calculate variances and standard deviations
- request that plots be produced on line printers and define special printing characters used for features
- suppress tables
- save statistics in an output data set

The following are the *options* that can be used with the PROC UNIVARIATE statement:

ALL
> requests all statistics and tables that the FREQ, MODES, NEXTRVAL=5, PLOT, and CIBASIC options generate. If the analysis variables are not weighted, this option also requests the statistics and tables generated by the CIPCTLDF, CIPCTLNORMAL, LOCCOUNT, NORMAL, ROBUSTSCALE, TRIMMED=.25, and WINSORIZED=.25 options. PROC UNIVARIATE also uses any values that you specify for ALPHA=, MU0=, NEXTRVAL=, CIBASIC, CIPCTLDF, CIPCTLNORMAL, TRIMMED=, or WINSORIZED= to produce the output.

ALPHA=α

specifies the level of significance α for $100(1 - \alpha)\%$ confidence intervals. The value α must be between 0 and 1; the default value is 0.05, which results in 95% confidence intervals.

Note that specialized ALPHA= options are available for a number of confidence interval options. For example, you can specify CIBASIC(ALPHA=0.10) to request a table of basic confidence limits at the 90% level. The default value of these options is the value of the ALPHA= option in the PROC statement.

ANNOTATE=*SAS-data-set*

ANNO=*SAS-data-set*

specifies an input data set that contains annotate variables as described in *SAS/GRAPH: Reference*. You can use this data set to add features to your traditional graphics. PROC UNIVARIATE adds the features in this data set to every graph that is produced in the procedure. PROC UNIVARIATE does not use the ANNOTATE= data set unless you create a traditional graph with a plot statement. The option does not apply to ODS Graphics output. Use the ANNOTATE= option in the plot statement if you want to add a feature to a specific graph produced by that statement.

CIBASIC < (< **TYPE**=*keyword* > < **ALPHA**=α >) >

requests confidence limits for the mean, standard deviation, and variance based on the assumption that the data are normally distributed. If you use the CIBASIC option, you must use the default value of VARDEF=, which is DF.

TYPE=*keyword*

specifies the type of confidence limit, where *keyword* is LOWER, UPPER, or TWOSIDED. The default value is TWOSIDED.

ALPHA=α

specifies the level of significance α for $100(1 - \alpha)\%$ confidence intervals. The value α must be between 0 and 1; the default value is 0.05, which results in 95% confidence intervals. The default value is the value of ALPHA= given in the PROC statement.

CIPCTLDF < (< **TYPE**=*keyword* > < **ALPHA**=α >) >

CIQUANTDF < (< **TYPE**=*keyword* > < **ALPHA**=α >) >

requests confidence limits for quantiles based on a method that is distribution-free. In other words, no specific parametric distribution such as the normal is assumed for the data. PROC UNIVARIATE uses order statistics (ranks) to compute the confidence limits as described by Hahn and Meeker (1991). This option does not apply if you use a WEIGHT statement.

TYPE=*keyword*

specifies the type of confidence limit, where *keyword* is LOWER, UPPER, SYMMETRIC, or ASYMMETRIC. The default value is SYMMETRIC.

ALPHA=α

specifies the level of significance α for $100(1 - \alpha)\%$ confidence intervals. The value α must be between 0 and 1; the default value is 0.05, which results in 95% confidence intervals. The default value is the value of ALPHA= given in the PROC statement.

CIPCTLNORMAL < (< TYPE=*keyword* > < ALPHA=α >) >

CIQUANTNORMAL < (< TYPE=*keyword* > < ALPHA=α >) >

>   requests confidence limits for quantiles based on the assumption that the data are normally distributed. The computational method is described in Section 4.4.1 of Hahn and Meeker (1991) and uses the noncentral *t* distribution as given by Odeh and Owen (1980). This option does not apply if you use a WEIGHT statement
>
>   **TYPE=***keyword*
>
>   >   specifies the type of confidence limit, where *keyword* is LOWER, UPPER, or TWOSIDED. The default is TWOSIDED.
>
>   **ALPHA=**α
>
>   >   specifies the level of significance α for $100(1 - \alpha)\%$ confidence intervals. The value α must be between 0 and 1; the default value is 0.05, which results in 95% confidence intervals. The default value is the value of ALPHA= given in the PROC statement.

DATA=*SAS-data-set*

>   specifies the input SAS data set to be analyzed. If the DATA= option is omitted, the procedure uses the most recently created SAS data set.

EXCLNPWGT

EXCLNPWGTS

>   excludes observations with nonpositive weight values (zero or negative) from the analysis. By default, PROC UNIVARIATE counts observations with negative or zero weights in the total number of observations. This option applies only when you use a WEIGHT statement.

FREQ

>   requests a frequency table that consists of the variable values, frequencies, cell percentages, and cumulative percentages.
>
>   If you specify the WEIGHT statement, PROC UNIVARIATE includes the weighted count in the table and uses this value to compute the percentages.

GOUT=*graphics-catalog*

>   specifies the SAS catalog that PROC UNIVARIATE uses to save traditional graphics output. If you omit the libref in the name of the *graphics-catalog*, PROC UNIVARIATE looks for the catalog in the temporary library called WORK and creates the catalog if it does not exist. The option does not apply to ODS Graphics output.

IDOUT

>   includes ID variables in the output data set created by an OUTPUT statement. The value of an ID variable in the output data set is its first value from the input data set or BY group. By default, ID variables are not included in OUTPUT statement data sets.

LOCCOUNT

>   requests a table that shows the number of observations greater than, not equal to, and less than the value of MU0=. PROC UNIVARIATE uses these values to construct the sign test and the signed rank test. This option does not apply if you use a WEIGHT statement.

MODES|MODE

requests a table of all possible modes. By default, when the data contain multiple modes, PROC UNIVARIATE displays the lowest mode in the table of basic statistical measures. When all the values are unique, PROC UNIVARIATE does not produce a table of modes.

MU0=values
LOCATION=values

specifies the value of the mean or location parameter (μ_0) in the null hypothesis for tests of location summarized in the table labeled "Tests for Location: Mu0=value." If you specify one value, PROC UNIVARIATE tests the same null hypothesis for all analysis variables. If you specify multiple values, a VAR statement is required, and PROC UNIVARIATE tests a different null hypothesis for each analysis variable, matching variables and location values by their order in the two lists. The default *value* is 0.

The following statement tests the hypothesis $\mu_0 = 0$ for the first variable and the hypothesis $\mu_0 = 0.5$ for the second variable.

```
proc univariate mu0=0 0.5;
```

NEXTROBS=n

specifies the number of extreme observations that PROC UNIVARIATE lists in the table of extreme observations. The table lists the *n* lowest observations and the *n* highest observations. The default value is 5. You can specify NEXTROBS=0 to suppress the table of extreme observations.

NEXTRVAL=n

specifies the number of extreme values that PROC UNIVARIATE lists in the table of extreme values. The table lists the *n* lowest unique values and the *n* highest unique values. By default, $n = 0$ and no table is displayed.

NOBYPLOT

suppresses side-by-side line printer box plots that are created by default when you use the BY statement and either the ALL option or the PLOT option in the PROC statement.

NOPRINT

suppresses all the tables of descriptive statistics that the PROC UNIVARIATE statement creates. NOPRINT does not suppress the tables that the HISTOGRAM statement creates. You can use the NOPRINT option in the HISTOGRAM statement to suppress the creation of its tables. Use NOPRINT when you want to create an OUT= or OUTTABLE= output data set only.

NORMAL
NORMALTEST

requests tests for normality that include a series of goodness-of-fit tests based on the empirical distribution function. The table provides test statistics and *p*-values for the Shapiro-Wilk test (provided the sample size is less than or equal to 2000), the Kolmogorov-Smirnov test, the Anderson-Darling test, and the Cramér-von Mises test. This option does not apply if you use a WEIGHT statement.

NOTABCONTENTS

suppresses the table of contents entries for tables of summary statistics produced by the PROC UNIVARIATE statement.

NOVARCONTENTS

suppresses grouping entries associated with analysis variables in the table of contents. By default, the table of contents lists results associated with an analysis variable in a group with the variable name.

OUTTABLE=SAS-data-set

creates an output data set that contains univariate statistics arranged in tabular form, with one observation per analysis variable. See the section "OUTTABLE= Output Data Set" on page 374 for details.

PCTLDEF=value

DEF=value

specifies the definition that PROC UNIVARIATE uses to calculate quantiles. The default value is 5. Values can be 1, 2, 3, 4, or 5. You cannot use PCTLDEF= when you compute weighted quantiles. See the section "Calculating Percentiles" on page 328 for details on quantile definitions.

PLOTS | PLOT

produces a stem-and-leaf plot (or a horizontal bar chart), a box plot, and a normal probability plot in line printer output. If you use a BY statement, side-by-side box plots that are labeled "Schematic Plots" appear after the univariate analysis for the last BY group.

PLOTSIZE=n

specifies the approximate number of rows used in line-printer plots requested with the PLOTS option. If n is larger than the value of the SAS system option PAGESIZE=, PROC UNIVARIATE uses the value of PAGESIZE=. If n is less than 8, PROC UNIVARIATE uses eight rows to draw the plots.

ROBUSTSCALE

produces a table with robust estimates of scale. The statistics include the interquartile range, Gini's mean difference, the median absolute deviation about the median (*MAD*), and two statistics proposed by Rousseeuw and Croux (1993), Q_n, and S_n. See the section "Robust Estimates of Scale" on page 335 for details. This option does not apply if you use a WEIGHT statement.

ROUND=units

specifies the units to use to round the analysis variables prior to computing statistics. If you specify one unit, PROC UNIVARIATE uses this unit to round all analysis variables. If you specify multiple units, a VAR statement is required, and each unit rounds the values of the corresponding analysis variable. If ROUND=0, no rounding occurs. The ROUND= option reduces the number of unique variable values, thereby reducing memory requirements for the procedure. For example, to make the rounding unit 1 for the first analysis variable and 0.5 for the second analysis variable, submit the statement

```
proc univariate round=1 0.5;
   var Yieldstrength tenstren;
run;
```

When a variable value is midway between the two nearest rounded points, the value is rounded to the nearest even multiple of the roundoff value. For example, with a roundoff value of 1, the variable values of −2.5, −2.2, and −1.5 are rounded to −2; the values of −0.5, 0.2, and 0.5 are rounded to 0; and the values of 0.6, 1.2, and 1.4 are rounded to 1.

SUMMARYCONTENTS='string'

specifies the table of contents entry used for grouping the summary statistics produced by the PROC UNIVARIATE statement. You can specify SUMMARYCONTENTS='' to suppress the grouping entry.

TRIMMED=values <(< **TYPE=**keyword > < **ALPHA=**α >)>

TRIM=values <(< **TYPE=**keyword > < **ALPHA=**α >)>

requests a table of trimmed means, where *value* specifies the number or the proportion of observations that PROC UNIVARIATE trims. If the *value* is the number n of trimmed observations, n must be between 0 and half the number of nonmissing observations. If *value* is a proportion p between 0 and ½, the number of observations that PROC UNIVARIATE trims is the smallest integer that is greater than or equal to np, where n is the number of observations. To include confidence limits for the mean and the Student's t test in the table, you must use the default value of VARDEF=, which is DF. For details concerning the computation of trimmed means, see the section "Trimmed Means" on page 335. The TRIMMED= option does not apply if you use a WEIGHT statement.

TYPE=keyword

specifies the type of confidence limit for the mean, where *keyword* is LOWER, UPPER, or TWOSIDED. The default value is TWOSIDED.

ALPHA=α

specifies the level of significance α for $100(1-\alpha)\%$ confidence intervals. The value α must be between 0 and 1; the default value is 0.05, which results in 95% confidence intervals.

VARDEF=divisor

specifies the divisor to use in the calculation of variances and standard deviation. By default, VARDEF=DF. Table 4.1 shows the possible values for *divisor* and associated divisors.

Table 4.1 Possible Values for VARDEF=

Value	Divisor	Formula for Divisor
DF	degrees of freedom	$n-1$
N	number of observations	n
WDF	sum of weights minus one	$(\Sigma_i w_i) - 1$
WEIGHT | WGT	sum of weights	$\Sigma_i w_i$

The procedure computes the variance as $\frac{CSS}{\text{divisor}}$ where CSS is the corrected sums of squares and equals $\sum_{i=1}^{n}(x_i - \bar{x})^2$. When you weight the analysis variables, $CSS = \sum_{i=1}^{n}(w_i x_i - \bar{x}_w)^2$ where $\bar{x}_w$ is the weighted mean.

The default value is DF. To compute the standard error of the mean, confidence limits, and Student's t test, use the default value of VARDEF=.

When you use the WEIGHT statement and VARDEF=DF, the variance is an estimate of s^2 where the variance of the ith observation is $var(x_i) = \frac{s^2}{w_i}$ and w_i is the weight for the ith observation. This yields an estimate of the variance of an observation with unit weight.

When you use the WEIGHT statement and VARDEF=WGT, the computed variance is asymptotically (for large n) an estimate of $\frac{s^2}{\bar{w}}$ where $\bar{w}$ is the average weight. This yields an asymptotic estimate of the variance of an observation with average weight.

WINSORIZED=values < (< **TYPE=**keyword > < **ALPHA=**α >) >

WINSOR=values < (< **TYPE=**keyword > < **ALPHA=**α >) >

requests of a table of Winsorized means, where *value* is the number or the proportion of observations that PROC UNIVARIATE uses to compute the Winsorized mean. If the *value* is the number n of Winsorized observations, n must be between 0 and half the number of nonmissing observations. If *value* is a proportion p between 0 and ½, the number of observations that PROC UNIVARIATE uses is equal to the smallest integer that is greater than or equal to np, where n is the number of observations. To include confidence limits for the mean and the Student t test in the table, you must use the default value of VARDEF=, which is DF. For details concerning the computation of Winsorized means, see the section "Winsorized Means" on page 334. The WINSORIZED= option does not apply if you use a WEIGHT statement.

TYPE=keyword

specifies the type of confidence limit for the mean, where *keyword* is LOWER, UPPER, or TWOSIDED. The default is TWOSIDED.

ALPHA=α

specifies the level of significance α for $100(1 - \alpha)\%$ confidence intervals. The value α must be between 0 and 1; the default value is 0.05, which results in 95% confidence intervals.

BY Statement

BY *variables* ;

You can specify a BY statement with PROC UNIVARIATE to obtain separate analyses for each BY group. The BY statement specifies the *variables* that the procedure uses to form BY groups. You can specify more than one *variable*. If you do not use the NOTSORTED option in the BY statement, the observations in the data set must either be sorted by all the *variables* that you specify or be indexed appropriately.

DESCENDING
>specifies that the data set is sorted in descending order by the variable that immediately follows the word DESCENDING in the BY statement.

NOTSORTED
>specifies that observations are not necessarily sorted in alphabetic or numeric order. The data are grouped in another way—for example, chronological order.
>
>The requirement for ordering or indexing observations according to the values of BY variables is suspended for BY-group processing when you use the NOTSORTED option. In fact, the procedure does not use an index if you specify NOTSORTED. The procedure defines a BY group as a set of contiguous observations that have the same values for all BY variables. If observations with the same values for the BY variables are not contiguous, the procedure treats each contiguous set as a separate BY group.

CDFPLOT Statement

>**CDFPLOT** < *variables* > < / *options* > ;

The CDFPLOT statement plots the observed cumulative distribution function (cdf) of a variable, defined as

$$F_N(x) = \text{percent of nonmissing values} \leq x$$
$$= \frac{\text{number of values} \leq x}{N} \times 100\%$$

where N is the number of nonmissing observations. The cdf is an increasing step function that has a vertical jump of $\frac{1}{N}$ at each value of x equal to an observed value. The cdf is also referred to as the empirical cumulative distribution function (ecdf).

You can use any number of CDFPLOT statements in the UNIVARIATE procedure. The components of the CDFPLOT statement are as follows.

variables
>specify variables for which to create cdf plots. If you specify a VAR statement, the *variables* must also be listed in the VAR statement. Otherwise, the *variables* can be any numeric variables in the input data set. If you do not specify a list of *variables*, then by default the procedure creates a cdf plot for each variable listed in the VAR statement, or for each numeric variable in the DATA= data set if you do not specify a VAR statement.
>
>For example, suppose a data set named Steel contains exactly three numeric variables: Length, Width, and Height. The following statements create a cdf plot for each of the three variables:
>
>```
>   proc univariate data=Steel;
>      cdfplot;
>   run;
>```
>
>The following statements create a cdf plot for Length and a cdf plot for Width:

```
proc univariate data=Steel;
   var Length Width;
   cdfplot;
run;
```

The following statements create a cdf plot for Width:

```
proc univariate data=Steel;
   var Length Width;
   cdfplot Width;
run;
```

options

specify the theoretical distribution for the plot or add features to the plot. If you specify more than one variable, the *options* apply equally to each variable. Specify all *options* after the slash (/) in the CDFPLOT statement. You can specify only one *option* that names a distribution in each CDFPLOT statement, but you can specify any number of other *options*. The distributions available are the beta, exponential, gamma, lognormal, normal, and three-parameter Weibull. By default, the procedure produces a plot for the normal distribution.

Table 4.2 through Table 4.10 list the CDFPLOT *options* by function. For complete descriptions, see the sections "Dictionary of Options" on page 246 and "Dictionary of Common Options" on page 317. *Options* can be any of the following:

- *primary options*
- *secondary options*
- *general options*

Distribution Options

Table 4.2 lists *primary options* for requesting a theoretical distribution.

Table 4.2 Primary Options for Theoretical Distributions

Option	Description
BETA(*beta-options*)	plots two-parameter beta distribution function, parameters θ and σ assumed known
EXPONENTIAL(*exponential-options*)	plots one-parameter exponential distribution function, parameter θ assumed known
GAMMA(*gamma-options*)	plots two-parameter gamma distribution function, parameter θ assumed known
LOGNORMAL(*lognormal-options*)	plots two-parameter lognormal distribution function, parameter θ assumed known
NORMAL(*normal-options*)	plots normal distribution function
WEIBULL(*Weibull-options*)	plots two-parameter Weibull distribution function, parameter θ assumed known

Table 4.3 through Table 4.9 list *secondary options* that specify distribution parameters and control the display of a theoretical distribution function. Specify these options in parentheses after the distribution keyword. For example, you can request a normal probability plot with a distribution reference line by specifying the NORMAL option as follows:

```
proc univariate;
   cdfplot / normal(mu=10 sigma=0.5 color=red);
run;
```

The COLOR= option specifies the color for the curve, and the *normal-options* MU= and SIGMA= specify the parameters $\mu = 10$ and $\sigma = 0.5$ for the distribution function. If you do not specify these parameters, maximum likelihood estimates are computed.

Table 4.3 Secondary Options Used with All Distributions

Option	Description
COLOR=	specifies color of theoretical distribution function
L=	specifies line type of theoretical distribution function
W=	specifies width of theoretical distribution function

Table 4.4 Secondary Beta-Options

Option	Description
ALPHA=	specifies first shape parameter α for beta distribution function
BETA=	specifies second shape parameter β for beta distribution function
SIGMA=	specifies scale parameter σ for beta distribution function
THETA=	specifies lower threshold parameter θ for beta distribution function

Table 4.5 Secondary Exponential-Options

Option	Description
SIGMA=	specifies scale parameter σ for exponential distribution function
THETA=	specifies threshold parameter θ for exponential distribution function

Table 4.6 Secondary Gamma-Options

Option	Description
ALPHA=	specifies shape parameter α for gamma distribution function
ALPHADELTA=	specifies change in successive estimates of α at which the Newton-Raphson approximation of $\hat{\alpha}$ terminates
ALPHAINITIAL=	specifies initial value for α in the Newton-Raphson approximation of $\hat{\alpha}$
MAXITER=	specifies maximum number of iterations in the Newton-Raphson approximation of $\hat{\alpha}$
SIGMA=	specifies scale parameter σ for gamma distribution function
THETA=	specifies threshold parameter θ for gamma distribution function

Table 4.7 Secondary Lognormal-Options

Option	Description
SIGMA=	specifies shape parameter σ for lognormal distribution function
THETA=	specifies threshold parameter θ for lognormal distribution function
ZETA=	specifies scale parameter ζ for lognormal distribution function

Table 4.8 Secondary Normal-Options

Option	Description
MU=	specifies mean μ for normal distribution function
SIGMA=	specifies standard deviation σ for normal distribution function

Table 4.9 Secondary Weibull-Options

Option	Description
C=	specifies shape parameter c for Weibull distribution function
CDELTA=	specifies change in successive estimates of c at which the Newton-Raphson approximation of $\hat{c}$ terminates
CINITIAL=	specifies initial value for c in the Newton-Raphson approximation of $\hat{c}$
MAXITER=	specifies maximum number of iterations in the Newton-Raphson approximation of $\hat{c}$
SIGMA=	specifies scale parameter σ for Weibull distribution function
THETA=	specifies threshold parameter θ for Weibull distribution function

General Options

Table 4.10 summarizes general options for enhancing cdf plots.

Table 4.10 General Graphics Options

Option	Description
ANNOKEY	applies annotation requested in ANNOTATE= data set to key cell only
ANNOTATE=	specifies annotate data set
CAXIS=	specifies color for axis
CFRAME=	specifies color for frame
CFRAMESIDE=	specifies color for filling row label frames
CFRAMETOP=	specifies color for filling column label frames
CHREF=	specifies color for HREF= lines
CONTENTS=	specifies table of contents entry for cdf plot grouping
CPROP=	specifies color for proportion of frequency bar
CTEXT=	specifies color for text
CTEXTSIDE=	specifies color for row labels
CTEXTTOP=	specifies color for column labels
CVREF=	specifies color for VREF= lines
DESCRIPTION=	specifies description for graphics catalog member
FONT=	specifies text font
HAXIS=	specifies AXIS statement for horizontal axis
HEIGHT=	specifies height of text used outside framed areas
HMINOR=	specifies number of horizontal axis minor tick marks
HREF=	specifies reference lines perpendicular to the horizontal axis
HREFLABELS=	specifies labels for HREF= lines
HREFLABPOS=	specifies position for HREF= line labels
INFONT=	specifies software font for text inside framed areas
INHEIGHT=	specifies height of text inside framed areas
INTERTILE=	specifies distance between tiles in comparative plot
LHREF=	specifies line style for HREF= lines
LVREF=	specifies line style for VREF= lines

Table 4.10 *continued*

Option	Description
NAME=	specifies name for plot in graphics catalog
NCOLS=	specifies number of columns in comparative plot
NOECDF	suppresses plot of empirical (observed) distribution function
NOFRAME	suppresses frame around plotting area
NOHLABEL	suppresses label for horizontal axis
NOVLABEL	suppresses label for vertical axis
NOVTICK	suppresses tick marks and tick mark labels for vertical axis
NROWS=	specifies number of rows in comparative plot
OVERLAY	overlays plots for different class levels (ODS Graphics only)
TURNVLABELS	turns and vertically strings out characters in labels for vertical axis
VAXIS=	specifies AXIS statement for vertical axis
VAXISLABEL=	specifies label for vertical axis
VMINOR=	specifies number of vertical axis minor tick marks
VREF=	specifies reference lines perpendicular to the vertical axis
VREFLABELS=	specifies labels for VREF= lines
VREFLABPOS=	specifies position for VREF= line labels
VSCALE=	specifies scale for vertical axis
WAXIS=	specifies line thickness for axes and frame

Dictionary of Options

The following entries provide detailed descriptions of the options specific to the CDFPLOT statement. See the section "Dictionary of Common Options" on page 317 for detailed descriptions of options common to all plot statements.

ALPHA=_value_

specifies the shape parameter α for distribution functions requested with the BETA and GAMMA options. Enclose the ALPHA= option in parentheses after the BETA or GAMMA keywords. If you do not specify a value for α, the procedure calculates a maximum likelihood estimate. For examples, see the entries for the BETA and GAMMA options.

BETA<(*beta-options*)>

displays a fitted beta distribution function on the cdf plot. The equation of the fitted cdf is

$$F(x) = \begin{cases} 0 & \text{for } x \leq \theta \\ I_{\frac{x-\theta}{\sigma}}(\alpha, \beta) & \text{for } \theta < x < \theta + \sigma \\ 1 & \text{for } x \geq \sigma + \theta \end{cases}$$

where $I_y(\alpha, \beta)$ is the incomplete beta function and

θ = lower threshold parameter (lower endpoint)

σ = scale parameter ($\sigma > 0$)

α = shape parameter ($\alpha > 0$)

β = shape parameter ($\beta > 0$)

The beta distribution is bounded below by the parameter θ and above by the value $\theta + \sigma$. You can specify θ and σ by using the THETA= and SIGMA= *beta-options*, as illustrated in the following statements, which fit a beta distribution bounded between 50 and 75. The default values for θ and σ are 0 and 1, respectively.

```
proc univariate;
    cdfplot / beta(theta=50 sigma=25);
run;
```

The beta distribution has two shape parameters: α and β. If these parameters are known, you can specify their values with the ALPHA= and BETA= *beta-options*. If you do not specify values for α and β, the procedure calculates maximum likelihood estimates.

The BETA option can appear only once in a CDFPLOT statement. Table 4.3 and Table 4.4 list options you can specify with the BETA distribution option.

BETA=value

B=value

specifies the second shape parameter β for beta distribution functions requested by the BETA option. Enclose the BETA= option in parentheses after the BETA keyword. If you do not specify a value for β, the procedure calculates a maximum likelihood estimate. For examples, see the preceding entry for the BETA option.

C=value

specifies the shape parameter c for Weibull distribution functions requested with the WEIBULL option. Enclose the C= option in parentheses after the WEIBULL keyword. If you do not specify a value for c, the procedure calculates a maximum likelihood estimate. You can specify the SHAPE= option as an alias for the C= option.

EXPONENTIAL< (*exponential-options*) >

EXP< (*exponential-options*) >

displays a fitted exponential distribution function on the cdf plot. The equation of the fitted cdf is

$$F(x) = \begin{cases} 0 & \text{for } x \leq \theta \\ 1 - \exp\left(-\frac{x-\theta}{\sigma}\right) & \text{for } x > \theta \end{cases}$$

where

θ = threshold parameter

σ = scale parameter ($\sigma > 0$)

The parameter θ must be less than or equal to the minimum data value. You can specify θ with the THETA= *exponential-option*. The default value for θ is 0. You can specify σ with the SIGMA= *exponential-option*. By default, a maximum likelihood estimate is computed for σ. For example, the following statements fit an exponential distribution with $\theta = 10$ and a maximum likelihood estimate for σ:

```
proc univariate;
    cdfplot / exponential(theta=10 l=2 color=green);
run;
```

The exponential curve is green and has a line type of 2.

The EXPONENTIAL option can appear only once in a CDFPLOT statement. Table 4.3 and Table 4.5 list the options you can specify with the EXPONENTIAL option.

GAMMA< (*gamma-options*) **>**

displays a fitted gamma distribution function on the cdf plot. The equation of the fitted cdf is

$$F(x) = \begin{cases} 0 & \text{for } x \leq \theta \\ \frac{1}{\Gamma(\alpha)\sigma} \int_{\theta}^{x} \left(\frac{t-\theta}{\sigma}\right)^{\alpha-1} \exp\left(-\frac{t-\theta}{\sigma}\right) dt & \text{for } x > \theta \end{cases}$$

where

θ = threshold parameter

σ = scale parameter ($\sigma > 0$)

α = shape parameter ($\alpha > 0$)

The parameter θ for the gamma distribution must be less than the minimum data value. You can specify θ with the THETA= *gamma-option*. The default value for θ is 0. In addition, the gamma distribution has a shape parameter α and a scale parameter σ. You can specify these parameters with the ALPHA= and SIGMA= *gamma-options*. By default, maximum likelihood estimates are computed for α and σ. For example, the following statements fit a gamma distribution function with $\theta = 4$ and maximum likelihood estimates for α and σ:

```
proc univariate;
    cdfplot / gamma(theta=4);
run;
```

Note that the maximum likelihood estimate of α is calculated iteratively using the Newton-Raphson approximation. The *gamma-options* ALPHADELTA=, ALPHAINITIAL=, and MAXITER= control the approximation.

The GAMMA option can appear only once in a CDFPLOT statement. Table 4.3 and Table 4.6 list the options you can specify with the GAMMA option.

LOGNORMAL< (*lognormal-options*) **>**

displays a fitted lognormal distribution function on the cdf plot. The equation of the fitted cdf is

$$F(x) = \begin{cases} 0 & \text{for } x \leq \theta \\ \Phi\left(\frac{\log(x-\theta)-\zeta}{\sigma}\right) & \text{for } x > \theta \end{cases}$$

where $\Phi(\cdot)$ is the standard normal cumulative distribution function and

θ = threshold parameter

ζ = scale parameter

σ = shape parameter ($\sigma > 0$)

The parameter θ for the lognormal distribution must be less than the minimum data value. You can specify θ with the THETA= *lognormal-option*. The default value for θ is 0. In addition, the lognormal distribution has a shape parameter σ and a scale parameter ζ. You can specify these parameters with the SIGMA= and ZETA= *lognormal-options*. By default, maximum likelihood estimates are computed for σ and ζ. For example, the following statements fit a lognormal distribution function with $\theta = 10$ and maximum likelihood estimates for σ and ζ:

```
proc univariate;
   cdfplot / lognormal(theta = 10);
run;
```

The LOGNORMAL option can appear only once in a CDFPLOT statement. Table 4.3 and Table 4.7 list options that you can specify with the LOGNORMAL option.

MU=*value*

specifies the parameter μ for normal distribution functions requested with the NORMAL option. Enclose the MU= option in parentheses after the NORMAL keyword. The default value is the sample mean.

NOECDF

suppresses the observed distribution function (the empirical cumulative distribution function) of the variable, which is drawn by default. This option enables you to create theoretical cdf plots without displaying the data distribution. The NOECDF option can be used only with a theoretical distribution (such as the NORMAL option).

NORMAL<(*normal-options***)>**

displays a fitted normal distribution function on the cdf plot. The equation of the fitted cdf is

$$F(x) = \Phi\left(\frac{x-\mu}{\sigma}\right) \quad \text{for } -\infty < x < \infty$$

where $\Phi(\cdot)$ is the standard normal cumulative distribution function and

μ = mean

σ = standard deviation ($\sigma > 0$)

You can specify known values for μ and σ with the MU= and SIGMA= *normal-options*, as shown in the following statements:

```
proc univariate;
   cdfplot / normal(mu=14 sigma=.05);
run;
```

By default, the sample mean and sample standard deviation are calculated for μ and σ. The NORMAL option can appear only once in a CDFPLOT statement. Table 4.3 and Table 4.8 list options that you can specify with the NORMAL option.

SIGMA=*value* **| EST**

specifies the parameter σ for distribution functions requested by the BETA, EXPONENTIAL, GAMMA, LOGNORMAL, NORMAL, and WEIBULL options. Enclose the SIGMA= option in parentheses after the distribution keyword. The following table summarizes the use of the SIGMA= option:

Distribution Option	SIGMA= Specifies	Default Value	Alias
BETA	scale parameter σ	1	SCALE=
EXPONENTIAL	scale parameter σ	maximum likelihood estimate	SCALE=
GAMMA	scale parameter σ	maximum likelihood estimate	SCALE=
LOGNORMAL	shape parameter σ	maximum likelihood estimate	SHAPE=
NORMAL	scale parameter σ	standard deviation	
WEIBULL	scale parameter σ	maximum likelihood estimate	SCALE=

THETA=value | **EST**

THRESHOLD=value | **EST**

specifies the lower threshold parameter θ for theoretical cumulative distribution functions requested with the BETA, EXPONENTIAL, GAMMA, LOGNORMAL, and WEIBULL options. Enclose the THETA= option in parentheses after the distribution keyword. The default value is 0.

VSCALE=PERCENT | PROPORTION

specifies the scale of the vertical axis. The value PERCENT scales the data in units of percent of observations per data unit. The value PROPORTION scales the data in units of proportion of observations per data unit. The default is PERCENT.

WEIBULL<(Weibull-options**)>**

displays a fitted Weibull distribution function on the cdf plot. The equation of the fitted cdf is

$$F(x) = \begin{cases} 0 & \text{for } x \leq \theta \\ 1 - \exp\left(-\left(\frac{x-\theta}{\sigma}\right)^c\right) & \text{for } x > \theta \end{cases}$$

where

θ = threshold parameter

σ = scale parameter ($\sigma > 0$)

c = shape parameter ($c > 0$)

The parameter θ must be less than the minimum data value. You can specify θ with the THETA= Weibull-option. The default value for θ is 0. In addition, the Weibull distribution has a shape parameter c and a scale parameter σ. You can specify these parameters with the SIGMA= and C= Weibull-options. By default, maximum likelihood estimates are computed for c and σ. For example, the following statements fit a Weibull distribution function with $\theta = 15$ and maximum likelihood estimates for σ and c:

```
proc univariate;
   cdfplot / weibull(theta=15);
run;
```

Note that the maximum likelihood estimate of c is calculated iteratively using the Newton-Raphson approximation. The Weibull-options CDELTA=, CINITIAL=, and MAXITER= control the approximation.

The WEIBULL option can appear only once in a CDFPLOT statement. Table 4.3 and Table 4.9 list options that you can specify with the WEIBULL option.

ZETA=value

specifies a value for the scale parameter ζ for a lognormal distribution function requested with the LOGNORMAL option. Enclose the ZETA= option in parentheses after the LOGNORMAL keyword. If you do not specify a value for ζ, a maximum likelihood estimate is computed. You can specify the SCALE= option as an alias for the ZETA= option.

CLASS Statement

CLASS variable-1 < (v-options) > < variable-2 < (v-options) > > < / KEYLEVEL= value1 | (value1 value2) > ;

The CLASS statement specifies one or two variables used to group the data into classification levels. Variables in a CLASS statement are referred to as *CLASS variables*. CLASS variables can be numeric or character. Class variables can have floating point values, but they typically have a few discrete values that define levels of the variable. You do not have to sort the data by CLASS variables. PROC UNIVARIATE uses the formatted values of the CLASS variables to determine the classification levels.

You can specify the following *v-options* enclosed in parentheses after the CLASS variable:

MISSING

specifies that missing values for the CLASS variable are to be treated as valid classification levels. Special missing values that represent numeric values ('.A' through '.Z' and '._') are each considered as a separate value. If you omit MISSING, PROC UNIVARIATE excludes the observations with a missing CLASS variable value from the analysis. Enclose this option in parentheses after the CLASS variable.

ORDER=DATA | FORMATTED | FREQ | INTERNAL

specifies the display order for the CLASS variable values. The default value is INTERNAL. You can specify the following values with the ORDER=*option*:

DATA   orders values according to their order in the input data set. When you use a plot statement, PROC UNIVARIATE displays the rows (columns) of the comparative plot from top to bottom (left to right) in the order that the CLASS variable values first appear in the input data set.

FORMATTED   orders values by their ascending formatted values. This order might depend on your operating environment. When you use a plot statement, PROC UNIVARIATE displays the rows (columns) of the comparative plot from top to bottom (left to right) in increasing order of the formatted CLASS variable values. For example, suppose a numeric CLASS variable DAY (with values 1, 2, and 3) has a user-defined format that assigns Wednesday to the value 1, Thursday to the value 2, and Friday to the value 3. The rows of the comparative plot will appear in alphabetical order (Friday, Thursday, Wednesday) from top to bottom.

If there are two or more distinct internal values with the same formatted value, then PROC UNIVARIATE determines the order by the internal

	value that occurs first in the input data set. For numeric variables without an explicit format, the levels are ordered by their internal values.
FREQ	orders values by descending frequency count so that levels with the most observations are listed first. If two or more values have the same frequency count, PROC UNIVARIATE uses the formatted values to determine the order.

When you use a plot statement, PROC UNIVARIATE displays the rows (columns) of the comparative plot from top to bottom (left to right) in order of decreasing frequency count for the CLASS variable values.

INTERNAL   orders values by their unformatted values, which yields the same order as PROC SORT. This order may depend on your operating environment.

When you use a plot statement, PROC UNIVARIATE displays the rows (columns) of the comparative plot from top to bottom (left to right) in increasing order of the internal (unformatted) values of the CLASS variable. The first CLASS variable is used to label the rows of the comparative plots (top to bottom). The second CLASS variable is used to label the columns of the comparative plots (left to right). For example, suppose a numeric CLASS variable DAY (with values 1, 2, and 3) has a user-defined format that assigns Wednesday to the value 1, Thursday to the value 2, and Friday to the value 3. The rows of the comparative plot will appear in day-of-the-week order (Wednesday, Thursday, Friday) from top to bottom.

You can specify the following *option* after the slash (/) in the CLASS statement.

KEYLEVEL=value | (value1 value2)

specifies the *key cells* in comparative plots. For each plot, PROC UNIVARIATE first determines the horizontal axis scaling for the key cell, and then extends the axis using the established tick interval to accommodate the data ranges for the remaining cells, if necessary. Thus, the choice of the key cell determines the uniform horizontal axis that PROC UNIVARIATE uses for all cells.

If you specify only one CLASS variable and use a plot statement, KEYLEVEL=*value* identifies the key cell as the level for which the CLASS variable is equal to *value*. By default, PROC UNIVARIATE sorts the levels in the order determined by the ORDER= option, and the key cell is the first occurrence of a level in this order. The cells display in order from top to bottom or left to right. Consequently, the key cell appears at the top (or left). When you specify a different key cell with the KEYLEVEL= option, this cell appears at the top (or left).

If you specify two CLASS variables, use KEYLEVEL= *(value1 value2)* to identify the key cell as the level for which CLASS variable *n* is equal to *valuen*. By default, PROC UNIVARIATE sorts the levels of the first CLASS variable in the order that is determined by its ORDER= option. Then, within each of these levels, it sorts the levels of the second CLASS variable in the order that is determined by its ORDER= option. The default key cell is the first occurrence of a combination of levels for the two variables in this order. The cells display in the order of the first CLASS variable from top to bottom and in the order of the second CLASS variable from left to right. Consequently, the default key cell appears at the upper left corner. When you specify a different key cell with the KEYLEVEL= option, this cell appears at the upper left corner.

The length of the KEYLEVEL= value cannot exceed 16 characters and you must specify a formatted value.

The KEYLEVEL= option has no effect unless you specify a plot statement.

NOKEYMOVE

specifies that the location of the key cell in a comparative plot be unchanged by the CLASS statement KEYLEVEL= option. By default, the key cell is positioned as the first cell in a comparative plot.

The NOKEYMOVE option has no effect unless you specify a plot statement.

FREQ Statement

FREQ *variable* ;

The FREQ statement specifies a numeric variable whose value represents the frequency of the observation. If you use the FREQ statement, the procedure assumes that each observation represents n observations, where n is the value of variable. If the variable is not an integer, the SAS System truncates it. If the variable is less than 1 or is missing, the procedure excludes that observation from the analysis. See Example 4.6.

NOTE: The FREQ statement affects the degrees of freedom, but the WEIGHT statement does not.

HISTOGRAM Statement

HISTOGRAM < *variables* > < / *options* > ;

The HISTOGRAM statement creates histograms and optionally superimposes estimated parametric and nonparametric probability density curves. You cannot use the WEIGHT statement with the HISTOGRAM statement. You can use any number of HISTOGRAM statements after a PROC UNIVARIATE statement. The components of the HISTOGRAM statement are follows.

variables

are the variables for which histograms are to be created. If you specify a VAR statement, the *variables* must also be listed in the VAR statement. Otherwise, the *variables* can be any numeric variables in the input data set. If you do not specify *variables* in a VAR statement or in the HISTOGRAM statement, then by default, a histogram is created for each numeric variable in the DATA= data set. If you use a VAR statement and do not specify any *variables* in the HISTOGRAM statement, then by default, a histogram is created for each variable listed in the VAR statement.

For example, suppose a data set named Steel contains exactly two numeric variables named Length and Width. The following statements create two histograms, one for Length and one for Width:

```
proc univariate data=Steel;
   histogram;
run;
```

Likewise, the following statements create histograms for Length and Width:

```
proc univariate data=Steel;
   var Length Width;
   histogram;
run;
```

The following statements create a histogram for Length only:

```
proc univariate data=Steel;
   var Length Width;
   histogram Length;
run;
```

options

add features to the histogram. Specify all *options* after the slash (/) in the HISTOGRAM statement. *Options* can be one of the following:

- *primary options* for fitted parametric distributions and kernel density estimates
- *secondary options* for fitted parametric distributions and kernel density estimates
- *general options* for graphics and output data sets

For example, in the following statements, the NORMAL option displays a fitted normal curve on the histogram, the MIDPOINTS= option specifies midpoints for the histogram, and the CTEXT= option specifies the color of the text:

```
proc univariate data=Steel;
   histogram Length / normal
                     midpoints = 5.6 5.8 6.0 6.2 6.4
                     ctext     = blue;
run;
```

Table 4.11 through Table 4.23 list the HISTOGRAM *options* by function. For complete descriptions, see the sections "Dictionary of Options" on page 261 and "Dictionary of Common Options" on page 317.

Parametric Density Estimation Options

Table 4.11 lists *primary options* that display parametric density estimates on the histogram. You can specify each primary option once in a given HISTOGRAM statement, and each primary option can display multiple curves from its family on the histogram.

Table 4.11 Primary Options for Parametric Fitted Distributions

Option	Description
BETA(*beta-options*)	fits beta distribution with threshold parameter θ, scale parameter σ, and shape parameters α and β
EXPONENTIAL(*exponential-options*)	fits exponential distribution with threshold parameter θ and scale parameter σ
GAMMA(*gamma-options*)	fits gamma distribution with threshold parameter θ, scale parameter σ, and shape parameter α
LOGNORMAL(*lognormal-options*)	fits lognormal distribution with threshold parameter θ, scale parameter ζ, and shape parameter σ
NORMAL(*normal-options*)	fits normal distribution with mean μ and standard deviation σ
SB(S_B-*options*)	fits Johnson S_B distribution with threshold parameter θ, scale parameter σ, and shape parameters δ and γ
SU(S_U-*options*)	fits Johnson S_U distribution with threshold parameter θ, scale parameter σ, and shape parameters δ and γ
WEIBULL(*Weibull-options*)	fits Weibull distribution with threshold parameter θ, scale parameter σ, and shape parameter c

Table 4.12 through Table 4.20 list *secondary options* that specify parameters for fitted parametric distributions and that control the display of fitted curves. Specify these *secondary options* in parentheses after the *primary distribution option*. For example, you can fit a normal curve by specifying the NORMAL option as follows:

```
proc univariate;
   histogram / normal(color=red mu=10 sigma=0.5);
run;
```

The COLOR= *normal-option* draws the curve in red, and the MU= and SIGMA= *normal-options* specify the parameters $\mu = 10$ and $\sigma = 0.5$ for the curve. Note that the sample mean and sample standard deviation are used to estimate μ and σ, respectively, when the MU= and SIGMA= *normal-options* are not specified.

You can specify lists of values for secondary options to display more than one fitted curve from the same distribution family on a histogram. Option values are matched by list position. You can specify the value EST in a list of distribution parameter values to use an estimate of the parameter.

For example, the following code displays two normal curves on a histogram:

```
proc univariate;
   histogram / normal(color=(red blue) mu=10 est sigma=0.5 est);
run;
```

The first curve is red, with $\mu = 10$ and $\sigma = 0.5$. The second curve is blue, with μ equal to the sample mean and σ equal to the sample standard deviation.

See the section "Formulas for Fitted Continuous Distributions" on page 348 for detailed information about the families of parametric distributions that you can fit with the HISTOGRAM statement.

Table 4.12 Secondary Options Used with All Parametric Distribution Options

Option	Description
COLOR=	specifies colors of density curves
CONTENTS=	specifies table of contents entry for density curve grouping
FILL	fills area under density curve
L=	specifies line types of density curves
MIDPERCENTS	prints table of midpoints of histogram intervals
NOPRINT	suppresses tables summarizing curves
PERCENTS=	lists percents for which quantiles calculated from data and quantiles estimated from curves are tabulated
W=	specifies widths of density curves

Table 4.13 Secondary Beta-Options

Option	Description
ALPHA=	specifies first shape parameter α for beta curve
BETA=	specifies second shape parameter β for beta curve
SIGMA=	specifies scale parameter σ for beta curve
THETA=	specifies lower threshold parameter θ for beta curve

Table 4.14 Secondary Exponential-Options

Option	Description
SIGMA=	specifies scale parameter σ for exponential curve
THETA=	specifies threshold parameter θ for exponential curve

Table 4.15 Secondary Gamma-Options

Option	Description
ALPHA=	specifies shape parameter α for gamma curve
ALPHADELTA=	specifies change in successive estimates of α at which the Newton-Raphson approximation of $\hat{\alpha}$ terminates
ALPHAINITIAL=	specifies initial value for α in the Newton-Raphson approximation of $\hat{\alpha}$
MAXITER=	specifies maximum number of iterations in the Newton-Raphson approximation of $\hat{\alpha}$
SIGMA=	specifies scale parameter σ for gamma curve
THETA=	specifies threshold parameter θ for gamma curve

Table 4.16 Secondary Lognormal-Options

Option	Description
SIGMA=	specifies shape parameter σ for lognormal curve
THETA=	specifies threshold parameter θ for lognormal curve
ZETA=	specifies scale parameter ζ for lognormal curve

Table 4.17 Secondary Normal-Options

Option	Description
MU=	specifies mean μ for normal curve
SIGMA=	specifies standard deviation σ for normal curve

Table 4.18 Secondary Johnson S_B-Options

Option	Description
DELTA=	specifies first shape parameter δ for Johnson S_B curve
FITINTERVAL=	specifies z-value for method of percentiles
FITMETHOD=	specifies method of parameter estimation
FITTOLERANCE=	specifies tolerance for method of percentiles
GAMMA=	specifies second shape parameter γ for Johnson S_B curve
SIGMA=	specifies scale parameter σ for Johnson S_B curve
THETA=	specifies lower threshold parameter θ for Johnson S_B curve

Table 4.19 Secondary Johnson S_U-Options

Option	Description
DELTA=	specifies first shape parameter δ for Johnson S_U curve
FITINTERVAL=	specifies z-value for method of percentiles
FITMETHOD=	specifies method of parameter estimation
FITTOLERANCE=	specifies tolerance for method of percentiles
GAMMA=	specifies second shape parameter γ for Johnson S_U curve
SIGMA=	specifies scale parameter σ for Johnson S_U curve
THETA=	specifies lower threshold parameter θ for Johnson S_U curve

Table 4.20 Secondary Weibull-Options

Option	Description
C=	specifies shape parameter c for Weibull curve
CDELTA=	specifies change in successive estimates of c at which the Newton-Raphson approximation of $\hat{c}$ terminates
CINITIAL=	specifies initial value for c in the Newton-Raphson approximation of $\hat{c}$
MAXITER=	specifies maximum number of iterations in the Newton-Raphson approximation of $\hat{c}$
SIGMA=	specifies scale parameter σ for Weibull curve
THETA=	specifies threshold parameter θ for Weibull curve

Nonparametric Density Estimation Options

Use the *option* KERNEL(*kernel-options*) to compute kernel density estimates. Specify the following *secondary options* in parentheses after the KERNEL option to control features of density estimates requested with the KERNEL option.

Table 4.21 Kernel-Options

Option	Description
C=	specifies standardized bandwidth parameter c
COLOR=	specifies color of the kernel density curve
FILL	fills area under kernel density curve
K=	specifies type of kernel function
L=	specifies line type used for kernel density curve
LOWER=	specifies lower bound for kernel density curve
UPPER=	specifies upper bound for kernel density curve
W=	specifies line width for kernel density curve

General Options

Table 4.22 summarizes *options* for enhancing histograms, and Table 4.23 summarizes options for requesting output data sets.

Table 4.22 General Graphics Options

Option	Description
ANNOKEY	applies annotation requested in ANNOTATE= data set to key cell only
ANNOTATE=	specifies annotate data set
BARLABEL=	produces labels above histogram bars
BARWIDTH=	specifies width for the bars
CAXIS=	specifies color for axis
CBARLINE=	specifies color for outlines of histogram bars
CFILL=	specifies color for filling under curve
CFRAME=	specifies color for frame
CFRAMESIDE=	specifies color for filling frame for row labels
CFRAMETOP=	specifies color for filling frame for column labels
CGRID=	specifies color for grid lines
CHREF=	specifies color for HREF= lines
CLIPREF	draws reference lines behind histogram bars
CONTENTS=	specifies table of contents entry for histogram grouping
CPROP=	specifies color for proportion of frequency bar
CTEXT=	specifies color for text
CTEXTSIDE=	specifies color for row labels of comparative histograms
CTEXTTOP=	specifies color for column labels of comparative histograms
CVREF=	specifies color for VREF= lines
DESCRIPTION=	specifies description for plot in graphics catalog
ENDPOINTS=	lists endpoints for histogram intervals
FONT=	specifies software font for text
FORCEHIST	forces creation of histogram
FRONTREF	draws reference lines in front of histogram bars
GRID	creates a grid
HANGING	constructs hanging histogram
HAXIS=	specifies AXIS statement for horizontal axis
HEIGHT=	specifies height of text used outside framed areas
HMINOR=	specifies number of horizontal minor tick marks
HOFFSET=	specifies offset for horizontal axis
HREF=	specifies reference lines perpendicular to the horizontal axis
HREFLABELS=	specifies labels for HREF= lines
HREFLABPOS=	specifies vertical position of labels for HREF= lines
INFONT=	specifies software font for text inside framed areas
INHEIGHT=	specifies height of text inside framed areas
INTERBAR=	specifies space between histogram bars
INTERTILE=	specifies distance between tiles
LGRID=	specifies a line type for grid lines
LHREF=	specifies line style for HREF= lines
LVREF=	specifies line style for VREF= lines

Table 4.22 *continued*

Option	Description
MAXNBIN=	specifies maximum number of bins to display
MAXSIGMAS=	limits the number of bins that display to within a specified number of standard deviations above and below mean of data in key cell
MIDPOINTS=	specifies midpoints for histogram intervals
NAME=	specifies name for plot in graphics catalog
NCOLS=	specifies number of columns in comparative histogram
NENDPOINTS=	specifies number of histogram interval endpoints
NMIDPOINTS=	specifies number of histogram interval midpoints
NOBARS	suppresses histogram bars
NOFRAME	suppresses frame around plotting area
NOHLABEL	suppresses label for horizontal axis
NOPLOT	suppresses plot
NOTABCONTENTS	suppresses table of contents entries for tables produced by HISTOGRAM statement
NOVLABEL	suppresses label for vertical axis
NOVTICK	suppresses tick marks and tick mark labels for vertical axis
NROWS=	specifies number of rows in comparative histogram
PFILL=	specifies pattern for filling under curve
RTINCLUDE	includes right endpoint in interval
TURNVLABELS	turns and vertically strings out characters in labels for vertical axis
VAXIS=	specifies AXIS statement or values for vertical axis
VAXISLABEL=	specifies label for vertical axis
VMINOR=	specifies number of vertical minor tick marks
VOFFSET=	specifies length of offset at upper end of vertical axis
VREF=	specifies reference lines perpendicular to the vertical axis
VREFLABELS=	specifies labels for VREF= lines
VREFLABPOS=	specifies horizontal position of labels for VREF= lines
VSCALE=	specifies scale for vertical axis
WAXIS=	specifies line thickness for axes and frame
WBARLINE=	specifies line thickness for bar outlines
WGRID=	specifies line thickness for grid

Table 4.23 Options for Requesting Output Data Sets

Option	Description
MIDPERCENTS	creates table of histogram intervals
OUTHISTOGRAM=	specifies information about histogram intervals
OUTKERNEL=	creates a data set containing kernel density estimates

Dictionary of Options

The following entries provide detailed descriptions of *options* in the HISTOGRAM statement. See the section "Dictionary of Common Options" on page 317 for detailed descriptions of options common to all plot statements.

ALPHA=*value-list*

specifies the shape parameter α for fitted curves requested with the BETA and GAMMA options. Enclose the ALPHA= option in parentheses after the BETA or GAMMA options. By default, or if you specify the value EST, the procedure calculates a maximum likelihood estimate for α. You can specify A= as an alias for ALPHA= if you use it as a *beta-option*. You can specify SHAPE= as an alias for ALPHA= if you use it as a *gamma-option*.

BARLABEL=COUNT | PERCENT | PROPORTION

displays labels above the histogram bars. If you specify BARLABEL=COUNT, the label shows the number of observations associated with a given bar. If you specify BARLABEL=PERCENT, the label shows the percentage of observations represented by that bar. If you specify BARLABEL=PROPORTION, the label displays the proportion of observations associated with the bar.

BARWIDTH=*value*

specifies the width of the histogram bars in percentage screen units. If both the BARWIDTH= and INTERBAR= options are specified, the INTERBAR= option takes precedence.

BETA < (*beta-options***) >**

displays fitted beta density curves on the histogram. The BETA option can occur only once in a HISTOGRAM statement, but it can request any number of beta curves. The beta distribution is bounded below by the parameter θ and above by the value $\theta + \sigma$. Use the THETA= and SIGMA= *beta-options* to specify these parameters. By default, THETA=0 and SIGMA=1. You can specify THETA=EST and SIGMA=EST to request maximum likelihood estimates for θ and σ.

The beta distribution has two shape parameters: α and β. If these parameters are known, you can specify their values with the ALPHA= and BETA= *beta-options*. By default, the procedure computes maximum likelihood estimates for α and β. **NOTE:** Three- and four-parameter maximum likelihood estimation may not always converge.

Table 4.12 and Table 4.13 list secondary options you can specify with the BETA option. See the section "Beta Distribution" on page 348 for details and Example 4.21 for an example that uses the BETA option.

BETA=*value-list*

B=*value-list*

specifies the second shape parameter β for beta density curves requested with the BETA option. Enclose the BETA= option in parentheses after the BETA option. By default, or if you specify the value EST, the procedure calculates a maximum likelihood estimate for β.

C=*value-list*

specifies the shape parameter c for Weibull density curves requested with the WEIBULL option. Enclose the C= *Weibull-option* in parentheses after the WEIBULL option. By default, or if you specify the value EST, the procedure calculates a maximum likelihood estimate for c. You can specify the SHAPE= *Weibull-option* as an alias for the C= *Weibull-option*.

C=*value-list*

specifies the standardized bandwidth parameter c for kernel density estimates requested with the KERNEL option. Enclose the C= *kernel-option* in parentheses after the KERNEL option. You can specify a list of values to request multiple estimates. You can specify the value MISE to produce the estimate with a bandwidth that minimizes the approximate mean integrated square error (MISE), or SJPI to select the bandwidth by using the Sheather-Jones plug-in method.

You can also use the C= *kernel-option* with the K= *kernel-option* (which specifies the kernel function) to compute multiple estimates. If you specify more kernel functions than bandwidths, the last bandwidth in the list is repeated for the remaining estimates. Similarly, if you specify more bandwidths than kernel functions, the last kernel function is repeated for the remaining estimates. If you do not specify the C= *kernel-option*, the bandwidth that minimizes the approximate MISE is used for all the estimates.

See the section "Kernel Density Estimates" on page 360 for more information about kernel density estimates.

CBARLINE=*color*

specifies the color for the outline of the histogram bars when producing traditional graphics. The option does not apply to ODS Graphics output.

CFILL=*color*

specifies the color to fill the bars of the histogram (or the area under a fitted density curve if you also specify the FILL option) when producing traditional graphics. See the entries for the FILL and PFILL= options for additional details. Refer to *SAS/GRAPH: Reference* for a list of colors. The option does not apply to ODS Graphics output.

CGRID=*color*

specifies the color for grid lines when a grid displays on the histogram in traditional graphics. This option also produces a grid if the GRID= option is not specified.

CLIPREF

draws reference lines requested with the HREF= and VREF= options behind the histogram bars. When the GSTYLE system option is in effect for traditional graphics, reference lines are drawn in front of the bars by default.

CONTENTS=

specifies the table of contents grouping entry for tables associated with a density curve. Enclose the CONTENTS= option in parentheses after the distribution option. You can specify CONTENTS='' to suppress the grouping entry.

DELTA=*value-list*

specifies the first shape parameter δ for Johnson S_B and Johnson S_U distribution functions requested with the SB and SU options. Enclose the DELTA= option in parentheses after the SB or SU option. If you do not specify a value for δ, or if you specify the value EST, the procedure calculates an estimate.

ENDPOINTS < =*values* | KEY | UNIFORM >
uses histogram bin endpoints as the tick mark values for the horizontal axis and determines how to compute the bin width of the histogram bars. The *values* specify both the left and right endpoint of each histogram interval. The width of the histogram bars is the difference between consecutive endpoints. The procedure uses the same values for all variables.

The range of endpoints must cover the range of the data. For example, if you specify

```
endpoints=2 to 10 by 2
```

then all of the observations must fall in the intervals [2,4) [4,6) [6,8) [8,10]. You also must use evenly spaced endpoints which you list in increasing order.

KEY determines the endpoints for the data in the key cell. The initial number of endpoints is based on the number of observations in the key cell by using the method of Terrell and Scott (1985). The procedure extends the endpoint list for the key cell in either direction as necessary until it spans the data in the remaining cells.

UNIFORM determines the endpoints by using all the observations as if there were no cells. In other words, the number of endpoints is based on the total sample size by using the method of Terrell and Scott (1985).

Neither KEY nor UNIFORM apply unless you use the CLASS statement.

If you omit ENDPOINTS, the procedure uses the histogram midpoints as horizontal axis tick values. If you specify ENDPOINTS, the procedure computes the endpoints by using an algorithm (Terrell and Scott 1985) that is primarily applicable to continuous data that are approximately normally distributed.

If you specify both MIDPOINTS= and ENDPOINTS, the procedure issues a warning message and uses the endpoints.

If you specify RTINCLUDE, the procedure includes the right endpoint of each histogram interval in that interval instead of including the left endpoint.

If you use a CLASS statement and specify ENDPOINTS, the procedure uses END-POINTS=KEY as the default. However if the key cell is empty, then the procedure uses ENDPOINTS=UNIFORM.

EXPONENTIAL < (*exponential-options*) >

EXP < (*exponential-options*) >
displays fitted exponential density curves on the histogram. The EXPONENTIAL option can occur only once in a HISTOGRAM statement, but it can request any number of exponential curves. The parameter θ must be less than or equal to the minimum data value. Use the THETA= *exponential-option* to specify θ. By default, THETA=0. You can specify THETA=EST to request the maximum likelihood estimate for θ. Use the SIGMA= *exponential-option* to specify σ. By default, the procedure computes a maximum likelihood estimate for σ. Table 4.12 and Table 4.14 list options you can specify with the EXPONENTIAL option. See the section "Exponential Distribution" on page 350 for details.

FILL

fills areas under the fitted density curve or the kernel density estimate with colors and patterns. The FILL option can occur with only one fitted curve. Enclose the FILL option in parentheses after a density curve option or the KERNEL option. The CFILL= and PFILL= options specify the color and pattern for the area under the curve when producing traditional graphics. For a list of available colors and patterns, see *SAS/GRAPH: Reference*.

FORCEHIST

forces the creation of a histogram if there is only one unique observation. By default, a histogram is not created if the standard deviation of the data is zero.

FRONTREF

draws reference lines requested with the HREF= and VREF= options in front of the histogram bars. When the NOGSTYLE system option is in effect for traditional graphics, reference lines are drawn behind the histogram bars by default, and they can be obscured by filled bars.

GAMMA < (*gamma-options*) >

displays fitted gamma density curves on the histogram. The GAMMA option can occur only once in a HISTOGRAM statement, but it can request any number of gamma curves. The parameter θ must be less than the minimum data value. Use the THETA= *gamma-option* to specify θ. By default, THETA=0. You can specify THETA=EST to request the maximum likelihood estimate for θ. Use the ALPHA= and the SIGMA= *gamma-options* to specify the shape parameter α and the scale parameter σ. By default, PROC UNIVARIATE computes maximum likelihood estimates for α and σ. The procedure calculates the maximum likelihood estimate of α iteratively by using the Newton-Raphson approximation. Table 4.12 and Table 4.15 list options you can specify with the GAMMA option. See the section "Gamma Distribution" on page 350 for details, and see Example 4.22 for an example that uses the GAMMA option.

GAMMA=*value-list*

specifies the second shape parameter γ for Johnson S_B and Johnson S_U distribution functions requested with the SB and SU options. Enclose the GAMMA= option in parentheses after the SB or SU option. If you do not specify a value for γ, or if you specify the value EST, the procedure calculates an estimate.

GRID

displays a grid on the histogram. Grid lines are horizontal lines that are positioned at major tick marks on the vertical axis.

HANGING

HANG

requests a hanging histogram, as illustrated in Figure 4.7.

Figure 4.7 Hanging Histogram

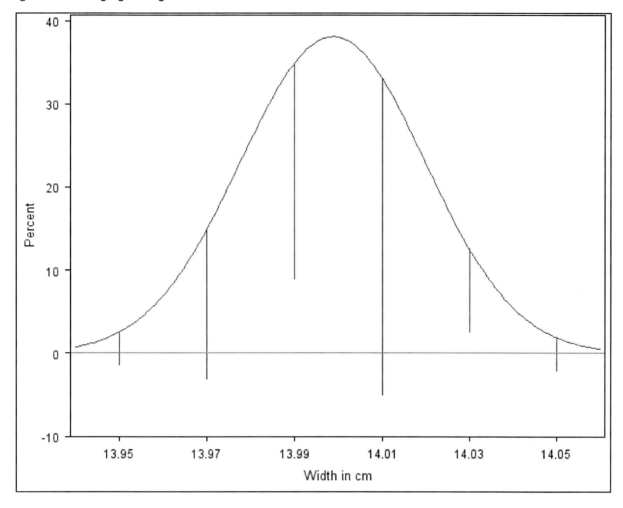

You can use the HANGING option only when exactly one fitted density curve is requested. A hanging histogram aligns the tops of the histogram bars (displayed as lines) with the fitted curve. The lines are positioned at the midpoints of the histogram bins. A hanging histogram is a goodness-of-fit diagnostic in the sense that the closer the lines are to the horizontal axis, the better the fit. Hanging histograms are discussed by Tukey (1977), Wainer (1974), and Velleman and Hoaglin (1981).

HOFFSET=value

specifies the offset, in percentage screen units, at both ends of the horizontal axis. You can use HOFFSET=0 to eliminate the default offset.

INTERBAR=value

specifies the space between histogram bars in percentage screen units. If both the INTERBAR= and BARWIDTH= options are specified, the INTERBAR= option takes precedence.

K=NORMAL | QUADRATIC | TRIANGULAR

specifies the kernel function (normal, quadratic, or triangular) used to compute a kernel density estimate. You can specify a list of values to request multiple estimates. You must enclose this option in parentheses after the KERNEL option. You can also use the K= *kernel-*

option with the C= *kernel-option*, which specifies standardized bandwidths. If you specify more kernel functions than bandwidths, the procedure repeats the last bandwidth in the list for the remaining estimates. Similarly, if you specify more bandwidths than kernel functions, the procedure repeats the last kernel function for the remaining estimates. By default, K=NORMAL.

KERNEL< (*kernel-options***) >**

superimposes kernel density estimates on the histogram. By default, the procedure uses the AMISE method to compute kernel density estimates. To request multiple kernel density estimates on the same histogram, specify a list of values for the C= *kernel-option* or K= *kernel-option*. Table 4.21 lists options you can specify with the KERNEL option. See the section "Kernel Density Estimates" on page 360 for more information about kernel density estimates, and see Example 4.23.

LGRID=*linetype*

specifies the line type for the grid when a grid displays on the histogram. This option also creates a grid if the GRID option is not specified.

LOGNORMAL< (*lognormal-options***) >**

displays fitted lognormal density curves on the histogram. The LOGNORMAL option can occur only once in a HISTOGRAM statement, but it can request any number of lognormal curves. The parameter θ must be less than the minimum data value. Use the THETA= *lognormal-option* to specify θ. By default, THETA=0. You can specify THETA=EST to request the maximum likelihood estimate for θ. Use the SIGMA= and ZETA= *lognormal-options* to specify σ and ζ. By default, the procedure computes maximum likelihood estimates for σ and ζ. Table 4.12 and Table 4.16 list options you can specify with the LOGNORMAL option. See the section "Lognormal Distribution" on page 351 for details, and see Example 4.22 and Example 4.24 for examples using the LOGNORMAL option.

LOWER=*value-list*

specifies lower bounds for kernel density estimates requested with the KERNEL option. Enclose the LOWER= option in parentheses after the KERNEL option. If you specify more kernel estimates than lower bounds, the last lower bound is repeated for the remaining estimates. The default is a missing value, indicating no lower bounds for fitted kernel density curves.

MAXNBIN=*n*

limits the number of bins displayed in the comparative histogram. This option is useful when the scales or ranges of the data distributions differ greatly from cell to cell. By default, the bin size and midpoints are determined for the key cell, and then the midpoint list is extended to accommodate the data ranges for the remaining cells. However, if the cell scales differ considerably, the resulting number of bins can be so great that each cell histogram is scaled into a narrow region. By using MAXNBIN= to limit the number of bins, you can narrow the window about the data distribution in the key cell. This option is not available unless you specify the CLASS statement. The MAXNBIN= option is an alternative to the MAXSIGMAS= option.

MAXSIGMAS=*value*

limits the number of bins displayed in the comparative histogram to a range of *value* standard deviations (of the data in the key cell) above and below the mean of the data in the key cell. This option is useful when the scales or ranges of the data distributions differ greatly from cell to cell. By default, the bin size and midpoints are determined for the key cell, and then the midpoint list is extended to accommodate the data ranges for the remaining cells. However, if the cell scales differ considerably, the resulting number of bins can be so great that each cell histogram is scaled into a narrow region. By using MAXSIGMAS= to limit the number of bins, you can narrow the window that surrounds the data distribution in the key cell. This option is not available unless you specify the CLASS statement.

MIDPERCENTS

requests a table listing the midpoints and percentage of observations in each histogram interval. If you specify MIDPERCENTS in parentheses after a density estimate option, the procedure displays a table that lists the midpoints, the observed percentage of observations, and the estimated percentage of the population in each interval (estimated from the fitted distribution). See Example 4.18.

MIDPOINTS=values | **KEY** | **UNIFORM**

specifies how to determine the midpoints for the histogram intervals, where values determines the width of the histogram bars as the difference between consecutive midpoints. The procedure uses the same values for all variables.

The range of midpoints, extended at each end by half of the bar width, must cover the range of the data. For example, if you specify

```
midpoints=2 to 10 by 0.5
```

then all of the observations should fall between 1.75 and 10.25. You must use evenly spaced midpoints listed in increasing order.

KEY determines the midpoints for the data in the key cell. The initial number of midpoints is based on the number of observations in the key cell that use the method of Terrell and Scott (1985). The procedure extends the midpoint list for the key cell in either direction as necessary until it spans the data in the remaining cells.

UNIFORM determines the midpoints by using all the observations as if there were no cells. In other words, the number of midpoints is based on the total sample size by using the method of Terrell and Scott (1985).

Neither KEY nor UNIFORM apply unless you use the CLASS statement. By default, if you use a CLASS statement, MIDPOINTS=KEY; however, if the key cell is empty then MIDPOINTS=UNIFORM. Otherwise, the procedure computes the midpoints by using an algorithm (Terrell and Scott 1985) that is primarily applicable to continuous data that are approximately normally distributed.

MU=value-list

specifies the parameter μ for normal density curves requested with the NORMAL option. Enclose the MU= option in parentheses after the NORMAL option. By default, or if you specify the value EST, the procedure uses the sample mean for μ.

NENDPOINTS=n
> uses histogram interval endpoints as the tick mark values for the horizontal axis and determines the number of bins.

NMIDPOINTS=n
> specifies the number of histogram intervals.

NOBARS
> suppresses drawing of histogram bars, which is useful for viewing fitted curves only.

NOPLOT

NOCHART
> suppresses the creation of a plot. Use this option when you only want to tabulate summary statistics for a fitted density or create an OUTHISTOGRAM= data set.

NOPRINT
> suppresses tables summarizing the fitted curve. Enclose the NOPRINT option in parentheses following the distribution option.

NORMAL<(normal-options**)>**
> displays fitted normal density curves on the histogram. The NORMAL option can occur only once in a HISTOGRAM statement, but it can request any number of normal curves. Use the MU= and SIGMA= normal-options to specify μ and σ. By default, the procedure uses the sample mean and sample standard deviation for μ and σ. Table 4.12 and Table 4.17 list options you can specify with the NORMAL option. See the section "Normal Distribution" on page 352 for details, and see Example 4.19 for an example that uses the NORMAL option.

NOTABCONTENTS
> suppresses the table of contents entries for tables produced by the HISTOGRAM statement.

OUTHISTOGRAM=SAS-data-set

OUTHIST=SAS-data-set
> creates a SAS data set that contains information about histogram intervals. Specifically, the data set contains the midpoints of the histogram intervals (or the lower endpoints of the intervals if you specify the ENDPOINTS option), the observed percentage of observations in each interval, and the estimated percentage of observations in each interval (estimated from each of the specified fitted curves).

PERCENTS=values

PERCENT=values
> specifies a list of percents for which quantiles calculated from the data and quantiles estimated from the fitted curve are tabulated. The percents must be between 0 and 100. Enclose the PERCENTS= option in parentheses after the curve option. The default percents are 1, 5, 10, 25, 50, 75, 90, 95, and 99.

PFILL=pattern
> specifies a pattern used to fill the bars of the histograms (or the areas under a fitted curve if you also specify the FILL option) when producing traditional graphics. See the entries for the CFILL= and FILL options for additional details. Refer to *SAS/GRAPH: Reference* for a list of pattern values. The option does not apply to ODS Graphics output.

RTINCLUDE
> includes the right endpoint of each histogram interval in that interval. By default, the left endpoint is included in the histogram interval.

SB<(S_B-options)>
> displays fitted Johnson S_B density curves on the histogram. The SB option can occur only once in a HISTOGRAM statement, but it can request any number of Johnson S_B curves. Use the THETA= and SIGMA= *normal-options* to specify θ and σ. By default, the procedure computes maximum likelihood estimates of θ and σ. Table 4.12 and Table 4.18 list options you can specify with the SB option. See the section "Johnson S_B Distribution" on page 353 for details.

SIGMA=value-list
> specifies the parameter σ for the fitted density curve when you request the BETA, EXPONENTIAL, GAMMA, LOGNORMAL, NORMAL, SB, SU, or WEIBULL options.
>
> See Table 4.24 for a summary of how to use the SIGMA= option. You must enclose this option in parentheses after the density curve option. You can specify the value EST to request a maximum likelihood estimate for σ.

Table 4.24   Uses of the SIGMA= Option

Distribution Keyword	SIGMA= Specifies	Default Value	Alias
BETA	scale parameter σ	1	SCALE=
EXPONENTIAL	scale parameter σ	maximum likelihood estimate	SCALE=
GAMMA	scale parameter σ	maximum likelihood estimate	SCALE=
LOGNORMAL	shape parameter σ	maximum likelihood estimate	SHAPE=
NORMAL	scale parameter σ	standard deviation	
SB	scale parameter σ	1	SCALE=
SU	scale parameter σ	percentile-based estimate	
WEIBULL	scale parameter σ	maximum likelihood estimate	SCALE=

SU<(S_U-options)>
> displays fitted Johnson S_U density curves on the histogram. The SU option can occur only once in a HISTOGRAM statement, but it can request any number of Johnson S_U curves. Use the THETA= and SIGMA= *normal-options* to specify θ and σ. By default, the procedure computes maximum likelihood estimates of θ and σ. Table 4.12 and Table 4.19 list options you can specify with the SU option. See the section "Johnson S_U Distribution" on page 354 for details.

THETA=value-list

THRESHOLD= value-list
> specifies the lower threshold parameter θ for curves requested with the BETA, EXPONENTIAL, GAMMA, LOGNORMAL, SB, SU, and WEIBULL options. Enclose the THETA= option in parentheses after the curve option. By default, THETA=0. If you specify the value EST, an estimate is computed for θ.

UPPER=*value-list*

specifies upper bounds for kernel density estimates requested with the KERNEL option. Enclose the UPPER= option in parentheses after the KERNEL option. If you specify more kernel estimates than upper bounds, the last upper bound is repeated for the remaining estimates. The default is a missing value, indicating no upper bounds for fitted kernel density curves.

VOFFSET=*value*

specifies the offset, in percentage screen units, at the upper end of the vertical axis.

VSCALE=COUNT | PERCENT | PROPORTION

specifies the scale of the vertical axis for a histogram. The value COUNT requests the data be scaled in units of the number of observations per data unit. The value PERCENT requests the data be scaled in units of percent of observations per data unit. The value PROPORTION requests the data be scaled in units of proportion of observations per data unit. The default is PERCENT.

WBARLINE=*n*

specifies the width of bar outlines when producing traditional graphics. The option does not apply to ODS Graphics output.

WEIBULL<(*Weibull-options*)>

displays fitted Weibull density curves on the histogram. The WEIBULL option can occur only once in a HISTOGRAM statement, but it can request any number of Weibull curves. The parameter θ must be less than the minimum data value. Use the THETA= *Weibull-option* to specify θ. By default, THETA=0. You can specify THETA=EST to request the maximum likelihood estimate for θ. Use the C= and SIGMA= *Weibull-options* to specify the shape parameter c and the scale parameter σ. By default, the procedure computes the maximum likelihood estimates for c and σ. Table 4.12 and Table 4.20 list options you can specify with the WEIBULL option. See the section "Weibull Distribution" on page 355 for details, and see Example 4.22 for an example that uses the WEIBULL option.

PROC UNIVARIATE calculates the maximum likelihood estimate of a iteratively by using the Newton-Raphson approximation. See also the C=, SIGMA=, and THETA= *Weibull-options*.

WGRID=*n*

specifies the line thickness for the grid when producing traditional graphics. The option does not apply to ODS Graphics output.

ZETA= *value-list*

specifies a value for the scale parameter ζ for lognormal density curves requested with the LOGNORMAL option. Enclose the ZETA= *lognormal-option* in parentheses after the LOGNORMAL option. By default, or if you specify the value EST, the procedure calculates a maximum likelihood estimate for ζ. You can specify the SCALE= option as an alias for the ZETA= option.

ID Statement

ID *variables* ;

The ID statement specifies one or more variables to include in the table of extreme observations. The corresponding values of the ID variables appear beside the *n* largest and *n* smallest observations, where *n* is the value of NEXTROBS= option. See Example 4.3.

You can also include ID variables in the output data set created by an OUTPUT statement by specifying the IDOUT option in the PROC UNIVARIATE statement.

INSET Statement

INSET *keywords* < / *options* > ;

An INSET statement places a box or table of summary statistics, called an *inset*, directly in a graph created with a CDFPLOT, HISTOGRAM, PPPLOT, PROBPLOT, or QQPLOT statement. The INSET statement must follow the plot statement that creates the plot that you want to augment. The inset appears in all the graphs that the preceding plot statement produces.

You can use multiple INSET statements after a plot statement to add more than one inset to a plot. See Example 4.17.

In an INSET statement, you specify one or more *keywords* that identify the information to display in the inset. The information is displayed in the order that you request the *keywords*. *Keywords* can be any of the following:

- *statistical keywords*
- *primary keywords*
- *secondary keywords*

The available *statistical keywords* are listed in Table 4.25 through Table 4.29.

Table 4.25 Descriptive Statistic Keywords

Keyword	Description
CSS	corrected sum of squares
CV	coefficient of variation
KURTOSIS	kurtosis
MAX	largest value
MEAN	sample mean
MIN	smallest value
MODE	most frequent value
N	sample size
NEXCL	number of observations excluded by MAXNBIN= or MAXSIGMAS= option
NMISS	number of missing values
NOBS	number of observations
RANGE	range
SKEWNESS	skewness
STD	standard deviation
STDMEAN	standard error of the mean
SUM	sum of the observations
SUMWGT	sum of the weights
USS	uncorrected sum of squares
VAR	variance

Table 4.26 Percentile Statistic Keywords

Keyword	Description
P1	1st percentile
P5	5th percentile
P10	10th percentile
Q1	lower quartile (25th percentile)
MEDIAN	median (50th percentile)
Q3	upper quartile (75th percentile)
P90	90th percentile
P95	95th percentile
P99	99th percentile
QRANGE	interquartile range (Q3 - Q1)

Table 4.27 Robust Statistics Keywords

Keyword	Description
GINI	Gini's mean difference
MAD	median absolute difference about the median
QN	Q_n, alternative to MAD
SN	S_n, alternative to MAD
STD_GINI	Gini's standard deviation
STD_MAD	MAD standard deviation
STD_QN	Q_n standard deviation
STD_QRANGE	interquartile range standard deviation
STD_SN	S_n standard deviation

Table 4.28 Hypothesis Testing Keywords

Keyword	Description
MSIGN	sign statistic
NORMALTEST	test statistic for normality
PNORMAL	probability value for the test of normality
SIGNRANK	signed rank statistic
PROBM	probability of greater absolute value for the sign statistic
PROBN	probability value for the test of normality
PROBS	probability value for the signed rank test
PROBT	probability value for the Student's t test
T	statistics for Student's t test

Table 4.29 Keyword for Reading Input Data Set

Keyword	Description
DATA=	(label, value) pairs from input data set

To create a completely customized inset, use a DATA= data set.

DATA=*SAS-data-set*

requests that PROC UNIVARIATE display customized statistics from a SAS data set in the inset table. The data set must contain two variables:

LABEL a character variable whose values provide labels for inset entries

VALUE a variable that is either character or numeric and whose values provide values for inset entries

The label and value from each observation in the data set occupy one line in the inset. The position of the DATA= keyword in the keyword list determines the position of its lines in the inset.

A *primary keyword* enables you to specify *secondary keywords* in parentheses immediately after the primary keyword. *Primary keywords* are BETA, EXPONENTIAL, GAMMA, KERNEL, KERNEL*n*, LOGNORMAL, NORMAL, SB, SU, WEIBULL, and WEIBULL2. If you specify a *primary keyword* but omit a *secondary keyword*, the inset displays a colored line and the distribution name as a key for the density curve.

By default, PROC UNIVARIATE identifies inset statistics with appropriate labels and prints numeric values with appropriate formats. To customize the label, specify the *keyword* followed by an equal sign (=) and the desired label in quotes. To customize the format, specify a numeric format in parentheses after the *keyword*. Labels can have up to 24 characters. If you specify both a label and a format for a statistic, the label must appear before the format. For example,

```
inset n='Sample Size' std='Std Dev' (5.2);
```

requests customized labels for two statistics and displays the standard deviation with a field width of 5 and two decimal places.

Table 4.30 and Table 4.31 list *primary keywords*.

Table 4.30 Parametric Density Primary Keywords

Keyword	Distribution	Plot Statement Availability
BETA	beta	all plot statements
EXPONENTIAL	exponential	all plot statements
GAMMA	gamma	all plot statements
LOGNORMAL	lognormal	all plot statements
NORMAL	normal	all plot statements
SB	Johnson S_B	HISTOGRAM
SU	Johnson S_U	HISTOGRAM
WEIBULL	Weibull(3-parameter)	all plot statements
WEIBULL2	Weibull(2-parameter)	PROBPLOT and QQPLOT

Table 4.31 Kernel Density Estimate Primary Keywords

Keyword	Description
KERNEL	displays statistics for all kernel estimates
KERNEL*n*	displays statistics for only the *n*th kernel density estimate $n = 1, 2, 3, 4,$ or 5

Table 4.32 through Table 4.41 list the *secondary keywords* available with *primary keywords* in Table 4.30 and Table 4.31.

Table 4.32 Secondary Keywords Available with the BETA Keyword

Secondary Keyword	Alias	Description
ALPHA	SHAPE1	first shape parameter α
BETA	SHAPE2	second shape parameter β
MEAN		mean of the fitted distribution
SIGMA	SCALE	scale parameter σ
STD		standard deviation of the fitted distribution
THETA	THRESHOLD	lower threshold parameter θ

Table 4.33 Secondary Keywords Available with the EXPONENTIAL Keyword

Secondary Keyword	Alias	Description
MEAN		mean of the fitted distribution
SIGMA	SCALE	scale parameter σ
STD		standard deviation of the fitted distribution
THETA	THRESHOLD	threshold parameter θ

Table 4.34 Secondary Keywords Available with the GAMMA Keyword

Secondary Keyword	Alias	Description
ALPHA	SHAPE	shape parameter α
MEAN		mean of the fitted distribution
SIGMA	SCALE	scale parameter σ
STD		standard deviation of the fitted distribution
THETA	THRESHOLD	threshold parameter θ

Table 4.35 Secondary Keywords Available with the LOGNORMAL Keyword

Secondary Keyword	Alias	Description
MEAN		mean of the fitted distribution
SIGMA	SHAPE	shape parameter σ
STD		standard deviation of the fitted distribution
THETA	THRESHOLD	threshold parameter θ
ZETA	SCALE	scale parameter ζ

Table 4.36 Secondary Keywords Available with the NORMAL Keyword

Secondary Keyword	Alias	Description
MU	MEAN	mean parameter μ
SIGMA	STD	scale parameter σ

Table 4.37 Secondary Keywords Available with the SB and SU Keywords

Secondary Keyword	Alias	Description
DELTA	SHAPE1	first shape parameter δ
GAMMA	SHAPE2	second shape parameter γ
MEAN		mean of the fitted distribution
SIGMA	SCALE	scale parameter σ
STD		standard deviation of the fitted distribution
THETA	THRESHOLD	lower threshold parameter θ

Table 4.38 Secondary Keywords Available with the WEIBULL

Secondary Keyword	Alias	Description
C	SHAPE	shape parameter c
MEAN		mean of the fitted distribution
SIGMA	SCALE	scale parameter σ
STD		standard deviation of the fitted distribution
THETA	THRESHOLD	threshold parameter θ

Table 4.39 Secondary Keywords Available with the WEIBULL2 Keyword

Secondary Keyword	Alias	Description
C	SHAPE	shape parameter c
MEAN		mean of the fitted distribution
SIGMA	SCALE	scale parameter σ
STD		standard deviation of the fitted distribution
THETA	THRESHOLD	known lower threshold θ_0

Table 4.40 Secondary Keywords Available with the KERNEL Keyword

Secondary Keyword	Description
AMISE	approximate mean integrated square error (MISE) for the kernel density
BANDWIDTH	bandwidth λ for the density estimate
BWIDTH	alias for BANDWIDTH
C	standardized bandwidth c for the density estimate: $c = \frac{\lambda}{Q} n^{\frac{1}{5}}$ where n = sample size, λ = bandwidth, and Q = interquartile range
TYPE	kernel type: normal, quadratic, or triangular

Table 4.41 Goodness-of-Fit Statistics for Fitted Curves

Secondary Keyword	Description
AD	Anderson-Darling EDF test statistic
ADPVAL	Anderson-Darling EDF test p-value
CVM	Cramér-von Mises EDF test statistic
CVMPVAL	Cramér-von Mises EDF test p-value
KSD	Kolmogorov-Smirnov EDF test statistic
KSDPVAL	Kolmogorov-Smirnov EDF test p-value

The inset statistics listed in Table 4.30 through Table 4.41 are not available unless you request a plot statement and options that calculate these statistics. For example, consider the following statements:

```
proc univariate data=score;
   histogram final / normal;
   inset mean std normal(ad adpval);
run;
```

The MEAN and STD *keywords* display the sample mean and standard deviation of final. The NORMAL *keyword* with the *secondary keywords* AD and ADPVAL display the Anderson-Darling goodness-of-fit test statistic and *p*-value. The statistics that are specified with the NORMAL *keyword* are available only because the NORMAL option is requested in the HISTOGRAM statement.

The KERNEL or KERNEL*n keyword* is available only if you request a kernel density estimate in a HISTOGRAM statement. The WEIBULL2 *keyword* is available only if you request a two-parameter Weibull distribution in the PROBPLOT or QQPLOT statement.

If you specify multiple kernel density estimates, you can request inset statistics for all the estimates with the KERNEL *keyword*. Alternatively, you can display inset statistics for individual curves with the KERNEL*n keyword*, where *n* is the curve number between 1 and 5.

Summary of Options

Table 4.42 lists INSET statement *options*, which are specified after the slash (/) in the INSET statement. For complete descriptions, see the section "Dictionary of Options" on page 278.

Table 4.42 INSET Options

Option	Description
CFILL=*color*	BLANK	specifies color of inset background
CFILLH=*color*	specifies color of header background
CFRAME=*color*	specifies color of frame
CHEADER=*color*	specifies color of header text
CSHADOW=*color*	specifies color of drop shadow
CTEXT=*color*	specifies color of inset text
DATA	specifies data units for POSITION=(x, y) coordinates

Table 4.42 *continued*

Option	Description
FONT=*font*	specifies font of text
FORMAT=*format*	specifies format of values in inset
HEADER='*string*'	specifies header text
HEIGHT=*value*	specifies height of inset text
NOFRAME	suppresses frame around inset
POSITION=*position*	specifies position of inset
REFPOINT=BR | BL | TR | TL	specifies reference point of inset positioned with POSITION=(*x*, *y*) coordinates

Dictionary of Options

The following entries provide detailed descriptions of options for the INSET statement.

CFILL=*color* **| BLANK**
> specifies the color of the background for traditional graphics. If you omit the CFILLH= option the header background is included. By default, the background is empty, which causes items that overlap the inset (such as curves or histogram bars) to show through the inset.
>
> If you specify a value for CFILL= option, then overlapping items no longer show through the inset. Use CFILL=BLANK to leave the background uncolored and to prevent items from showing through the inset.

CFILLH=*color*
> specifies the color of the header background for traditional graphics. The default value is the CFILL= color.

CFRAME=*color*
> specifies the color of the frame for traditional graphics. The default value is the same color as the axis of the plot.

CHEADER=*color*
> specifies the color of the header text for traditional graphics. The default value is the CTEXT= color.

CSHADOW=*color*
> specifies the color of the drop shadow for traditional graphics. By default, if a CSHADOW= option is not specified, a drop shadow is not displayed.

CTEXT=*color*
> specifies the color of the text for traditional graphics. The default value is the same color as the other text on the plot.

DATA
> specifies that data coordinates are to be used in positioning the inset with the POSITION= option. The DATA option is available only when you specify POSITION=(x,y). You must place

DATA immediately after the coordinates (x,y). **NOTE:** Positioning insets with coordinates is not supported for ODS Graphics output.

FONT=font

specifies the font of the text for traditional graphics. By default, if you locate the inset in the interior of the plot, then the font is SIMPLEX. If you locate the inset in the exterior of the plot, then the font is the same as the other text on the plot.

FORMAT=format

specifies a format for all the values in the inset. If you specify a format for a particular statistic, then this format overrides FORMAT= format. For more information about SAS formats, see *SAS Language Reference: Dictionary*

HEADER=string

specifies the header text. The *string* cannot exceed 40 characters. By default, no header line appears in the inset. If all the keywords that you list in the INSET statement are secondary keywords that correspond to a fitted curve on a histogram, PROC UNIVARIATE displays a default header that indicates the distribution and identifies the curve.

HEIGHT=value

specifies the height of the text for traditional graphics.

NOFRAME

suppresses the frame drawn around the text.

POSITION=position

POS=position

determines the position of the inset. The position is a compass point keyword, a margin keyword, or a pair of coordinates (x,y). You can specify coordinates in axis percent units or axis data units. The default value is NW, which positions the inset in the upper left (northwest) corner of the display. See the section "Positioning Insets" on page 344.

NOTE: Positioning insets with coordinates is not supported for ODS Graphics output.

REFPOINT=BR | BL | TR | TL

specifies the reference point for an inset that PROC UNIVARIATE positions by a pair of coordinates with the POSITION= option. The REFPOINT= option specifies which corner of the inset frame that you want to position at coordinates (x,y). The *keywords* are BL, BR, TL, and TR, which correspond to bottom left, bottom right, top left, and top right. The default value is BL. You must use REFPOINT= with POSITION=(x,y) coordinates. The option does not apply to ODS Graphics output.

OUTPUT Statement

OUTPUT < *OUT=SAS-data-set* < *keyword1=names* ... *keywordk=names* > < *percentile-options* > > ;

The OUTPUT statement saves statistics and BY variables in an output data set. When you use a BY statement, each observation in the OUT= data set corresponds to one of the BY groups. Otherwise, the OUT= data set contains only one observation.

You can use any number of OUTPUT statements in the UNIVARIATE procedure. Each OUTPUT statement creates a new data set containing the statistics specified in that statement. You must use the VAR statement with the OUTPUT statement. The OUTPUT statement must contain a specification of the form *keyword=names* or the PCTLPTS= and PCTLPRE= specifications. See Example 4.7 and Example 4.8.

OUT=SAS-data-set

identifies the output data set. If *SAS-data-set* does not exist, PROC UNIVARIATE creates it. If you omit OUT=, the data set is named DATA*n*, where *n* is the smallest integer that makes the name unique.

keyword=names

specifies the statistics to include in the output data set and gives names to the new variables that contain the statistics. Specify a *keyword* for each desired statistic, followed by an equal sign, followed by the *names* of the variables to contain the statistic. In the output data set, the first variable listed after a keyword in the OUTPUT statement contains the statistic for the first variable listed in the VAR statement, the second variable contains the statistic for the second variable in the VAR statement, and so on. If the list of *names* following the equal sign is shorter than the list of variables in the VAR statement, the procedure uses the *names* in the order in which the variables are listed in the VAR statement. The available keywords are listed in the following tables:

Table 4.43 Descriptive Statistic Keywords

Keyword	Description
CSS	corrected sum of squares
CV	coefficient of variation
KURTOSIS	kurtosis
MAX	largest value
MEAN	sample mean
MIN	smallest value
MODE	most frequent value
N	sample size
NMISS	number of missing values
NOBS	number of observations
RANGE	range
SKEWNESS	skewness
STD	standard deviation
STDMEAN	standard error of the mean
SUM	sum of the observations
SUMWGT	sum of the weights
USS	uncorrected sum of squares
VAR	variance

Table 4.44 Quantile Statistic Keywords

Keyword	Description
P1	1st percentile
P5	5th percentile
P10	10th percentile
Q1	lower quartile (25th percentile)
MEDIAN	median (50th percentile)
Q3	upper quartile (75th percentile)
P90	90th percentile
P95	95th percentile
P99	99th percentile
QRANGE	interquartile range (Q3 - Q1)

Table 4.45 Robust Statistics Keywords

Keyword	Description
GINI	Gini's mean difference
MAD	median absolute difference about the median
QN	Q_n, alternative to MAD
SN	S_n, alternative to MAD
STD_GINI	Gini's standard deviation
STD_MAD	MAD standard deviation
STD_QN	Q_n standard deviation
STD_QRANGE	interquartile range standard deviation
STD_SN	S_n standard deviation

Table 4.46 Hypothesis Testing Keywords

Keyword	Description
MSIGN	sign statistic
NORMALTEST	test statistic for normality
SIGNRANK	signed rank statistic
PROBM	probability of a greater absolute value for the sign statistic
PROBN	probability value for the test of normality
PROBS	probability value for the signed rank test
PROBT	probability value for the Student's t test
T	statistic for the Student's t test

The UNIVARIATE procedure automatically computes the 1st, 5th, 10th, 25th, 50th, 75th, 90th, 95th, and 99th percentiles for the data. These can be saved in an output data set by using *keyword=names* specifications. For additional percentiles, you can use the following *percentile-options*.

PCTLPTS=*percentiles*

specifies one or more percentiles that are not automatically computed by the UNIVARIATE procedure. The PCTLPRE= and PCTLPTS= options must be used together. You can specify percentiles with an expression of the form *start* TO *stop* BY *increment* where *start* is a starting number, *stop* is an ending number, and *increment* is a number to increment by. The PCTLPTS= option generates additional percentiles and outputs them to a data set. These additional percentiles are not printed.

To compute the 50th, 95th, 97.5th, and 100th percentiles, submit the statement

```
output pctlpre=P_ pctlpts=50,95 to 100 by 2.5;
```

PROC UNIVARIATE computes the requested percentiles based on the method that you specify with the PCTLDEF= option in the PROC UNIVARIATE statement. You must use PCTLPRE=, and optionally PCTLNAME=, to specify variable names for the percentiles. For example, the following statements create an output data set named Pctls that contains the 20th and 40th percentiles of the analysis variables PreTest and PostTest:

```
proc univariate data=Score;
   var PreTest PostTest;
   output out=Pctls pctlpts=20 40 pctlpre=PreTest_ PostTest_
          pctlname=P20 P40;
run;
```

PROC UNIVARIATE saves the 20th and 40th percentiles for PreTest and PostTest in the variables PreTest_P20, PostTest_P20, PreTest_P40, and PostTest_P40.

PCTLPRE=prefixes

specifies one or more prefixes to create the variable names for the variables that contain the PCTLPTS= percentiles. To save the same percentiles for more than one analysis variable, specify a list of prefixes. The order of the prefixes corresponds to the order of the analysis variables in the VAR statement. The PCTLPRE= and PCTLPTS= options must be used together.

The procedure generates new variable names by using the *prefix* and the percentile values. If the specified percentile is an integer, the variable name is simply the *prefix* followed by the value. If the specified value is not an integer, an underscore replaces the decimal point in the variable name, and decimal values are truncated to one decimal place. For example, the following statements create the variables pwid20, pwid33_3, pwid66_6, and pwid80 for the 20th, 33.33rd, 66.67th, and 80th percentiles of Width, respectively:

```
proc univariate noprint;
   var Width;
   output pctlpts=20 33.33 66.67 80 pctlpre=pwid;
run;
```

If you request percentiles for more than one variable, you should list prefixes in the same order in which the variables appear in the VAR statement. If combining the *prefix* and percentile value results in a name longer than 32 characters, the prefix is truncated so that the variable name is 32 characters.

PCTLNAME=suffixes

specifies one or more suffixes to create the names for the variables that contain the PCTLPTS= percentiles. PROC UNIVARIATE creates a variable name by combining the PCTLPRE= value and suffix name. Because the suffix names are associated with the percentiles that are requested, list the suffix names in the same order as the PCTLPTS= percentiles. If you specify *n suffixes* with the PCTLNAME= option and *m* percentile values with the PCTLPTS= option where $m > n$, the *suffixes* are used to name the first *n* percentiles and the default names are used for the remaining $m - n$ percentiles. For example, consider the following statements:

```
proc univariate;
   var Length Width Height;
   output pctlpts  = 20 40
          pctlpre  = pl pw ph
          pctlname = twenty;
run;
```

The value `twenty` in the PCTLNAME= option is used for only the first percentile in the PCTLPTS= list. This suffix is appended to the values in the PCTLPRE= option to generate the new variable names pltwenty, pwtwenty, and phtwenty, which contain the 20th percentiles for Length, Width, and Height, respectively. Because a second PCTLNAME= suffix is not specified, variable names for the 40th percentiles for Length, Width, and Height are generated using the prefixes and percentile values. Thus, the output data set contains the variables pltwenty, pl40, pwtwenty, pw40, phtwenty, and ph40.

You must specify PCTLPRE= to supply prefix names for the variables that contain the PCTLPTS= percentiles.

If the number of PCTLNAME= values is fewer than the number of percentiles or if you omit PCTLNAME=, PROC UNIVARIATE uses the percentile as the suffix to create the name of the variable that contains the percentile. For an integer percentile, PROC UNIVARIATE uses the percentile. Otherwise, PROC UNIVARIATE truncates decimal values of percentiles to two decimal places and replaces the decimal point with an underscore.

If either the prefix and suffix name combination or the prefix and percentile name combination is longer than 32 characters, PROC UNIVARIATE truncates the prefix name so that the variable name is 32 characters.

PPPLOT Statement

PPPLOT < variables > < / options > ;

The PPPLOT statement creates a probability-probability plot (also referred to as a P-P plot or percent plot), which compares the empirical cumulative distribution function (ecdf) of a variable with a specified theoretical cumulative distribution function such as the normal. If the two distributions match, the points on the plot form a linear pattern that passes through the origin and has unit slope. Thus, you can use a P-P plot to determine how well a theoretical distribution models a set of measurements.

You can specify one of the following theoretical distributions with the PPPLOT statement:

- beta
- exponential
- gamma
- lognormal
- normal
- Weibull

NOTE: Probability-probability plots should not be confused with probability plots, which compare a set of ordered measurements with *percentiles* from a specified distribution. You can create probability plots with the PROBPLOT statement.

You can use any number of PPPLOT statements in the UNIVARIATE procedure. The components of the PPPLOT statement are as follows.

variables

are the process variables for which P-P plots are created. If you specify a VAR statement, the *variables* must also be listed in the VAR statement. Otherwise, the *variables* can be any numeric variables in the input data set. If you do not specify a list of *variables*, then by default, the procedure creates a P-P plot for each variable listed in the VAR statement or for each numeric variable in the input data set if you do not specify a VAR statement. For example, if data set measures contains two numeric variables, length and width, the following two PPPLOT statements each produce a P-P plot for each of those variables:

```
proc univariate data=measures;
   var length width;
   ppplot;
run;

proc univariate data=measures;
   ppplot length width;
run;
```

options

specify the theoretical distribution for the plot or add features to the plot. If you specify more than one variable, the options apply equally to each variable. Specify all *options* after the slash (/) in the PPPLOT statement. You can specify only one option that names a distribution, but you can specify any number of other options. By default, the procedure produces a P-P plot based on the normal distribution.

In the following example, the NORMAL, MU=, and SIGMA= options request a P-P plot based on the normal distribution with mean 10 and standard deviation 0.3. The SQUARE option displays the plot in a square frame, and the CTEXT= option specifies the text color.

```
proc univariate data=measures;
   ppplot length width / normal(mu=10 sigma=0.3)
                        square
                        ctext=blue;
run;
```

Table 4.47 through Table 4.55 list the PPPLOT *options* by function. For complete descriptions, see the sections "Dictionary of Options" on page 289 and "Dictionary of Common Options" on page 317. *Options* can be any of the following:

- *primary options*
- *secondary options*
- *general options*

Distribution Options

Table 4.47 summarizes the options for requesting a specific theoretical distribution.

Table 4.47 Options for Specifying the Theoretical Distribution

Option	Description
BETA(*beta-options*)	specifies beta P-P plot
EXPONENTIAL(*exponential-options*)	specifies exponential P-P plot
GAMMA(*gamma-options*)	specifies gamma P-P plot
LOGNORMAL(*lognormal-options*)	specifies lognormal P-P plot
NORMAL(*normal-options*)	specifies normal P-P plot
WEIBULL(*Weibull-options*)	specifies Weibull P-P plot

Table 4.48 through Table 4.54 summarize options that specify distribution parameters and control the display of the diagonal distribution reference line. Specify these options in parentheses after the distribution option. For example, the following statements use the NORMAL option to request a normal P-P plot:

```
proc univariate data=measures;
   ppplot length / normal(mu=10 sigma=0.3 color=red);
run;
```

The MU= and SIGMA= *normal-options* specify μ and σ for the normal distribution, and the COLOR= *normal-option* specifies the color for the line.

Table 4.48 Distribution Reference Line Options

Option	Description
COLOR=	specifies color of distribution reference line
L=	specifies line type of distribution reference line
NOLINE	suppresses the distribution reference line
W=	specifies width of distribution reference line

Table 4.49 Secondary Beta-Options

Option	Description
ALPHA=	specifies shape parameter α
BETA=	specifies shape parameter β
SIGMA=	specifies scale parameter σ
THETA=	specifies lower threshold parameter θ

Table 4.50 Secondary Exponential-Options

Option	Description
SIGMA=	specifies scale parameter σ
THETA=	specifies threshold parameter θ

Table 4.51 Secondary Gamma-Options

Option	Description
ALPHA=	specifies shape parameter α
ALPHADELTA=	specifies change in successive estimates of α at which the Newton-Raphson approximation of $\hat{\alpha}$ terminates
ALPHAINITIAL=	specifies initial value for α in the Newton-Raphson approximation of $\hat{\alpha}$
MAXITER=	specifies maximum number of iterations in the Newton-Raphson approximation of $\hat{\alpha}$
SIGMA=	specifies scale parameter σ
THETA=	specifies threshold parameter θ

Table 4.52 Secondary Lognormal-Options

Option	Description
SIGMA=	specifies shape parameter σ
THETA=	specifies threshold parameter θ
ZETA=	specifies scale parameter ζ

Table 4.53 Secondary Normal-Options

Option	Description
MU=	specifies mean μ
SIGMA=	specifies standard deviation σ

Table 4.54 Secondary Weibull-Options

Option	Description
C=	specifies shape parameter c
CDELTA=	specifies change in successive estimates of c at which the Newton-Raphson approximation of $\hat{c}$ terminates
CINITIAL=	specifies initial value for c in the Newton-Raphson approximation of $\hat{c}$
MAXITER=	specifies maximum number of iterations in the Newton-Raphson approximation of $\hat{c}$
SIGMA=	specifies scale parameter σ
THETA=	specifies threshold parameter θ

General Options

Table 4.55 lists options that control the appearance of the plots. For complete descriptions, see the sections "Dictionary of Options" on page 289 and "Dictionary of Common Options" on page 317.

Table 4.55 General Graphics Options

Option	Description
ANNOKEY	applies annotation requested in ANNOTATE= data set to key cell only
ANNOTATE=	provides an annotate data set
CAXIS=	specifies color for axis
CFRAME=	specifies color for frame
CFRAMESIDE=	specifies color for filling row label frames
CFRAMETOP=	specifies color for filling column label frames
CHREF=	specifies color for HREF= lines
CONTENTS=	specifies table of contents entry for P-P plot grouping
CPROP=	specifies color for proportion of frequency bar
CTEXT=	specifies color for text
CTEXTSIDE=	specifies color for row labels
CTEXTTOP=	specifies color for column labels
CVREF=	specifies color for VREF= lines
DESCRIPTION=	specifies description for plot in graphics catalog
FONT=	specifies software font for text
HAXIS=	specifies AXIS statement for horizontal axis
HEIGHT=	specifies height of text used outside framed areas
HMINOR=	specifies number of minor tick marks on horizontal axis
HREF=	specifies reference lines perpendicular to the horizontal axis
HREFLABELS=	specifies line labels for HREF= lines
HREFLABPOS=	specifies position for HREF= line labels
INFONT=	specifies software font for text inside framed areas
INHEIGHT=	specifies height of text inside framed areas
INTERTILE=	specifies distance between tiles in comparative plot
LHREF=	specifies line type for HREF= lines
LVREF=	specifies line type for VREF= lines
NAME=	specifies name for plot in graphics catalog
NCOLS=	specifies number of columns in comparative plot
NOFRAME	suppresses frame around plotting area
NOHLABEL	suppresses label for horizontal axis
NOVLABEL	suppresses label for vertical axis
NOVTICK	suppresses tick marks and tick mark labels for vertical axis
NROWS=	specifies number of rows in comparative plot
OVERLAY	overlays plots for different class levels (ODS Graphics only)
SQUARE	displays P-P plot in square format
TURNVLABELS	turns and vertically strings out characters in labels for vertical axis
VAXIS=	specifies AXIS statement for vertical axis
VAXISLABEL=	specifies label for vertical axis
VMINOR=	specifies number of minor tick marks on vertical axis
VREF=	specifies reference lines perpendicular to the vertical axis
VREFLABELS=	specifies line labels for VREF= lines
VREFLABPOS=	specifies position for VREF= line labels
WAXIS=	specifies line thickness for axes and frame

Dictionary of Options

The following entries provide detailed descriptions of options for the PPPLOT statement. See the section "Dictionary of Common Options" on page 317 for detailed descriptions of options common to all plot statements.

ALPHA=value

specifies the shape parameter α ($\alpha > 0$) for P-P plots requested with the BETA and GAMMA options. For examples, see the entries for the BETA and GAMMA options.

BETA< (beta-options) **>**

creates a beta P-P plot. To create the plot, the n nonmissing observations are ordered from smallest to largest:

$$x_{(1)} \leq x_{(2)} \leq \cdots \leq x_{(n)}$$

The y-coordinate of the ith point is the empirical cdf value $\frac{i}{n}$. The x-coordinate is the theoretical beta cdf value

$$B_{\alpha\beta}\left(\frac{x_{(i)} - \theta}{\sigma}\right) = \int_{\theta}^{x_{(i)}} \frac{(t - \theta)^{\alpha-1}(\theta + \sigma - t)^{\beta-1}}{B(\alpha, \beta)\sigma^{(\alpha+\beta-1)}} dt$$

where $B_{\alpha\beta}(\cdot)$ is the normalized incomplete beta function, $B(\alpha, \beta) = \frac{\Gamma(\alpha)\Gamma(\beta)}{\Gamma(\alpha+\beta)}$, and

$\theta =$ lower threshold parameter

$\sigma =$ scale parameter ($\sigma > 0$)

$\alpha =$ first shape parameter ($\alpha > 0$)

$\beta =$ second shape parameter ($\beta > 0$)

You can specify α, β, σ, and θ with the ALPHA=, BETA=, SIGMA=, and THETA= beta-options, as illustrated in the following example:

```
proc univariate data=measures;
   ppplot width / beta(theta=1 sigma=2 alpha=3 beta=4);
run;
```

If you do not specify values for these parameters, then by default, $\theta = 0$, $\sigma = 1$, and maximum likelihood estimates are calculated for α and β.

IMPORTANT: If the default unit interval (0,1) does not adequately describe the range of your data, then you should specify THETA=θ and SIGMA=σ so that your data fall in the interval $(\theta, \theta + \sigma)$.

If the data are beta distributed with parameters α, β, σ, and θ, then the points on the plot for ALPHA=α, BETA=β, SIGMA=σ, and THETA=θ tend to fall on or near the diagonal line $y = x$, which is displayed by default. Agreement between the diagonal line and the point pattern is evidence that the specified beta distribution is a good fit. You can specify the SCALE= option as an alias for the SIGMA= option and the THRESHOLD= option as an alias for the THETA= option.

BETA=_value_

specifies the shape parameter β ($\beta > 0$) for P-P plots requested with the BETA distribution option. See the preceding entry for the BETA distribution option for an example.

C=_value_

specifies the shape parameter c ($c > 0$) for P-P plots requested with the WEIBULL option. See the entry for the WEIBULL option for examples.

EXPONENTIAL< (_exponential-options_) **>**

EXP< (_exponential-options_) **>**

creates an exponential P-P plot. To create the plot, the n nonmissing observations are ordered from smallest to largest:

$$x_{(1)} \leq x_{(2)} \leq \cdots \leq x_{(n)}$$

The y-coordinate of the ith point is the empirical cdf value $\frac{i}{n}$. The x-coordinate is the theoretical exponential cdf value

$$F(x_{(i)}) = 1 - \exp\left(-\frac{x_{(i)} - \theta}{\sigma}\right)$$

where

θ = threshold parameter

σ = scale parameter ($\sigma > 0$)

You can specify σ and θ with the SIGMA= and THETA= _exponential-options_, as illustrated in the following example:

```
proc univariate data=measures;
   ppplot width / exponential(theta=1 sigma=2);
run;
```

If you do not specify values for these parameters, then by default, $\theta = 0$ and a maximum likelihood estimate is calculated for σ.

IMPORTANT: Your data must be greater than or equal to the lower threshold θ. If the default $\theta = 0$ is not an adequate lower bound for your data, specify θ with the THETA= option.

If the data are exponentially distributed with parameters σ and θ, the points on the plot for SIGMA=σ and THETA=θ tend to fall on or near the diagonal line $y = x$, which is displayed by default. Agreement between the diagonal line and the point pattern is evidence that the specified exponential distribution is a good fit. You can specify the SCALE= option as an alias for the SIGMA= option and the THRESHOLD= option as an alias for the THETA= option.

GAMMA< (_gamma-options_) **>**

creates a gamma P-P plot. To create the plot, the n nonmissing observations are ordered from smallest to largest:

$$x_{(1)} \leq x_{(2)} \leq \cdots \leq x_{(n)}$$

The y-coordinate of the ith point is the empirical cdf value $\frac{i}{n}$. The x-coordinate is the theoretical gamma cdf value

$$G_\alpha\left(\frac{x_{(i)} - \theta}{\sigma}\right) = \int_\theta^{x_{(i)}} \frac{1}{\sigma\Gamma(\alpha)} \left(\frac{t - \theta}{\sigma}\right)^{\alpha-1} \exp\left(-\frac{t - \theta}{\sigma}\right) dt$$

where $G_\alpha(\cdot)$ is the normalized incomplete gamma function and

θ = threshold parameter

σ = scale parameter ($\sigma > 0$)

α = shape parameter ($\alpha > 0$)

You can specify α, σ, and θ with the ALPHA=, SIGMA=, and THETA= *gamma-options*, as illustrated in the following example:

```
proc univariate data=measures;
   ppplot width / gamma(alpha=1 sigma=2 theta=3);
run;
```

If you do not specify values for these parameters, then by default, $\theta = 0$ and maximum likelihood estimates are calculated for α and σ.

IMPORTANT: Your data must be greater than or equal to the lower threshold θ. If the default $\theta = 0$ is not an adequate lower bound for your data, specify θ with the THETA= option.

If the data are gamma distributed with parameters α, σ, and θ, the points on the plot for ALPHA=α, SIGMA=σ, and THETA=θ tend to fall on or near the diagonal line $y = x$, which is displayed by default. Agreement between the diagonal line and the point pattern is evidence that the specified gamma distribution is a good fit. You can specify the SHAPE= option as an alias for the ALPHA= option, the SCALE= option as an alias for the SIGMA= option, and the THRESHOLD= option as an alias for the THETA= option.

LOGNORMAL< (*lognormal-options*) >
LNORM< (*lognormal-options*) >
creates a lognormal P-P plot. To create the plot, the n nonmissing observations are ordered from smallest to largest:

$$x_{(1)} \leq x_{(2)} \leq \cdots \leq x_{(n)}$$

The y-coordinate of the ith point is the empirical cdf value $\frac{i}{n}$. The x-coordinate is the theoretical lognormal cdf value

$$\Phi\left(\frac{\log(x_{(i)} - \theta) - \zeta}{\sigma}\right)$$

where $\Phi(\cdot)$ is the cumulative standard normal distribution function and

θ = threshold parameter

ζ = scale parameter

σ = shape parameter ($\sigma > 0$)

You can specify θ, ζ, and σ with the THETA=, ZETA=, and SIGMA= *lognormal-options*, as illustrated in the following example:

```
proc univariate data=measures;
   ppplot width / lognormal(theta=1 zeta=2);
run;
```

If you do not specify values for these parameters, then by default, $\theta = 0$ and maximum likelihood estimates are calculated for σ and ζ.

IMPORTANT: Your data must be greater than the lower threshold θ. If the default $\theta = 0$ is not an adequate lower bound for your data, specify θ with the THETA= option.

If the data are lognormally distributed with parameters σ, θ, and ζ, the points on the plot for SIGMA=σ, THETA=θ, and ZETA=ζ tend to fall on or near the diagonal line $y = x$, which is displayed by default. Agreement between the diagonal line and the point pattern is evidence that the specified lognormal distribution is a good fit. You can specify the SHAPE= option as an alias for the SIGMA=option, the SCALE= option as an alias for the ZETA= option, and the THRESHOLD= option as an alias for the THETA= option.

MU=*value*

specifies the mean μ for a normal P-P plot requested with the NORMAL option. By default, the sample mean is used for μ. See Example 4.36.

NOLINE

suppresses the diagonal reference line.

NORMAL<(*normal-options***)>**

NORM<(*normal-options***)>**

creates a normal P-P plot. By default, if you do not specify a distribution option, the procedure displays a normal P-P plot. To create the plot, the n nonmissing observations are ordered from smallest to largest:

$$x_{(1)} \leq x_{(2)} \leq \cdots \leq x_{(n)}$$

The y-coordinate of the ith point is the empirical cdf value $\frac{i}{n}$. The x-coordinate is the theoretical normal cdf value

$$\Phi\left(\frac{x_{(i)} - \mu}{\sigma}\right) = \int_{-\infty}^{x_{(i)}} \frac{1}{\sigma\sqrt{2\pi}} \exp\left(-\frac{(t - \mu)^2}{2\sigma^2}\right) dt$$

where $\Phi(\cdot)$ is the cumulative standard normal distribution function and

μ = location parameter or mean

σ = scale parameter or standard deviation ($\sigma > 0$)

You can specify μ and σ with the MU= and SIGMA= *normal-options*, as illustrated in the following example:

```
proc univariate data=measures;
   ppplot width / normal(mu=1 sigma=2);
run;
```

By default, the sample mean and sample standard deviation are used for μ and σ.

If the data are normally distributed with parameters μ and σ, the points on the plot for MU=μ and SIGMA=σ tend to fall on or near the diagonal line $y = x$, which is displayed by default. Agreement between the diagonal line and the point pattern is evidence that the specified normal distribution is a good fit. See Example 4.36.

SIGMA=*value*

specifies the parameter σ, where $\sigma > 0$. When used with the BETA, EXPONENTIAL, GAMMA, NORMAL, and WEIBULL options, the SIGMA= option specifies the scale parameter. When used with the LOGNORMAL option, the SIGMA= option specifies the shape parameter. See Example 4.36.

SQUARE

displays the P-P plot in a square frame. The default is a rectangular frame. See Example 4.36.

THETA=*value*

THRESHOLD=*value*

specifies the lower threshold parameter θ for plots requested with the BETA, EXPONENTIAL, GAMMA, LOGNORMAL, and WEIBULL options.

WEIBULL<(*Weibull-options***)>**

WEIB<(*Weibull-options***)>**

creates a Weibull P-P plot. To create the plot, the n nonmissing observations are ordered from smallest to largest:

$$x_{(1)} \leq x_{(2)} \leq \cdots \leq x_{(n)}$$

The y-coordinate of the ith point is the empirical cdf value $\frac{i}{n}$. The x-coordinate is the theoretical Weibull cdf value

$$F(x_{(i)}) = 1 - \exp\left(-\left(\frac{x_{(i)} - \theta}{\sigma}\right)^c\right)$$

where

θ = threshold parameter

σ = scale parameter ($\sigma > 0$)

c = shape parameter ($c > 0$)

You can specify c, σ, and θ with the C=, SIGMA=, and THETA= *Weibull-options*, as illustrated in the following example:

```
proc univariate data=measures;
   ppplot width / weibull(theta=1 sigma=2);
run;
```

If you do not specify values for these parameters, then by default $\theta = 0$ and maximum likelihood estimates are calculated for σ and c.

IMPORTANT: Your data must be greater than or equal to the lower threshold θ. If the default $\theta = 0$ is not an adequate lower bound for your data, you should specify θ with the THETA= option.

If the data are Weibull distributed with parameters c, σ, and θ, the points on the plot for C=c, SIGMA=σ, and THETA=θ tend to fall on or near the diagonal line $y = x$, which is displayed by default. Agreement between the diagonal line and the point pattern is evidence that the specified Weibull distribution is a good fit. You can specify the SHAPE= option as an alias for the C= option, the SCALE= option as an alias for the SIGMA= option, and the THRESHOLD= option as an alias for the THETA= option.

ZETA=value

specifies a value for the scale parameter ζ for lognormal P-P plots requested with the LOG-NORMAL option.

PROBPLOT Statement

PROBPLOT < variables > < / options > ;

The PROBPLOT statement creates a probability plot, which compares ordered variable values with the percentiles of a specified theoretical distribution. If the data distribution matches the theoretical distribution, the points on the plot form a linear pattern. Consequently, you can use a probability plot to determine how well a theoretical distribution models a set of measurements.

Probability plots are similar to Q-Q plots, which you can create with the QQPLOT statement. Probability plots are preferable for graphical estimation of percentiles, whereas Q-Q plots are preferable for graphical estimation of distribution parameters.

You can use any number of PROBPLOT statements in the UNIVARIATE procedure. The components of the PROBPLOT statement are as follows.

variables

are the variables for which probability plots are created. If you specify a VAR statement, the *variables* must also be listed in the VAR statement. Otherwise, the *variables* can be any numeric variables in the input data set. If you do not specify a list of *variables*, then by default the procedure creates a probability plot for each variable listed in the VAR statement, or for each numeric variable in the DATA= data set if you do not specify a VAR statement. For example, each of the following PROBPLOT statements produces two probability plots, one for Length and one for Width:

```
proc univariate data=Measures;
   var Length Width;
   probplot;

proc univariate data=Measures;
   probplot Length Width;
run;
```

options

specify the theoretical distribution for the plot or add features to the plot. If you specify more than one variable, the *options* apply equally to each variable. Specify all *options* after the slash (/) in the PROBPLOT statement. You can specify only one *option* that names a distribution in each PROBPLOT statement, but you can specify any number of other *options*. The distributions available are the beta, exponential, gamma, lognormal, normal, two-parameter Weibull, and three-parameter Weibull. By default, the procedure produces a plot for the normal distribution.

In the following example, the NORMAL option requests a normal probability plot for each variable, while the MU= and SIGMA= *normal-options* request a distribution reference line corresponding to the normal distribution with $\mu = 10$ and $\sigma = 0.3$. The SQUARE option displays the plot in a square frame, and the CTEXT= option specifies the text color.

```
proc univariate data=Measures;
   probplot Length1 Length2 / normal(mu=10 sigma=0.3)
                              square ctext=blue;
run;
```

Table 4.56 through Table 4.65 list the PROBPLOT *options* by function. For complete descriptions, see the sections "Dictionary of Options" on page 300 and "Dictionary of Common Options" on page 317. *Options* can be any of the following:

- *primary options*
- *secondary options*
- *general options*

Distribution Options

Table 4.56 lists *options* for requesting a theoretical distribution.

Table 4.56 Primary Options for Theoretical Distributions

Option	Description
BETA(*beta-options*)	specifies beta probability plot for shape parameters α and β specified with mandatory ALPHA= and BETA= *beta-options*
EXPONENTIAL(*exponential-options*)	specifies exponential probability plot
GAMMA(*gamma-options*)	specifies gamma probability plot for shape parameter α specified with mandatory ALPHA= *gamma-option*
LOGNORMAL(*lognormal-options*)	specifies lognormal probability plot for shape parameter σ specified with mandatory SIGMA= *lognormal-option*
NORMAL(*normal-options*)	specifies normal probability plot
WEIBULL(*Weibull-options*)	specifies three-parameter Weibull probability plot for shape parameter c specified with mandatory C= *Weibull-option*
WEIBULL2(*Weibull2-options*)	specifies two-parameter Weibull probability plot

Table 4.57 through Table 4.64 list *secondary options* that specify distribution parameters and control the display of a distribution reference line. Specify these options in parentheses after the distribution keyword. For example, you can request a normal probability plot with a distribution reference line by specifying the NORMAL option as follows:

```
proc univariate;
   probplot Length / normal(mu=10 sigma=0.3 color=red);
run;
```

The MU= and SIGMA= *normal-options* display a distribution reference line that corresponds to the normal distribution with mean $\mu_0 = 10$ and standard deviation $\sigma_0 = 0.3$, and the COLOR= *normal-option* specifies the color for the line.

Table 4.57 Secondary Reference Line Options Used with All Distributions

Option	Description
COLOR=	specifies color of distribution reference line
L=	specifies line type of distribution reference line
W=	specifies width of distribution reference line

Table 4.58 Secondary Beta-Options

Option	Description
ALPHA=	specifies mandatory shape parameter α
BETA=	specifies mandatory shape parameter β
SIGMA=	specifies σ_0 for distribution reference line
THETA=	specifies θ_0 for distribution reference line

Table 4.59 Secondary Exponential-Options

Option	Description
SIGMA=	specifies σ_0 for distribution reference line
THETA=	specifies θ_0 for distribution reference line

Table 4.60 Secondary Gamma-Options

Option	Description
ALPHA=	specifies mandatory shape parameter α
ALPHADELTA=	specifies change in successive estimates of α at which the Newton-Raphson approximation of $\hat{\alpha}$ terminates
ALPHAINITIAL=	specifies initial value for α in the Newton-Raphson approximation of $\hat{\alpha}$
MAXITER=	specifies maximum number of iterations in the Newton-Raphson approximation of $\hat{\alpha}$
SIGMA=	specifies σ_0 for distribution reference line
THETA=	specifies θ_0 for distribution reference line

Table 4.61 Secondary Lognormal-Options

Option	Description
SIGMA=	specifies mandatory shape parameter σ
SLOPE=	specifies slope of distribution reference line
THETA=	specifies θ_0 for distribution reference line
ZETA=	specifies ζ_0 for distribution reference line (slope is $\exp(\zeta_0)$)

Table 4.62 Secondary Normal-Options

Option	Description
MU=	specifies μ_0 for distribution reference line
SIGMA=	specifies σ_0 for distribution reference line

Table 4.63 Secondary Weibull-Options

Option	Description
C=	specifies mandatory shape parameter c
CDELTA=	specifies change in successive estimates of c at which the Newton-Raphson approximation of $\hat{c}$ terminates
CINITIAL=	specifies initial value for c in the Newton-Raphson approximation of $\hat{c}$
MAXITER=	specifies maximum number of iterations in the Newton-Raphson approximation of $\hat{c}$
SIGMA=	specifies σ_0 for distribution reference line
THETA=	specifies θ_0 for distribution reference line

Table 4.64 Secondary Weibull2-Options

Option	Description
C=	specifies c_0 for distribution reference line (slope is $1/c_0$)
CDELTA=	specifies change in successive estimates of c at which the Newton-Raphson approximation of $\hat{c}$ terminates
CINITIAL=	specifies initial value for c in the Newton-Raphson approximation of $\hat{c}$
MAXITER=	specifies maximum number of iterations in the Newton-Raphson approximation of $\hat{c}$
SIGMA=	specifies σ_0 for distribution reference line (intercept is $\log(\sigma_0)$)
SLOPE=	specifies slope of distribution reference line
THETA=	specifies known lower threshold θ_0

General Graphics Options

Table 4.65 summarizes the general options for enhancing probability plots.

Table 4.65 General Graphics Options

Option	Description
ANNOKEY	applies annotation requested in ANNOTATE= data set to key cell only
ANNOTATE=	specifies annotate data set
CAXIS=	specifies color for axis
CFRAME=	specifies color for frame
CFRAMESIDE=	specifies color for filling frame for row labels
CFRAMETOP=	specifies color for filling frame for column labels
CGRID=	specifies color for grid lines
CHREF=	specifies color for HREF= lines
CONTENTS=	specifies table of contents entry for probability plot grouping
CPROP=	specifies color for proportion of frequency bar
CTEXT=	specifies color for text

Table 4.65 (continued)

Option	Description
CTEXTSIDE=	specifies color for row labels
CTEXTTOP=	specifies color for column labels
CVREF=	specifies color for VREF= lines
DESCRIPTION=	specifies description for plot in graphics catalog
FONT=	specifies software font for text
GRID	creates a grid
HAXIS=	specifies AXIS statement for horizontal axis
HEIGHT=	specifies height of text used outside framed areas
HMINOR=	specifies number of horizontal minor tick marks
HREF=	specifies reference lines perpendicular to the horizontal axis
HREFLABELS=	specifies labels for HREF= lines
HREFLABPOS=	specifies position for HREF= line labels
INFONT=	specifies software font for text inside framed areas
INHEIGHT=	specifies height of text inside framed areas
INTERTILE=	specifies distance between tiles
LGRID=	specifies a line type for grid lines
LHREF=	specifies line style for HREF= lines
LVREF=	specifies line style for VREF= lines
NADJ=	adjusts sample size when computing percentiles
NAME=	specifies name for plot in graphics catalog
NCOLS=	specifies number of columns in comparative probability plot
NOFRAME	suppresses frame around plotting area
NOHLABEL	suppresses label for horizontal axis
NOVLABEL	suppresses label for vertical axis
NOVTICK	suppresses tick marks and tick mark labels for vertical axis
NROWS=	specifies number of rows in comparative probability plot
OVERLAY	overlays plots for different class levels (ODS Graphics only)
PCTLMINOR	requests minor tick marks for percentile axis
PCTLORDER=	specifies tick mark labels for percentile axis
RANKADJ=	adjusts ranks when computing percentiles
ROTATE	switches horizontal and vertical axes
SQUARE	displays plot in square format
TURNVLABELS	turns and vertically strings out characters in labels for vertical axis
VAXIS=	specifies AXIS statement for vertical axis
VAXISLABEL=	specifies label for vertical axis
VMINOR=	specifies number of vertical minor tick marks
VREF=	specifies reference lines perpendicular to the vertical axis
VREFLABELS=	specifies labels for VREF= lines
VREFLABPOS=	specifies horizontal position of labels for VREF= lines
WAXIS=	specifies line thickness for axes and frame
WGRID=	specifies line thickness for grid

Dictionary of Options

The following entries provide detailed descriptions of *options* in the PROBPLOT statement. See the section "Dictionary of Common Options" on page 317 for detailed descriptions of options common to all plot statements.

ALPHA=*value-list* | **EST**

specifies the mandatory shape parameter α for probability plots requested with the BETA and GAMMA options. Enclose the ALPHA= option in parentheses after the BETA or GAMMA options. If you specify ALPHA=EST, a maximum likelihood estimate is computed for α.

BETA(ALPHA=*value* | **EST BETA=***value* | **EST** < *beta-options* >**)**

creates a beta probability plot for each combination of the required shape parameters α and β specified by the required ALPHA= and BETA= *beta-options*. If you specify ALPHA=EST and BETA=EST, the procedure creates a plot based on maximum likelihood estimates for α and β. You can specify the SCALE= *beta-option* as an alias for the SIGMA= *beta-option* and the THRESHOLD= *beta-option* as an alias for the THETA= *beta-option*. To create a plot that is based on maximum likelihood estimates for α and β, specify ALPHA=EST and BETA=EST.

To obtain graphical estimates of α and β, specify lists of values in the ALPHA= and BETA= *beta-options*, and select the combination of α and β that most nearly linearizes the point pattern. To assess the point pattern, you can add a diagonal distribution reference line corresponding to lower threshold parameter θ_0 and scale parameter σ_0 with the THETA= and SIGMA= *beta-options*. Alternatively, you can add a line that corresponds to estimated values of θ_0 and σ_0 with the *beta-options* THETA=EST and SIGMA=EST. Agreement between the reference line and the point pattern indicates that the beta distribution with parameters α, β, θ_0, and σ_0 is a good fit.

BETA=*value-list* | **EST**

B=*value-list* | **EST**

specifies the mandatory shape parameter β for probability plots requested with the BETA option. Enclose the BETA= option in parentheses after the BETA option. If you specify BETA=EST, a maximum likelihood estimate is computed for β.

C=*value-list* | **EST**

specifies the shape parameter c for probability plots requested with the WEIBULL and WEIBULL2 options. Enclose this option in parentheses after the WEIBULL or WEIBULL2 option. C= is a required *Weibull-option* in the WEIBULL option; in this situation, it accepts a list of values, or if you specify C=EST, a maximum likelihood estimate is computed for c. You can optionally specify C=*value* or C=EST as a *Weibull2-option* with the WEIBULL2 option to request a distribution reference line; in this situation, you must also specify *Weibull2-option* SIGMA=*value* or SIGMA=EST.

CGRID=*color*

specifies the color for grid lines when a grid displays on the plot. This option also produces a grid.

EXPONENTIAL<(*exponential-options***)>**

EXP<(*exponential-options***)>**

creates an exponential probability plot. To assess the point pattern, add a diagonal distribution reference line corresponding to θ_0 and σ_0 with the THETA= and SIGMA= *exponential-options*. Alternatively, you can add a line corresponding to estimated values of the threshold parameter θ_0 and the scale parameter σ with the *exponential-options* THETA=EST and SIGMA=EST. Agreement between the reference line and the point pattern indicates that the exponential distribution with parameters θ_0 and σ_0 is a good fit. You can specify the SCALE= *exponential-option* as an alias for the SIGMA= *exponential-option* and the THRESHOLD= *exponential-option* as an alias for the THETA= *exponential-option*.

GAMMA(ALPHA=*value* **| EST <** *gamma-options* **>)**

creates a gamma probability plot for each value of the shape parameter α given by the mandatory ALPHA= *gamma-option*. If you specify ALPHA=EST, the procedure creates a plot based on a maximum likelihood estimate for α. To obtain a graphical estimate of α, specify a list of values for the ALPHA= *gamma-option* and select the value that most nearly linearizes the point pattern. To assess the point pattern, add a diagonal distribution reference line corresponding to θ_0 and σ_0 with the THETA= and SIGMA= *gamma-options*. Alternatively, you can add a line corresponding to estimated values of the threshold parameter θ_0 and the scale parameter σ with the *gamma-options* THETA=EST and SIGMA=EST. Agreement between the reference line and the point pattern indicates that the gamma distribution with parameters α, θ_0, and σ_0 is a good fit. You can specify the SCALE= *gamma-option* as an alias for the SIGMA= *gamma-option* and the THRESHOLD= *gamma-option* as an alias for the THETA= *gamma-option*.

GRID

displays a grid. Grid lines are reference lines that are perpendicular to the percentile axis at major tick marks.

LGRID=*linetype*

specifies the line type for the grid requested by the GRID= option. By default, LGRID=1, which produces a solid line.

LOGNORMAL(SIGMA=*value* **| EST <** *lognormal-options* **>)**

LNORM(SIGMA=*value* **| EST <** *lognormal-options* **>)**

creates a lognormal probability plot for each value of the shape parameter σ given by the mandatory SIGMA= *lognormal-option*. If you specify SIGMA=EST, the procedure creates a plot based on a maximum likelihood estimate for σ. To obtain a graphical estimate of σ, specify a list of values for the SIGMA= *lognormal-option* and select the value that most nearly linearizes the point pattern. To assess the point pattern, add a diagonal distribution reference line corresponding to θ_0 and ζ_0 with the THETA= and ZETA= *lognormal-options*. Alternatively, you can add a line corresponding to estimated values of the threshold parameter θ_0 and the scale parameter ζ_0 with the *lognormal-options* THETA=EST and ZETA=EST. Agreement between the reference line and the point pattern indicates that the lognormal distribution with parameters σ, θ_0, and ζ_0 is a good fit. You can specify the THRESHOLD= *lognormal-option* as an alias for the THETA= *lognormal-option* and the SCALE= *lognormal-option* as an alias for the ZETA= *lognormal-option*. See Example 4.26.

MU=value | **EST**

specifies the mean μ_0 for a normal probability plot requested with the NORMAL option. Enclose the MU= *normal-option* in parentheses after the NORMAL option. The MU= *normal-option* must be specified with the SIGMA= *normal-option*, and they request a distribution reference line. You can specify MU=EST to request a distribution reference line with μ_0 equal to the sample mean.

NADJ=value

specifies the adjustment value added to the sample size in the calculation of theoretical percentiles. By default, NADJ=$\frac{1}{4}$. Refer to Chambers et al. (1983).

NORMAL< (*normal-options*) **>**

creates a normal probability plot. This is the default if you omit a distribution option. To assess the point pattern, you can add a diagonal distribution reference line corresponding to μ_0 and σ_0 with the MU= and SIGMA= *normal-options*. Alternatively, you can add a line corresponding to estimated values of μ_0 and σ_0 with the *normal-options* MU=EST and SIGMA=EST; the estimates of the mean μ_0 and the standard deviation σ_0 are the sample mean and sample standard deviation. Agreement between the reference line and the point pattern indicates that the normal distribution with parameters μ_0 and σ_0 is a good fit.

PCTLMINOR

requests minor tick marks for the percentile axis. The HMINOR option overrides the minor tick marks requested by the PCTLMINOR option.

PCTLORDER=values

specifies the tick marks that are labeled on the theoretical percentile axis. Because the values are percentiles, the labels must be between 0 and 100, exclusive. The values must be listed in increasing order and must cover the plotted percentile range. Otherwise, the default values of 1, 5, 10, 25, 50, 75, 90, 95, and 99 are used.

RANKADJ=value

specifies the adjustment value added to the ranks in the calculation of theoretical percentiles. By default, RANKADJ=$-\frac{3}{8}$, as recommended by Blom (1958). Refer to Chambers et al. (1983) for additional information.

ROTATE

switches the horizontal and vertical axes so that the theoretical percentiles are plotted vertically while the data are plotted horizontally. Regardless of whether the plot has been rotated, horizontal axis options (such as HAXIS=) still refer to the horizontal axis, and vertical axis options (such as VAXIS=) still refer to the vertical axis. All other options that depend on axis placement adjust to the rotated axes.

SIGMA=value-list | **EST**

specifies the parameter σ, where $\sigma > 0$. Alternatively, you can specify SIGMA=EST to request a maximum likelihood estimate for σ_0. The interpretation and use of the SIGMA= option depend on the distribution option with which it is used. See Table 4.66 for a summary of how to use the SIGMA= option. You must enclose this option in parentheses after the distribution option.

Table 4.66 Uses of the SIGMA= Option

Distribution Option	Use of the SIGMA= Option
BETA EXPONENTIAL GAMMA WEIBULL	THETA=θ_0 and SIGMA=σ_0 request a distribution reference line corresponding to θ_0 and σ_0.
LOGNORMAL	SIGMA=$\sigma_1 \ldots \sigma_n$ requests n probability plots with shape parameters $\sigma_1 \ldots \sigma_n$. The SIGMA= option must be specified.
NORMAL	MU=μ_0 and SIGMA=σ_0 request a distribution reference line corresponding to μ_0 and σ_0. SIGMA=EST requests a line with σ_0 equal to the sample standard deviation.
WEIBULL2	SIGMA=σ_0 and C=c_0 request a distribution reference line corresponding to σ_0 and c_0.

SLOPE=value | **EST**

specifies the slope for a distribution reference line requested with the LOGNORMAL and WEIBULL2 options. Enclose the SLOPE= option in parentheses after the distribution option. When you use the SLOPE= *lognormal-option* with the LOGNORMAL option, you must also specify a threshold parameter value θ_0 with the THETA= *lognormal-option* to request the line. The SLOPE= *lognormal-option* is an alternative to the ZETA= *lognormal-option* for specifying ζ_0, because the slope is equal to $\exp(\zeta_0)$.

When you use the SLOPE= *Weibull2-option* with the WEIBULL2 option, you must also specify a scale parameter value σ_0 with the SIGMA= *Weibull2-option* to request the line. The SLOPE= *Weibull2-option* is an alternative to the C= *Weibull2-option* for specifying c_0, because the slope is equal to $\frac{1}{c_0}$.

For example, the first and second PROBPLOT statements produce the same probability plots and the third and fourth PROBPLOT statements produce the same probability plots:

```
proc univariate data=Measures;
   probplot Width / lognormal(sigma=2 theta=0 zeta=0);
   probplot Width / lognormal(sigma=2 theta=0 slope=1);
   probplot Width / weibull2(sigma=2 theta=0 c=.25);
   probplot Width / weibull2(sigma=2 theta=0 slope=4);
run;
```

SQUARE

displays the probability plot in a square frame. By default, the plot is in a rectangular frame.

THETA=value | **EST**

THRESHOLD=value | **EST**

specifies the lower threshold parameter θ for plots requested with the BETA, EXPONENTIAL, GAMMA, LOGNORMAL, WEIBULL, and WEIBULL2 options. Enclose the THETA= option in parentheses after a distribution option. When used with the WEIBULL2 option, the THETA= option specifies the known lower threshold θ_0, for which the default is 0. When used with the other distribution options, the THETA= option specifies θ_0 for a distribution reference line; alternatively in this situation, you can specify THETA=EST to request a maximum likelihood estimate for θ_0. To request the line, you must also specify a scale parameter.

WEIBULL(C=*value* **| EST** < *Weibull-options* >**)**

WEIB(C=*value* **| EST** < *Weibull-options* >**)**

creates a three-parameter Weibull probability plot for each value of the required shape parameter c specified by the mandatory C= *Weibull-option*. To create a plot that is based on a maximum likelihood estimate for c, specify C=EST. To obtain a graphical estimate of c, specify a list of values in the C= *Weibull-option* and select the value that most nearly linearizes the point pattern. To assess the point pattern, add a diagonal distribution reference line corresponding to θ_0 and σ_0 with the THETA= and SIGMA= *Weibull-options*. Alternatively, you can add a line corresponding to estimated values of θ_0 and σ_0 with the *Weibull-options* THETA=EST and SIGMA=EST. Agreement between the reference line and the point pattern indicates that the Weibull distribution with parameters c, θ_0, and σ_0 is a good fit. You can specify the SCALE= *Weibull-option* as an alias for the SIGMA= *Weibull-option* and the THRESHOLD= *Weibull-option* as an alias for the THETA= *Weibull-option*.

WEIBULL2<(*Weibull2-options***) >**

W2<(*Weibull2-options***) >**

creates a two-parameter Weibull probability plot. You should use the WEIBULL2 option when your data have a *known* lower threshold θ_0, which is 0 by default. To specify the threshold value θ_0, use the THETA= *Weibull2-option*. By default, THETA=0. An advantage of the two-parameter Weibull plot over the three-parameter Weibull plot is that the parameters c and σ can be estimated from the slope and intercept of the point pattern. A disadvantage is that the two-parameter Weibull distribution applies only in situations where the threshold parameter is known. To obtain a graphical estimate of θ_0, specify a list of values for the THETA= *Weibull2-option* and select the value that most nearly linearizes the point pattern. To assess the point pattern, add a diagonal distribution reference line corresponding to σ_0 and c_0 with the SIGMA= and C= *Weibull2-options*. Alternatively, you can add a distribution reference line corresponding to estimated values of σ_0 and c_0 with the *Weibull2-options* SIGMA=EST and C=EST. Agreement between the reference line and the point pattern indicates that the Weibull distribution with parameters c_0, θ_0, and σ_0 is a good fit. You can specify the SCALE= *Weibull2-option* as an alias for the SIGMA= *Weibull2-option* and the SHAPE= *Weibull2-option* as an alias for the C= *Weibull2-option*.

WGRID=*n*

specifies the line thickness for the grid when producing traditional graphics. The option does not apply to ODS Graphics output.

ZETA=*value* **| EST**

specifies a value for the scale parameter ζ for the lognormal probability plots requested with the LOGNORMAL option. Enclose the ZETA= *lognormal-option* in parentheses after the LOGNORMAL option. To request a distribution reference line with intercept θ_0 and slope $\exp(\zeta_0)$, specify the THETA=θ_0 and ZETA=ζ_0.

QQPLOT Statement

QQPLOT < *variables* > < / *options* > ;

The QQPLOT statement creates quantile-quantile plots (Q-Q plots) and compares ordered variable values with quantiles of a specified theoretical distribution. If the data distribution matches the theoretical distribution, the points on the plot form a linear pattern. Thus, you can use a Q-Q plot to determine how well a theoretical distribution models a set of measurements.

Q-Q plots are similar to probability plots, which you can create with the PROBPLOT statement. Q-Q plots are preferable for graphical estimation of distribution parameters, whereas probability plots are preferable for graphical estimation of percentiles.

You can use any number of QQPLOT statements in the UNIVARIATE procedure. The components of the QQPLOT statement are as follows.

variables

are the variables for which Q-Q plots are created. If you specify a VAR statement, the *variables* must also be listed in the VAR statement. Otherwise, the *variables* can be any numeric variables in the input data set. If you do not specify a list of *variables*, then by default the procedure creates a Q-Q plot for each variable listed in the VAR statement, or for each numeric variable in the DATA= data set if you do not specify a VAR statement. For example, each of the following QQPLOT statements produces two Q-Q plots, one for Length and one for Width:

```
proc univariate data=Measures;
   var Length Width;
   qqplot;

proc univariate data=Measures;
   qqplot Length Width;
run;
```

options

specify the theoretical distribution for the plot or add features to the plot. If you specify more than one variable, the *options* apply equally to each variable. Specify all *options* after the slash (/) in the QQPLOT statement. You can specify only one *option* that names the distribution in each QQPLOT statement, but you can specify any number of other *options*. The distributions available are the beta, exponential, gamma, lognormal, normal, two-parameter Weibull, and three-parameter Weibull. By default, the procedure produces a plot for the normal distribution.

In the following example, the NORMAL option requests a normal Q-Q plot for each variable. The MU= and SIGMA= *normal-options* request a distribution reference line with intercept 10 and slope 0.3 for each plot, corresponding to a normal distribution with mean $\mu = 10$ and standard deviation $\sigma = 0.3$. The SQUARE option displays the plot in a square frame, and the CTEXT= option specifies the text color.

```
proc univariate data=measures;
   qqplot length1 length2 / normal(mu=10 sigma=0.3)
                            square ctext=blue;
run;
```

Table 4.67 through Table 4.76 list the QQPLOT *options* by function. For complete descriptions, see the sections "Dictionary of Options" on page 310 and "Dictionary of Common Options" on page 317.

Options can be any of the following:

- *primary options*
- *secondary options*
- *general options*

Distribution Options

Table 4.67 lists *primary options* for requesting a theoretical distribution. See the section "Distributions for Probability and Q-Q Plots" on page 364 for detailed descriptions of these distributions.

Table 4.67 Primary Options for Theoretical Distributions

Option	Description
BETA(*beta-options*)	specifies beta Q-Q plot for shape parameters α and β specified with mandatory ALPHA= and BETA= *beta-options*
EXPONENTIAL(*exponential-options*)	specifies exponential Q-Q plot
GAMMA(*gamma-options*)	specifies gamma Q-Q plot for shape parameter α specified with mandatory ALPHA= *gamma-option*
LOGNORMAL(*lognormal-options*)	specifies lognormal Q-Q plot for shape parameter σ specified with mandatory SIGMA= *lognormal-option*
NORMAL(*normal-options*)	specifies normal Q-Q plot
WEIBULL(*Weibull-options*)	specifies three-parameter Weibull Q-Q plot for shape parameter c specified with mandatory C= *Weibull-option*
WEIBULL2(*Weibull2-options*)	specifies two-parameter Weibull Q-Q plot

Table 4.68 through Table 4.75 list *secondary options* that specify distribution parameters and control the display of a distribution reference line. Specify these options in parentheses after the distribution keyword. For example, you can request a normal Q-Q plot with a distribution reference line by specifying the NORMAL option as follows:

```
proc univariate;
   qqplot Length / normal(mu=10 sigma=0.3 color=red);
run;
```

The MU= and SIGMA= *normal-options* display a distribution reference line that corresponds to the normal distribution with mean $\mu_0 = 10$ and standard deviation $\sigma_0 = 0.3$, and the COLOR= *normal-option* specifies the color for the line.

Table 4.68   Secondary Reference Line Options Used with All Distributions

Option	Description
COLOR=	specifies color of distribution reference line
L=	specifies line type of distribution reference line
W=	specifies width of distribution reference line

Table 4.69   Secondary Beta-Options

Option	Description
ALPHA=	specifies mandatory shape parameter α
BETA=	specifies mandatory shape parameter β
SIGMA=	specifies σ_0 for distribution reference line
THETA=	specifies θ_0 for distribution reference line

Table 4.70   Secondary Exponential-Options

Option	Description
SIGMA=	specifies σ_0 for distribution reference line
THETA=	specifies θ_0 for distribution reference line

Table 4.71   Secondary Gamma-Options

Option	Description
ALPHA=	specifies mandatory shape parameter α
ALPHADELTA=	specifies change in successive estimates of α at which the Newton-Raphson approximation of $\hat{\alpha}$ terminates
ALPHAINITIAL=	specifies initial value for α in the Newton-Raphson approximation of $\hat{\alpha}$
MAXITER=	specifies maximum number of iterations in the Newton-Raphson approximation of $\hat{\alpha}$
SIGMA=	specifies σ_0 for distribution reference line
THETA=	specifies θ_0 for distribution reference line

Table 4.72 Secondary Lognormal-Options

Option	Description
SIGMA=	specifies mandatory shape parameter σ
SLOPE=	specifies slope of distribution reference line
THETA=	specifies θ_0 for distribution reference line
ZETA=	specifies ζ_0 for distribution reference line (slope is $\exp(\zeta_0)$)

Table 4.73 Secondary Normal-Options

Option	Description
MU=	specifies μ_0 for distribution reference line
SIGMA=	specifies σ_0 for distribution reference line

Table 4.74 Secondary Weibull-Options

Option	Description
C=	specifies mandatory shape parameter c
SIGMA=	specifies σ_0 for distribution reference line
THETA=	specifies θ_0 for distribution reference line

Table 4.75 Secondary Weibull2-Options

Option	Description
C=	specifies c_0 for distribution reference line (slope is $1/c_0$)
SIGMA=	specifies σ_0 for distribution reference line (intercept is $\log(\sigma_0)$)
SLOPE=	specifies slope of distribution reference line
THETA=	specifies known lower threshold θ_0

General Options

Table 4.76 summarizes *general options* for enhancing Q-Q plots.

Table 4.76 General Graphics Options

Option	Description
ANNOKEY	applies annotation requested in ANNOTATE= data set to key cell only
ANNOTATE=	specifies annotate data set
CAXIS=	specifies color for axis
CFRAME=	specifies color for frame
CFRAMESIDE=	specifies color for filling frame for row labels

Table 4.76 (continued)

Option	Description
CFRAMETOP=	specifies color for filling frame for column labels
CGRID=	specifies color for grid lines
CHREF=	specifies color for HREF= lines
CONTENTS=	specifies table of contents entry for Q-Q plot grouping
CTEXT=	specifies color for text
CVREF=	specifies color for VREF= lines
DESCRIPTION=	specifies description for plot in graphics catalog
FONT=	specifies software font for text
GRID	creates a grid
HEIGHT=	specifies height of text used outside framed areas
HMINOR=	specifies number of horizontal minor tick marks
HREF=	specifies reference lines perpendicular to the horizontal axis
HREFLABELS=	specifies labels for HREF= lines
HREFLABPOS=	specifies vertical position of labels for HREF= lines
INFONT=	specifies software font for text inside framed areas
INHEIGHT=	specifies height of text inside framed areas
INTERTILE=	specifies distance between tiles
LGRID=	specifies a line type for grid lines
LHREF=	specifies line style for HREF= lines
LVREF=	specifies line style for VREF= lines
NADJ=	adjusts sample size when computing percentiles
NAME=	specifies name for plot in graphics catalog
NCOLS=	specifies number of columns in comparative Q-Q plot
NOFRAME	suppresses frame around plotting area
NOHLABEL	suppresses label for horizontal axis
NOVLABEL	suppresses label for vertical axis
NOVTICK	suppresses tick marks and tick mark labels for vertical axis
NROWS=	specifies number of rows in comparative Q-Q plot
PCTLAXIS	displays a nonlinear percentile axis
PCTLMINOR	requests minor tick marks for percentile axis
PCTLSCALE	replaces theoretical quantiles with percentiles
RANKADJ=	adjusts ranks when computing percentiles
ROTATE	switches horizontal and vertical axes
SQUARE	displays plot in square format
VAXIS=	specifies AXIS statement for vertical axis
VAXISLABEL=	specifies label for vertical axis
VMINOR=	specifies number of vertical minor tick marks
VREF=	specifies reference lines perpendicular to the vertical axis
VREFLABELS=	specifies labels for VREF= lines
VREFLABPOS=	specifies horizontal position of labels for VREF= lines
WAXIS=	specifies line thickness for axes and frame
WGRID=	specifies line thickness for grid

Dictionary of Options

The following entries provide detailed descriptions of *options* in the QQPLOT statement. See the section "Dictionary of Common Options" on page 317 for detailed descriptions of options common to all plot statements.

ALPHA=*value-list* **| EST**

specifies the mandatory shape parameter α for quantile plots requested with the BETA and GAMMA options. Enclose the ALPHA= option in parentheses after the BETA or GAMMA options. If you specify ALPHA=EST, a maximum likelihood estimate is computed for α.

BETA(ALPHA=*value* **| EST BETA=***value* **| EST** < *beta-options* >**)**

creates a beta quantile plot for each combination of the required shape parameters α and β specified by the required ALPHA= and BETA= *beta-options*. If you specify ALPHA=EST and BETA=EST, the procedure creates a plot based on maximum likelihood estimates for α and β. You can specify the SCALE= *beta-option* as an alias for the SIGMA= *beta-option* and the THRESHOLD= *beta-option* as an alias for the THETA= *beta-option*. To create a plot that is based on maximum likelihood estimates for α and β, specify ALPHA=EST and BETA=EST. See the section "Beta Distribution" on page 365 for details.

To obtain graphical estimates of α and β, specify lists of values in the ALPHA= and BETA= *beta-options* and select the combination of α and β that most nearly linearizes the point pattern. To assess the point pattern, you can add a diagonal distribution reference line corresponding to lower threshold parameter θ_0 and scale parameter σ_0 with the THETA= and SIGMA= *beta-options*. Alternatively, you can add a line that corresponds to estimated values of θ_0 and σ_0 with the *beta-options* THETA=EST and SIGMA=EST. Agreement between the reference line and the point pattern indicates that the beta distribution with parameters α, β, θ_0, and σ_0 is a good fit.

BETA=*value-list* **| EST**

B=*value* **| EST**

specifies the mandatory shape parameter β for quantile plots requested with the BETA option. Enclose the BETA= option in parentheses after the BETA option. If you specify BETA=EST, a maximum likelihood estimate is computed for β.

C=*value-list* **| EST**

specifies the shape parameter c for quantile plots requested with the WEIBULL and WEIBULL2 options. Enclose this option in parentheses after the WEIBULL or WEIBULL2 option. C= is a required *Weibull-option* in the WEIBULL option; in this situation, it accepts a list of values, or if you specify C=EST, a maximum likelihood estimate is computed for c. You can optionally specify C=*value* or C=EST as a *Weibull2-option* with the WEIBULL2 option to request a distribution reference line; in this situation, you must also specify *Weibull2-option* SIGMA=*value* or SIGMA=EST.

CGRID=*color*

specifies the color for grid lines when a grid displays on the plot. This option also produces a grid.

EXPONENTIAL< (*exponential-options* **) >**

EXP< (*exponential-options* **) >**

creates an exponential quantile plot. To assess the point pattern, add a diagonal distribution reference line corresponding to θ_0 and σ_0 with the THETA= and SIGMA= *exponential-options*. Alternatively, you can add a line corresponding to estimated values of the threshold parameter θ_0 and the scale parameter σ with the *exponential-options* THETA=EST and SIGMA=EST. Agreement between the reference line and the point pattern indicates that the exponential distribution with parameters θ_0 and σ_0 is a good fit. You can specify the SCALE= *exponential-option* as an alias for the SIGMA= *exponential-option* and the THRESHOLD= *exponential-option* as an alias for the THETA= *exponential-option*. See the section "Exponential Distribution" on page 365 for details.

GAMMA(ALPHA=*value* **| EST <** *gamma-options* **>)**

creates a gamma quantile plot for each value of the shape parameter α given by the mandatory ALPHA= *gamma-option*. If you specify ALPHA=EST, the procedure creates a plot based on a maximum likelihood estimate for α. To obtain a graphical estimate of α, specify a list of values for the ALPHA= *gamma-option* and select the value that most nearly linearizes the point pattern. To assess the point pattern, add a diagonal distribution reference line corresponding to θ_0 and σ_0 with the THETA= and SIGMA= *gamma-options*. Alternatively, you can add a line corresponding to estimated values of the threshold parameter θ_0 and the scale parameter σ with the *gamma-options* THETA=EST and SIGMA=EST. Agreement between the reference line and the point pattern indicates that the gamma distribution with parameters α, θ_0, and σ_0 is a good fit. You can specify the SCALE= *gamma-option* as an alias for the SIGMA= *gamma-option* and the THRESHOLD= *gamma-option* as an alias for the THETA= *gamma-option*. See the section "Gamma Distribution" on page 365 for details.

GRID

displays a grid of horizontal lines positioned at major tick marks on the vertical axis.

LGRID=*linetype*

specifies the line type for the grid requested by the GRID option. By default, LGRID=1, which produces a solid line. The LGRID= option also produces a grid.

LOGNORMAL(SIGMA=*value* **| EST <** *lognormal-options* **>)**

LNORM(SIGMA=*value* **| EST <** *lognormal-options* **>)**

creates a lognormal quantile plot for each value of the shape parameter σ given by the mandatory SIGMA= *lognormal-option*. If you specify SIGMA=EST, the procedure creates a plot based on a maximum likelihood estimate for σ. To obtain a graphical estimate of σ, specify a list of values for the SIGMA= *lognormal-option* and select the value that most nearly linearizes the point pattern. To assess the point pattern, add a diagonal distribution reference line corresponding to θ_0 and ζ_0 with the THETA= and ZETA= *lognormal-options*. Alternatively, you can add a line corresponding to estimated values of the threshold parameter θ_0 and the scale parameter ζ_0 with the *lognormal-options* THETA=EST and ZETA=EST. Agreement between the reference line and the point pattern indicates that the lognormal distribution with parameters σ, θ_0, and ζ_0 is a good fit. You can specify the THRESHOLD= *lognormal-option* as an alias for the THETA= *lognormal-option* and the SCALE= *lognormal-option* as an alias for the ZETA= *lognormal-option*. See the section "Lognormal Distribution" on page 366 for details, and see Example 4.31 through Example 4.33 for examples that use the LOGNORMAL option.

MU=value | **EST**

specifies the mean μ_0 for a normal quantile plot requested with the NORMAL option. Enclose the MU= *normal-option* in parentheses after the NORMAL option. The MU= *normal-option* must be specified with the SIGMA= *normal-option*, and they request a distribution reference line. You can specify MU=EST to request a distribution reference line with μ_0 equal to the sample mean.

NADJ=value

specifies the adjustment value added to the sample size in the calculation of theoretical percentiles. By default, NADJ=$\frac{1}{4}$. Refer to Chambers et al. (1983) for additional information.

NORMAL< *(normal-options)* **>**

creates a normal quantile plot. This is the default if you omit a distribution option. To assess the point pattern, you can add a diagonal distribution reference line corresponding to μ_0 and σ_0 with the MU= and SIGMA= *normal-options*. Alternatively, you can add a line corresponding to estimated values of μ_0 and σ_0 with the *normal-options* MU=EST and SIGMA=EST; the estimates of the mean μ_0 and the standard deviation σ_0 are the sample mean and sample standard deviation. Agreement between the reference line and the point pattern indicates that the normal distribution with parameters μ_0 and σ_0 is a good fit. See the section "Normal Distribution" on page 366 for details, and see Example 4.28 and Example 4.30 for examples that use the NORMAL option.

PCTLAXIS< *(axis-options)* **>**

adds a nonlinear percentile axis along the frame of the Q-Q plot opposite the theoretical quantile axis. The added axis is identical to the axis for probability plots produced with the PROBPLOT statement. When using the PCTLAXIS option, you must specify HREF= values in quantile units, and you cannot use the NOFRAME option. You can specify the following *axis-options*:

Table 4.77 PCTLAXIS Axis Options

Option	Description
CGRID=	specifies color for grid lines
GRID	draws grid lines at major percentiles
LABEL='*string*'	specifies label for percentile axis
LGRID=*linetype*	specifies line type for grid
WGRID=*n*	specifies line thickness for grid

PCTLMINOR

requests minor tick marks for the percentile axis when you specify PCTLAXIS. The HMINOR option overrides the PCTLMINOR option.

PCTLSCALE

requests scale labels for the theoretical quantile axis in percentile units, resulting in a nonlinear axis scale. Tick marks are drawn uniformly across the axis based on the quantile scale. In all other respects, the plot remains the same, and you must specify HREF= values in quantile units. For a true nonlinear axis, use the PCTLAXIS option or use the PROBPLOT statement.

RANKADJ=value

specifies the adjustment value added to the ranks in the calculation of theoretical percentiles. By default, RANKADJ=$-\frac{3}{8}$, as recommended by Blom (1958). Refer to Chambers et al. (1983) for additional information.

ROTATE

switches the horizontal and vertical axes so that the theoretical quantiles are plotted vertically while the data are plotted horizontally. Regardless of whether the plot has been rotated, horizontal axis options (such as HAXIS=) still refer to the horizontal axis, and vertical axis options (such as VAXIS=) still refer to the vertical axis. All other options that depend on axis placement adjust to the rotated axes.

SIGMA=value | **EST**

specifies the parameter σ, where $\sigma > 0$. Alternatively, you can specify SIGMA=EST to request a maximum likelihood estimate for σ_0. The interpretation and use of the SIGMA= option depend on the distribution option with which it is used, as summarized in Table 4.78. Enclose this option in parentheses after the distribution option.

Table 4.78 Uses of the SIGMA= Option

Distribution Option	Use of the SIGMA= Option
BETA EXPONENTIAL GAMMA WEIBULL	THETA=θ_0 and SIGMA=σ_0 request a distribution reference line corresponding to θ_0 and σ_0.
LOGNORMAL	SIGMA=$\sigma_1 \ldots \sigma_n$ requests n quantile plots with shape parameters $\sigma_1 \ldots \sigma_n$. The SIGMA= option must be specified.
NORMAL	MU=μ_0 and SIGMA=σ_0 request a distribution reference line corresponding to μ_0 and σ_0. SIGMA=EST requests a line with σ_0 equal to the sample standard deviation.
WEIBULL2	SIGMA=σ_0 and C=c_0 request a distribution reference line corresponding to σ_0 and c_0.

SLOPE=value | **EST**

specifies the slope for a distribution reference line requested with the LOGNORMAL and WEIBULL2 options. Enclose the SLOPE= option in parentheses after the distribution option. When you use the SLOPE= *lognormal-option* with the LOGNORMAL option, you must also specify a threshold parameter value θ_0 with the THETA= *lognormal-option* to request the line. The SLOPE= *lognormal-option* is an alternative to the ZETA= *lognormal-option* for specifying ζ_0, because the slope is equal to $\exp(\zeta_0)$.

When you use the SLOPE= *Weibull2-option* with the WEIBULL2 option, you must also specify a scale parameter value σ_0 with the SIGMA= *Weibull2-option* to request the line. The SLOPE= *Weibull2-option* is an alternative to the C= *Weibull2-option* for specifying c_0, because the slope is equal to $\frac{1}{c_0}$.

For example, the first and second QQPLOT statements produce the same quantile plots and the third and fourth QQPLOT statements produce the same quantile plots:

```
proc univariate data=Measures;
   qqplot Width / lognormal(sigma=2 theta=0 zeta=0);
   qqplot Width / lognormal(sigma=2 theta=0 slope=1);
   qqplot Width / weibull2(sigma=2 theta=0 c=.25);
   qqplot Width / weibull2(sigma=2 theta=0 slope=4);
```

SQUARE

displays the quantile plot in a square frame. By default, the frame is rectangular.

THETA=_value_ **| EST**

THRESHOLD=_value_ **| EST**

specifies the lower threshold parameter θ for plots requested with the BETA, EXPONENTIAL, GAMMA, LOGNORMAL, WEIBULL, and WEIBULL2 options. Enclose the THETA= option in parentheses after a distribution option. When used with the WEIBULL2 option, the THETA= option specifies the known lower threshold θ_0, for which the default is 0. When used with the other distribution options, the THETA= option specifies θ_0 for a distribution reference line; alternatively in this situation, you can specify THETA=EST to request a maximum likelihood estimate for θ_0. To request the line, you must also specify a scale parameter.

WEIBULL(C=_value_ **| EST < **_Weibull-options_** >)**

WEIB(C=_value_ **| EST < **_Weibull-options_** >)**

creates a three-parameter Weibull quantile plot for each value of the required shape parameter c specified by the mandatory C= _Weibull-option_. To create a plot that is based on a maximum likelihood estimate for c, specify C=EST. To obtain a graphical estimate of c, specify a list of values in the C= _Weibull-option_ and select the value that most nearly linearizes the point pattern. To assess the point pattern, add a diagonal distribution reference line corresponding to θ_0 and σ_0 with the THETA= and SIGMA= _Weibull-options_. Alternatively, you can add a line corresponding to estimated values of θ_0 and σ_0 with the _Weibull-options_ THETA=EST and SIGMA=EST. Agreement between the reference line and the point pattern indicates that the Weibull distribution with parameters c, θ_0, and σ_0 is a good fit. You can specify the SCALE= _Weibull-option_ as an alias for the SIGMA= _Weibull-option_ and the THRESHOLD= _Weibull-option_ as an alias for the THETA= _Weibull-option_. See Example 4.34.

WEIBULL2< (_Weibull2-options_**) >**

W2< (_Weibull2-options_**) >**

creates a two-parameter Weibull quantile plot. You should use the WEIBULL2 option when your data have a _known_ lower threshold θ_0, which is 0 by default. To specify the threshold value θ_0, use the THETA= _Weibull2-option_. By default, THETA=0. An advantage of the two-parameter Weibull plot over the three-parameter Weibull plot is that the parameters c and σ can be estimated from the slope and intercept of the point pattern. A disadvantage is that the two-parameter Weibull distribution applies only in situations where the threshold parameter is known. To obtain a graphical estimate of θ_0, specify a list of values for the THETA= _Weibull2-option_ and select the value that most nearly linearizes the point pattern. To assess the point pattern, add a diagonal distribution reference line corresponding to σ_0

and c_0 with the SIGMA= and C= *Weibull2-options*. Alternatively, you can add a distribution reference line corresponding to estimated values of σ_0 and c_0 with the *Weibull2-options* SIGMA=EST and C=EST. Agreement between the reference line and the point pattern indicates that the Weibull distribution with parameters c_0, θ_0, and σ_0 is a good fit. You can specify the SCALE= *Weibull2-option* as an alias for the SIGMA= *Weibull2-option* and the SHAPE= *Weibull2-option* as an alias for the C= *Weibull2-option*. See Example 4.34.

WGRID=n

specifies the line thickness for the grid when producing traditional graphics. The option does not apply to ODS Graphics output.

ZETA=value | **EST**

specifies a value for the scale parameter ζ for the lognormal quantile plots requested with the LOGNORMAL option. Enclose the ZETA= *lognormal-option* in parentheses after the LOGNORMAL option. To request a distribution reference line with intercept θ_0 and slope $\exp(\zeta_0)$, specify the THETA=θ_0 and ZETA=ζ_0.

VAR Statement

VAR *variables* ;

The VAR statement specifies the analysis variables and their order in the results. By default, if you omit the VAR statement, PROC UNIVARIATE analyzes all numeric variables that are not listed in the other statements.

Using the Output Statement with the VAR Statement

You must provide a VAR statement when you use an OUTPUT statement. To store the same statistic for several analysis variables in the OUT= data set, you specify a list of names in the OUTPUT statement. PROC UNIVARIATE makes a one-to-one correspondence between the order of the analysis variables in the VAR statement and the list of names that follow a statistic keyword.

WEIGHT Statement

WEIGHT *variable* ;

The WEIGHT statement specifies numeric weights for analysis variables in the statistical calculations. The UNIVARIATE procedure uses the values w_i of the WEIGHT variable to modify the computation of a number of summary statistics by assuming that the variance of the ith value x_i of the analysis variable is equal to σ^2/w_i, where σ is an unknown parameter. The values of the

WEIGHT variable do not have to be integers and are typically positive. By default, observations with nonpositive or missing values of the WEIGHT variable are handled as follows:[1]

- If the value is zero, the observation is counted in the total number of observations.
- If the value is negative, it is converted to zero, and the observation is counted in the total number of observations.
- If the value is missing, the observation is excluded from the analysis.

To exclude observations that contain negative and zero weights from the analysis, use EXCLNPWGT. Note that most SAS/STAT procedures, such as PROC GLM, exclude negative and zero weights by default. The weight variable does not change how the procedure determines the range, mode, extreme values, extreme observations, or number of missing values. When you specify a WEIGHT statement, the procedure also computes a weighted standard error and a weighted version of Student's t test. The Student's t test is the only test of location that PROC UNIVARIATE computes when you weight the analysis variables.

When you specify a WEIGHT variable, the procedure uses its values, w_i, to compute weighted versions of the statistics[2] provided in the Moments table. For example, the procedure computes a weighted mean $\overline{x}_w$ and a weighted variance s_w^2 as

$$\overline{x}_w = \frac{\sum_i w_i x_i}{\sum_i w_i}$$

and

$$s_w^2 = \frac{1}{d} \sum_i w_i (x_i - \overline{x}_w)^2$$

where x_i is the ith variable value. The divisor d is controlled by the VARDEF= option in the PROC UNIVARIATE statement.

The WEIGHT statement does not affect the determination of the mode, extreme values, extreme observations, or the number of missing values of the analysis variables. However, the weights w_i are used to compute weighted percentiles.[3] The WEIGHT variable has no effect on graphical displays produced with the plot statements.

The CIPCTLDF, CIPCTLNORMAL, LOCCOUNT, NORMAL, ROBUSTSCALE, TRIMMED=, and WINSORIZED= options are not available with the WEIGHT statement.

To compute weighted skewness or kurtosis, use VARDEF=DF or VARDEF=N in the PROC statement.

You cannot specify the HISTOGRAM, PROBPLOT, or QQPLOT statements with the WEIGHT statement.

When you use the WEIGHT statement, consider which value of the VARDEF= option is appropriate. See VARDEF= and the calculation of weighted statistics for more information.

---

[1] In SAS 6.12 and earlier releases, observations were used in the analysis if and only if the WEIGHT variable value was greater than zero.

[2] In SAS 6.12 and earlier releases, weighted skewness and kurtosis were not computed.

[3] In SAS 6.12 and earlier releases, the weights did not affect the computation of percentiles and the procedure did not exclude the observations with missing weights from the count of observations.

Dictionary of Common Options

The following entries provide detailed descriptions of *options* that are common to all the plot statements: CDFPLOT, HISTOGRAM, PPPLOT, PROBPLOT, and QQPLOT.

ALPHADELTA=value

specifies the change in successive estimates of $\hat{\alpha}$ at which iteration terminates in the Newton-Raphson approximation of the maximum likelihood estimate of α for gamma distributions requested with the GAMMA option. Enclose the ALPHADELTA= option in parentheses after the GAMMA keyword. Iteration continues until the change in α is less than the value specified or the number of iterations exceeds the value of the MAXITER= option. The default value is 0.00001.

ALPHAINITIAL=value

specifies the initial value for $\hat{\alpha}$ in the Newton-Raphson approximation of the maximum likelihood estimate of α for gamma distributions requested with the GAMMA option. Enclose the ALPHAINITIAL= option in parentheses after the GAMMA keyword. The default value is Thom's approximation of the estimate of α (refer to Johnson, Kotz, and Balakrishnan (1995)).

ANNOKEY

applies the annotation requested with the ANNOTATE= option only to the key cell of a comparative plot. By default, the procedure applies annotation to all of the cells. This option is not available unless you use the CLASS statement. You can use the KEYLEVEL= option in the CLASS statement to specify the key cell.

ANNOTATE=SAS-data-set

ANNO=SAS-data-set

specifies an input data set that contains annotate variables as described in *SAS/GRAPH: Reference*. The ANNOTATE= data set you specify in the plot statement is used for all plots created by the statement. You can also specify an ANNOTATE= data set in the PROC UNIVARIATE statement to enhance all plots created by the procedure (see the section "ANNOTATE= Data Sets" on page 370).

CAXIS=color

CAXES=color

CA=color

specifies the color for the axes and tick marks. This option overrides any COLOR= specifications in an AXIS statement.

CDELTA=value

specifies the change in successive estimates of c at which iterations terminate in the Newton-Raphson approximation of the maximum likelihood estimate of c for Weibull distributions requested by the WEIBULL option. Enclose the CDELTA= option in parentheses after the WEIBULL keyword. Iteration continues until the change in c between consecutive steps is less than the *value* specified or until the number of iterations exceeds the value of the MAXITER= option. The default value is 0.00001.

CFRAME=_color_
> specifies the color for the area that is enclosed by the axes and frame. The area is not filled by default.

CFRAMESIDE=_color_
> specifies the color to fill the frame area for the row labels that display along the left side of a comparative plot. This color also fills the frame area for the label of the corresponding CLASS variable (if you associate a label with the variable). By default, these areas are not filled. This option is not available unless you use the CLASS statement.

CFRAMETOP=_color_
> specifies the color to fill the frame area for the column labels that display across the top of a comparative plot. This color also fills the frame area for the label of the corresponding CLASS variable (if you associate a label with the variable). By default, these areas are not filled. This option is not available unless you use the CLASS statement.

CHREF=_color_

CH=_color_
> specifies the color for horizontal axis reference lines requested by the HREF= option.

CINITIAL=_value_
> specifies the initial value for $\hat{c}$ in the Newton-Raphson approximation of the maximum likelihood estimate of c for Weibull distributions requested with the WEIBULL or WEIBULL2 option. The default value is 1.8 (see Johnson, Kotz, and Balakrishnan (1995).

COLOR=_color_

COLOR=_color-list_
> specifies the color of the curve or reference line associated with a distribution or kernel density estimate. Enclose the COLOR= option in parentheses after a distribution option or the KERNEL option. In a HISTOGRAM statement, you can specify a list of colors in parentheses for multiple density curves.

CONTENTS='_string_'
> specifies the table of contents grouping entry for output produced by the plot statement. You can specify CONTENTS='' to suppress the grouping entry.

CPROP=_color_ | **EMPTY**

CPROP
> specifies the color for a horizontal bar whose length (relative to the width of the tile) indicates the proportion of the total frequency that is represented by the corresponding cell in a comparative plot. By default, no proportion bars are displayed. This option is not available unless you use the CLASS statement. You can specify the keyword EMPTY to display empty bars. See Example 4.20.
>
> For ODS Graphics and traditional graphics with the GSTYLE system option in effect, you can specify CPROP with no argument to produce proportion bars using an appropriate color from the ODS style.

CTEXT=*color*

CT=*color*

> specifies the color for tick mark values and axis labels. The default is the color specified for the CTEXT= option in the GOPTIONS statement.

CTEXTSIDE=*color*

> specifies the color for the row labels that display along the left side of a comparative plot. By default, the color specified by the CTEXT= option is used. If you omit the CTEXT= option, the color specified in the GOPTIONS statement is used. This option is not available unless you use the CLASS statement. You can specify the CFRAMESIDE= option to change the background color for the row labels.

CTEXTTOP=*color*

> specifies the color for the column labels that display along the left side of a comparative plot. By default, the color specified by the CTEXT= option is used. If you omit the CTEXT= option, the color specified in the GOPTIONS statement is used. This option is not available unless you specify the CLASS statement. You can use the CFRAMETOP= option to change the background color for the column labels.

CVREF=*color*

CV=*color*

> specifies the color for lines requested with the VREF= option.

DESCRIPTION=*'string'*

DES=*'string'*

> specifies a description, up to 256 characters long, that appears in the PROC GREPLAY master menu for a traditional graphics chart. The default value is the analysis variable name.

FITINTERVAL=*value*

> specifies the value of z for the method of percentiles when this method is used to fit a Johnson S_B or Johnson S_U distribution. The FITINTERVAL= option is specified in parentheses after the SB or SU option. The default of z is 0.524.

FITMETHOD=PERCENTILE | MLE | MOMENTS

> specifies the method used to estimate the parameters of a Johnson S_B or Johnson S_U distribution. The FITMETHOD= option is specified in parentheses after the SB or SU option. By default, the method of percentiles is used.

FITTOLERANCE=*value*

> specifies the tolerance value for the ratio criterion when the method of percentiles is used to fit a Johnson S_B or Johnson S_U distribution. The FITTOLERANCE= option is specified in parentheses after the SB or SU option. The default value is 0.01.

FONT=*font*

> specifies a software font for reference line and axis labels. You can also specify fonts for axis labels in an AXIS statement. The FONT= font takes precedence over the FTEXT= font specified in the GOPTIONS statement.

HAXIS=*value*

> specifies the name of an AXIS statement describing the horizontal axis.

HEIGHT=_value_

specifies the height, in percentage screen units, of text for axis labels, tick mark labels, and legends. This option takes precedence over the HTEXT= option in the GOPTIONS statement.

HMINOR=_n_

HM=_n_

specifies the number of minor tick marks between each major tick mark on the horizontal axis. Minor tick marks are not labeled. By default, HMINOR=0.

HREF=_values_

draws reference lines that are perpendicular to the horizontal axis at the values that you specify. Also see the CHREF= and LHREF= options.

HREFLABELS='_label1_' ... '_labeln_'

HREFLABEL='_label1_' ... '_labeln_'

HREFLAB='_label1_' ... '_labeln_'

specifies labels for the lines requested by the HREF= option. The number of labels must equal the number of lines. Enclose each label in quotes. Labels can have up to 16 characters.

HREFLABPOS=_n_

specifies the vertical position of HREFLABELS= labels, as described in the following table.

n	Position
1	along top of plot
2	staggered from top to bottom of plot
3	along bottom of plot
4	staggered from bottom to top of plot

By default, HREFLABPOS=1. NOTE: HREFLABPOS=2 and HREFLABPOS=4 are not supported for ODS Graphics output.

INFONT=_font_

specifies a software font to use for text inside the framed areas of the plot. The INFONT= option takes precedence over the FTEXT= option in the GOPTIONS statement. For a list of fonts, see _SAS/GRAPH: Reference_.

INHEIGHT=_value_

specifies the height, in percentage screen units, of text used inside the framed areas of the histogram. By default, the height specified by the HEIGHT= option is used. If you do not specify the HEIGHT= option, the height specified with the HTEXT= option in the GOPTIONS statement is used.

INTERTILE=_value_

specifies the distance in horizontal percentage screen units between the framed areas, called _tiles_, of a comparative plot. By default, INTERTILE=0.75 percentage screen units. This option is not available unless you use the CLASS statement. You can specify INTERTILE=0 to create contiguous tiles.

L=_linetype_

L=_linetype-list_

> specifies the line type of the curve or reference line associated with a distribution or kernel density estimate. Enclose the L= option in parentheses after the distribution option or the KERNEL option. In a HISTOGRAM statement, you can specify a list of line types in parentheses for multiple density curves.

LHREF=_linetype_

LH=_linetype_

> specifies the line type for the reference lines that you request with the HREF= option. By default, LHREF=2, which produces a dashed line.

LVREF=_linetype_

LV=_linetype_

> specifies the line type for lines requested with the VREF= option. By default, LVREF=2, which produces a dashed line.

MAXITER=_n_

> specifies the maximum number of iterations in the Newton-Raphson approximation of the maximum likelihood estimate of α for gamma distributions requested with the GAMMA option and c for Weibull distributions requested with the WEIBULL and WEIBULL2 options. Enclose the MAXITER= option in parentheses after the GAMMA, WEIBULL, or WEIBULL2 keywords. The default value of n is 20.

NAME='_string_'

> specifies a name for the plot, up to eight characters long, that appears in the PROC GREPLAY master menu for a traditional graphics chart. The default value is 'UNIVAR'.

NCOLS=_n_

NCOL=_n_

> specifies the number of columns per panel in a comparative plot. This option is not available unless you use the CLASS statement. By default, NCOLS=1 if you specify only one CLASS variable, and NCOLS=2 if you specify two CLASS variables. If you specify two CLASS variables, you can use the NCOLS= option with the NROWS= option.

NOFRAME

> suppresses the frame around the subplot area.

NOHLABEL

> suppresses the label for the horizontal axis. You can use this option to reduce clutter.

NOVLABEL

> suppresses the label for the vertical axis. You can use this option to reduce clutter.

NOVTICK

> suppresses the tick marks and tick mark labels for the vertical axis. This option also suppresses the label for the vertical axis.

NROWS=_n_

NROW=_n_

specifies the number of rows per panel in a comparative plot. This option is not available unless you use the CLASS statement. By default, NROWS=2. If you specify two CLASS variables, you can use the NCOLS= option with the NROWS= option.

OVERLAY

specifies that plots associated with different levels of a CLASS variable be overlaid onto a single plot, rather than displayed as separate cells in a comparative plot. If you specify the OVERLAY option with one CLASS variable, the output associated with each level of the CLASS variable is overlaid on a single plot. If you specify the OVERLAY option with two CLASS variables, a comparative plot based on the first CLASS variable's levels is produced. Each cell in this comparative plot contains overlaid output associated with the levels of the second CLASS variable.

The OVERLAY option applies only to ODS Graphics output and it is not available in the HISTOGRAM statement.

SCALE=_value_

is an alias for the SIGMA= option for distributions requested by the BETA, EXPONENTIAL, GAMMA, SB, SU, WEIBULL, and WEIBULL2 options and for the ZETA= option for distributions requested by the LOGNORMAL option.

SHAPE=_value_

is an alias for the ALPHA= option for distributions requested by the GAMMA option, for the SIGMA= option for distributions requested by the LOGNORMAL option, and for the C= option for distributions requested by the WEIBULL and WEIBULL2 options.

TURNVLABELS

TURNVLABEL

turns the characters in the vertical axis labels so that they display vertically. This happens by default when you use a hardware font.

VAXIS=_name_

VAXIS=_value-list_

specifies the name of an AXIS statement describing the vertical axis. In a HISTOGRAM statement, you can alternatively specify a _value-list_ for the vertical axis.

VAXISLABEL='_label_**'**

specifies a label for the vertical axis. Labels can have up to 40 characters.

VMINOR=_n_

VM=_n_

specifies the number of minor tick marks between each major tick mark on the vertical axis. Minor tick marks are not labeled. The default is zero.

VREF=_value-list_

draws reference lines perpendicular to the vertical axis at the values specified. Also see the CVREF= and LVREF= options.

VREFLABELS='*label1*'…'*labeln*'
VREFLABEL='*label1*'…'*labeln*'
VREFLAB='*label1*'…'*labeln*'

    specifies labels for the lines requested by the VREF= option. The number of labels must equal the number of lines. Enclose each label in quotes. Labels can have up to 16 characters.

VREFLABPOS=*n*

    specifies the horizontal position of VREFLABELS= labels. If you specify VREFLABPOS=1, the labels are positioned at the left of the plot. If you specify VREFLABPOS=2, the labels are positioned at the right of the plot. By default, VREFLABPOS=1.

W=*value*

W=*value-list*

    specifies the width in pixels of the curve or reference line associated with a distribution or kernel density estimate. Enclose the W= option in parentheses after the distribution option or the KERNEL option. In a HISTOGRAM statement, you can specify a list of widths in parentheses for multiple density curves.

WAXIS=*n*

    specifies the line thickness, in pixels, for the axes and frame.

Details: UNIVARIATE Procedure

Missing Values

PROC UNIVARIATE excludes missing values for an analysis variable before calculating statistics. Each analysis variable is treated individually; a missing value for an observation in one variable does not affect the calculations for other variables. The statements handle missing values as follows:

- If a BY or an ID variable value is missing, PROC UNIVARIATE treats it like any other BY or ID variable value. The missing values form a separate BY group.

- If the FREQ variable value is missing or nonpositive, PROC UNIVARIATE excludes the observation from the analysis.

- If the WEIGHT variable value is missing, PROC UNIVARIATE excludes the observation from the analysis.

PROC UNIVARIATE tabulates the number of missing values and reports this information in the ODS table named "Missing Values." See the section "ODS Table Names" on page 376. Before the number of missing values is tabulated, PROC UNIVARIATE excludes observations when either of the following conditions exist:

- you use the FREQ statement and the frequencies are nonpositive
- you use the WEIGHT statement and the weights are missing or nonpositive (you must specify the EXCLNPWGT option)

Rounding

When you specify ROUND=u, PROC UNIVARIATE rounds a variable by using the rounding unit to divide the number line into intervals with midpoints of the form ui, where u is the nonnegative rounding unit and i is an integer. The interval width is u. Any variable value that falls in an interval is rounded to the midpoint of that interval. A variable value that is midway between two midpoints, and is therefore on the boundary of two intervals, rounds to the even midpoint. Even midpoints occur when i is an even integer $(0, \pm 2, \pm 4, \ldots)$.

When ROUND=1 and the analysis variable values are between -2.5 and 2.5, the intervals are as follows:

Table 4.79 Intervals for Rounding When ROUND=1

i	Interval	Midpoint	Left endpt rounds to	Right endpt rounds to
-2	$[-2.5, -1.5]$	-2	-2	-2
-1	$[-1.5, -0.5]$	-1	-2	0
0	$[-0.5, 0.5]$	0	0	0
1	$[0.5, 1.5]$	1	0	2
2	$[1.5, 2.5]$	2	2	2

When ROUND=0.5 and the analysis variable values are between -1.25 and 1.25, the intervals are as follows:

Table 4.80 Intervals for Rounding When ROUND=0.5

i	Interval	Midpoint	Left endpt rounds to	Right endpt rounds to
-2	$[-1.25, -0.75]$	-1.0	-1	-1
-1	$[-0.75, -0.25]$	-0.5	-1	0
0	$[-0.25, 0.25]$	0.0	0	0
1	$[0.25, 0.75]$	0.5	0	1
2	$[0.75, 1.25]$	1.0	1	1

As the rounding unit increases, the interval width also increases. This reduces the number of unique values and decreases the amount of memory that PROC UNIVARIATE needs.

Descriptive Statistics

This section provides computational details for the descriptive statistics that are computed with the PROC UNIVARIATE statement. These statistics can also be saved in the OUT= data set by specifying the keywords listed in Table 4.43 in the OUTPUT statement.

Standard algorithms (Fisher 1973) are used to compute the moment statistics. The computational methods used by the UNIVARIATE procedure are consistent with those used by other SAS procedures for calculating descriptive statistics.

The following sections give specific details on a number of statistics calculated by the UNIVARIATE procedure.

Mean

The sample mean is calculated as

$$\bar{x}_w = \frac{\sum_{i=1}^{n} w_i x_i}{\sum_{i=1}^{n} w_i}$$

where n is the number of nonmissing values for a variable, x_i is the ith value of the variable, and w_i is the weight associated with the ith value of the variable. If there is no WEIGHT variable, the formula reduces to

$$\bar{x} = \frac{1}{n} \sum_{i=1}^{n} x_i$$

Sum

The sum is calculated as $\sum_{i=1}^{n} w_i x_i$, where n is the number of nonmissing values for a variable, x_i is the ith value of the variable, and w_i is the weight associated with the ith value of the variable. If there is no WEIGHT variable, the formula reduces to $\sum_{i=1}^{n} x_i$.

Sum of the Weights

The sum of the weights is calculated as $\sum_{i=1}^{n} w_i$, where n is the number of nonmissing values for a variable and w_i is the weight associated with the ith value of the variable. If there is no WEIGHT variable, the sum of the weights is n.

Variance

The variance is calculated as

$$\frac{1}{d} \sum_{i=1}^{n} w_i (x_i - \bar{x}_w)^2$$

where n is the number of nonmissing values for a variable, x_i is the ith value of the variable, $\bar{x}_w$ is the weighted mean, w_i is the weight associated with the ith value of the variable, and d is the divisor controlled by the VARDEF= option in the PROC UNIVARIATE statement:

$$d = \begin{cases} n-1 & \text{if VARDEF=DF (default)} \\ n & \text{if VARDEF=N} \\ (\sum_i w_i) - 1 & \text{if VARDEF=WDF} \\ \sum_i w_i & \text{if VARDEF=WEIGHT | WGT} \end{cases}$$

If there is no WEIGHT variable, the formula reduces to

$$\frac{1}{d} \sum_{i=1}^{n} (x_i - \bar{x})^2$$

Standard Deviation

The standard deviation is calculated as

$$s_w = \sqrt{\frac{1}{d} \sum_{i=1}^{n} w_i (x_i - \bar{x}_w)^2}$$

where n is the number of nonmissing values for a variable, x_i is the ith value of the variable, $\bar{x}_w$ is the weighted mean, w_i is the weight associated with the ith value of the variable, and d is the divisor controlled by the VARDEF= option in the PROC UNIVARIATE statement. If there is no WEIGHT variable, the formula reduces to

$$s = \sqrt{\frac{1}{d} \sum_{i=1}^{n} (x_i - \bar{x})^2}$$

Skewness

The sample skewness, which measures the tendency of the deviations to be larger in one direction than in the other, is calculated as follows depending on the VARDEF= option:

Table 4.81 Formulas for Skewness

VARDEF	Formula
DF (default)	$\dfrac{n}{(n-1)(n-2)} \sum_{i=1}^{n} w_i^{3/2} \left(\dfrac{x_i - \bar{x}_w}{s_w} \right)^3$
N	$\dfrac{1}{n} \sum_{i=1}^{n} w_i^{3/2} \left(\dfrac{x_i - \bar{x}_w}{s_w} \right)^3$
WDF	missing
WEIGHT | WGT	missing

where n is the number of nonmissing values for a variable, x_i is the ith value of the variable, $\bar{x}_w$ is the sample average, s is the sample standard deviation, and w_i is the weight associated with the ith value of the variable. If VARDEF=DF, then n must be greater than 2. If there is no WEIGHT variable, then $w_i = 1$ for all $i = 1, \ldots, n$.

The sample skewness can be positive or negative; it measures the asymmetry of the data distribution and estimates the theoretical skewness $\sqrt{\beta_1} = \mu_3 \mu_2^{-\frac{3}{2}}$, where μ_2 and μ_3 are the second and third central moments. Observations that are normally distributed should have a skewness near zero.

Kurtosis

The sample kurtosis, which measures the heaviness of tails, is calculated as follows depending on the VARDEF= option:

Table 4.82 Formulas for Kurtosis

VARDEF	Formula
DF (default)	$\dfrac{n(n+1)}{(n-1)(n-2)(n-3)} \sum_{i=1}^{n} w_i^2 \left(\dfrac{x_i - \bar{x}_w}{s_w} \right)^4 - \dfrac{3(n-1)^2}{(n-2)(n-3)}$
N	$\dfrac{1}{n} \sum_{i=1}^{n} w_i^2 \left(\dfrac{x_i - \bar{x}_w}{s_w} \right)^4 - 3$
WDF	missing
WEIGHT | WGT	missing

where n is the number of nonmissing values for a variable, x_i is the ith value of the variable, $\bar{x}_w$ is the sample average, s_w is the sample standard deviation, and w_i is the weight associated with the ith value of the variable. If VARDEF=DF, then n must be greater than 3. If there is no WEIGHT variable, then $w_i = 1$ for all $i = 1, \ldots, n$.

The sample kurtosis measures the heaviness of the tails of the data distribution. It estimates the adjusted theoretical kurtosis denoted as $\beta_2 - 3$, where $\beta_2 = \frac{\mu_4}{\mu_2^2}$, and μ_4 is the fourth central moment. Observations that are normally distributed should have a kurtosis near zero.

Coefficient of Variation (CV)

The coefficient of variation is calculated as

$$CV = \frac{100 \times s_w}{\bar{x}_w}$$

Calculating the Mode

The mode is the value that occurs most often in the data. PROC UNIVARIATE counts repetitions of the values of the analysis variables or, if you specify the ROUND= option, the rounded values. If a tie occurs for the most frequent value, the procedure reports the lowest mode in the table labeled "Basic Statistical Measures" in the statistical output. To list all possible modes, use the MODES option in the PROC UNIVARIATE statement. When no repetitions occur in the data (as with truly continuous data), the procedure does not report the mode. The WEIGHT statement has no effect on the mode. See Example 4.2.

Calculating Percentiles

The UNIVARIATE procedure automatically computes the 1st, 5th, 10th, 25th, 50th, 75th, 90th, 95th, and 99th percentiles (quantiles), as well as the minimum and maximum of each analysis variable. To compute percentiles other than these default percentiles, use the PCTLPTS= and PCTLPRE= options in the OUTPUT statement.

You can specify one of five definitions for computing the percentiles with the PCTLDEF= option. Let n be the number of nonmissing values for a variable, and let $x_1, x_2, \ldots, x_n$ represent the ordered values of the variable. Let the tth percentile be y, set $p = \frac{t}{100}$, and let

$$\begin{aligned} np &= j + g \quad \text{when PCTLDEF=1, 2, 3, or 5} \\ (n+1)p &= j + g \quad \text{when PCTLDEF=4} \end{aligned}$$

where j is the integer part of np, and g is the fractional part of np. Then the PCTLDEF= option defines the tth percentile, y, as described in the Table 4.83.

Table 4.83 Percentile Definitions

PCTLDEF	Description	Formula
1	weighted average at x_{np}	$y = (1-g)x_j + gx_{j+1}$ where x_0 is taken to be x_1
2	observation numbered closest to np	$y = x_j$ if $g < \frac{1}{2}$ $y = x_j$ if $g = \frac{1}{2}$ and j is even $y = x_{j+1}$ if $g = \frac{1}{2}$ and j is odd $y = x_{j+1}$ if $g > \frac{1}{2}$
3	empirical distribution function	$y = x_j$ if $g = 0$ $y = x_{j+1}$ if $g > 0$
4	weighted average aimed at $x_{(n+1)p}$	$y = (1-g)x_j + gx_{j+1}$ where x_{n+1} is taken to be x_n
5	empirical distribution function with averaging	$y = \frac{1}{2}(x_j + x_{j+1})$ if $g = 0$ $y = x_{j+1}$ if $g > 0$

Weighted Percentiles

When you use a WEIGHT statement, the percentiles are computed differently. The $100p$th weighted percentile y is computed from the empirical distribution function with averaging:

$$y = \begin{cases} x_1 & \text{if } w_1 > pW \\ \frac{1}{2}(x_i + x_{i+1}) & \text{if } \sum_{j=1}^{i} w_j = pW \\ x_{i+1} & \text{if } \sum_{j=1}^{i} w_j < pW < \sum_{j=1}^{i+1} w_j \end{cases}$$

where w_i is the weight associated with x_i and $W = \sum_{i=1}^{n} w_i$ is the sum of the weights.

Note that the PCTLDEF= option is not applicable when a WEIGHT statement is used. However, in this case, if all the weights are identical, the weighted percentiles are the same as the percentiles that would be computed without a WEIGHT statement and with PCTLDEF=5.

Confidence Limits for Percentiles

You can use the CIPCTLNORMAL option to request confidence limits for percentiles, assuming the data are normally distributed. These limits are described in Section 4.4.1 of Hahn and Meeker (1991). When $0 < p < \frac{1}{2}$, the two-sided $100(1-\alpha)\%$ confidence limits for the $100p$th percentile are

lower limit $= \bar{X} - g'(\frac{\alpha}{2}; 1-p, n)s$
upper limit $= \bar{X} - g'(1-\frac{\alpha}{2}; p, n)s$

where n is the sample size. When $\frac{1}{2} \leq p < 1$, the two-sided $100(1-\alpha)\%$ confidence limits for the $100p$th percentile are

$$\text{lower limit} = \bar{X} + g'(\tfrac{\alpha}{2}; 1-p, n)s$$
$$\text{upper limit} = \bar{X} + g'(1-\tfrac{\alpha}{2}; p, n)s$$

One-sided $100(1-\alpha)\%$ confidence bounds are computed by replacing $\frac{\alpha}{2}$ by α in the appropriate preceding equation. The factor $g'(\gamma, p, n)$ is related to the noncentral t distribution and is described in Owen and Hua (1977) and Odeh and Owen (1980). See Example 4.10.

You can use the CIPCTLDF option to request distribution-free confidence limits for percentiles. In particular, it is not necessary to assume that the data are normally distributed. These limits are described in Section 5.2 of Hahn and Meeker (1991). The two-sided $100(1-\alpha)\%$ confidence limits for the $100p$th percentile are

$$\text{lower limit} = X_{(l)}$$
$$\text{upper limit} = X_{(u)}$$

where $X_{(j)}$ is the jth order statistic when the data values are arranged in increasing order:

$$X_{(1)} \leq X_{(2)} \leq \ldots \leq X_{(n)}$$

The lower rank l and upper rank u are integers that are symmetric (or nearly symmetric) around $[np] + 1$, where $[np]$ is the integer part of np and n is the sample size. Furthermore, l and u are chosen so that $X_{(l)}$ and $X_{(u)}$ are as close to $X_{[n+1]p}$ as possible while satisfying the coverage probability requirement,

$$Q(u-1; n, p) - Q(l-1; n, p) \geq 1 - \alpha$$

where $Q(k; n, p)$ is the cumulative binomial probability,

$$Q(k; n, p) = \sum_{i=0}^{k} \binom{n}{i} p^i (1-p)^{n-i}$$

In some cases, the coverage requirement cannot be met, particularly when n is small and p is near 0 or 1. To relax the requirement of symmetry, you can specify CIPCTLDF(TYPE = ASYMMETRIC). This option requests symmetric limits when the coverage requirement can be met, and asymmetric limits otherwise.

If you specify CIPCTLDF(TYPE = LOWER), a one-sided $100(1-\alpha)\%$ lower confidence bound is computed as $X_{(l)}$, where l is the largest integer that satisfies the inequality

$$1 - Q(l-1; n, p) \geq 1 - \alpha$$

with $0 < l \leq n$. Likewise, if you specify CIPCTLDF(TYPE = UPPER), a one-sided $100(1-\alpha)\%$ lower confidence bound is computed as $X_{(u)}$, where u is the largest integer that satisfies the inequality

$$Q(u-1; n, p) \geq 1-\alpha \quad \text{where } 0 < u \leq n$$

Note that confidence limits for percentiles are not computed when a WEIGHT statement is specified. See Example 4.10.

Tests for Location

PROC UNIVARIATE provides three tests for location: Student's t test, the sign test, and the Wilcoxon signed rank test. All three tests produce a test statistic for the null hypothesis that the mean or median is equal to a given value μ_0 against the two-sided alternative that the mean or median is not equal to μ_0. By default, PROC UNIVARIATE sets the value of μ_0 to zero. You can use the MU0= option in the PROC UNIVARIATE statement to specify the value of μ_0. Student's t test is appropriate when the data are from an approximately normal population; otherwise, use nonparametric tests such as the sign test or the signed rank test. For large sample situations, the t test is asymptotically equivalent to a z test. If you use the WEIGHT statement, PROC UNIVARIATE computes only one weighted test for location, the t test. You must use the default value for the VARDEF= option in the PROC statement (VARDEF=DF). See Example 4.12.

You can also use these tests to compare means or medians of *paired data*. Data are said to be paired when subjects or units are matched in pairs according to one or more variables, such as pairs of subjects with the same age and gender. Paired data also occur when each subject or unit is measured at two times or under two conditions. To compare the means or medians of the two times, create an analysis variable that is the difference between the two measures. The test that the mean or the median difference of the variables equals zero is equivalent to the test that the means or medians of the two original variables are equal. Note that you can also carry out these tests by using the PAIRED statement in the TTEST procedure; see Chapter 92, "The TTEST Procedure" (*SAS/STAT User's Guide*). Also see Example 4.13.

Student's *t* Test

PROC UNIVARIATE calculates the t statistic as

$$t = \frac{\bar{x} - \mu_0}{s/\sqrt{n}}$$

where $\bar{x}$ is the sample mean, n is the number of nonmissing values for a variable, and s is the sample standard deviation. The null hypothesis is that the population mean equals μ_0. When the data values are approximately normally distributed, the probability under the null hypothesis of a t statistic that is as extreme, or more extreme, than the observed value (the *p*-value) is obtained from the t distribution with $n-1$ degrees of freedom. For large n, the t statistic is asymptotically equivalent to a z test. When you use the WEIGHT statement and the default value of VARDEF=, which is DF, the t statistic is calculated as

$$t_w = \frac{\bar{x}_w - \mu_0}{s_w/\sqrt{\sum_{i=1}^{n} w_i}}$$

where $\bar{x}_w$ is the weighted mean, s_w is the weighted standard deviation, and w_i is the weight for ith observation. The t_w statistic is treated as having a Student's t distribution with $n-1$ degrees of freedom. If you specify the EXCLNPWGT option in the PROC statement, n is the number of nonmissing observations when the value of the WEIGHT variable is positive. By default, n is the number of nonmissing observations for the WEIGHT variable.

Sign Test

PROC UNIVARIATE calculates the sign test statistic as

$$M = (n^+ - n^-)/2$$

where n^+ is the number of values that are greater than μ_0, and n^- is the number of values that are less than μ_0. Values equal to μ_0 are discarded. Under the null hypothesis that the population median is equal to μ_0, the p-value for the observed statistic M_{obs} is

$$\Pr(|M_{obs}| \geq |M|) = 0.5^{(n_t-1)} \sum_{j=0}^{min(n^+,n^-)} \binom{n_t}{i}$$

where $n_t = n^+ + n^-$ is the number of x_i values not equal to μ_0.

NOTE: If n^+ and n^- are equal, the p-value is equal to one.

Wilcoxon Signed Rank Test

The signed rank statistic S is computed as

$$S = \sum_{i:|x_i-\mu_0|>0} r_i^+ - \frac{n_t(n_t+1)}{4}$$

where r_i^+ is the rank of $|x_i - \mu_0|$ after discarding values of $x_i = \mu_0$, and n_t is the number of x_i values not equal to μ_0. Average ranks are used for tied values.

If $n_t \leq 20$, the significance of S is computed from the exact distribution of S, where the distribution is a convolution of scaled binomial distributions. When $n_t > 20$, the significance of S is computed by treating

$$S\sqrt{\frac{n_t - 1}{n_t V - S^2}}$$

as a Student's t variate with $n_t - 1$ degrees of freedom. V is computed as

$$V = \frac{1}{24}n_t(n_t+1)(2n_t+1) - \frac{1}{48}\sum t_i(t_i+1)(t_i-1)$$

where the sum is over groups tied in absolute value and where t_i is the number of values in the ith group (Iman 1974; Conover 1999). The null hypothesis tested is that the mean (or median) is μ_0, assuming that the distribution is symmetric. Refer to Lehmann (1998).

Confidence Limits for Parameters of the Normal Distribution

The two-sided $100(1-\alpha)\%$ confidence interval for the mean has upper and lower limits

$$\bar{x} \pm t_{1-\frac{\alpha}{2};n-1}\frac{s}{\sqrt{n}}$$

where $s^2 = \frac{1}{n-1}\sum(x_i - \bar{x})^2$ and $t_{1-\frac{\alpha}{2};n-1}$ is the $(1-\frac{\alpha}{2})$ percentile of the t distribution with $n-1$ degrees of freedom. The one-sided upper $100(1-\alpha)\%$ confidence limit is computed as $\bar{x} + \frac{s}{\sqrt{n}}t_{1-\alpha;n-1}$ and the one-sided lower $100(1-\alpha)\%$ confidence limit is computed as $\bar{x} - \frac{s}{\sqrt{n}}t_{1-\alpha;n-1}$. See Example 4.9.

The two-sided $100(1-\alpha)\%$ confidence interval for the standard deviation has lower and upper limits,

$$s\sqrt{\frac{n-1}{\chi^2_{1-\frac{\alpha}{2};n-1}}} \quad \text{and} \quad s\sqrt{\frac{n-1}{\chi^2_{\frac{\alpha}{2};n-1}}}$$

respectively, where $\chi^2_{1-\frac{\alpha}{2};n-1}$ and $\chi^2_{\frac{\alpha}{2};n-1}$ are the $(1-\frac{\alpha}{2})$ and $\frac{\alpha}{2}$ percentiles of the chi-square distribution with $n-1$ degrees of freedom. A one-sided $100(1-\alpha)\%$ confidence limit has lower and upper limits,

$$s\sqrt{\frac{n-1}{\chi^2_{1-\alpha;n-1}}} \quad \text{and} \quad s\sqrt{\frac{n-1}{\chi^2_{\alpha;n-1}}}$$

respectively. The $100(1-\alpha)\%$ confidence interval for the variance has upper and lower limits equal to the squares of the corresponding upper and lower limits for the standard deviation. When you use the WEIGHT statement and specify VARDEF=DF in the PROC statement, the $100(1-\alpha)\%$ confidence interval for the weighted mean is

$$\bar{x}_w \pm t_{1-\frac{\alpha}{2}}\frac{s_w}{\sqrt{\sum_{i=1}^{n}w_i}}$$

where $\bar{x}_w$ is the weighted mean, s_w is the weighted standard deviation, w_i is the weight for ith observation, and $t_{1-\frac{\alpha}{2}}$ is the $(1-\frac{\alpha}{2})$ percentile for the t distribution with $n-1$ degrees of freedom.

Robust Estimators

A statistical method is robust if it is insensitive to moderate or even large departures from the assumptions that justify the method. PROC UNIVARIATE provides several methods for robust estimation of location and scale. See Example 4.11.

Winsorized Means

The Winsorized mean is a robust estimator of the location that is relatively insensitive to outliers. The k-times Winsorized mean is calculated as

$$\bar{x}_{wk} = \frac{1}{n}\left((k+1)x_{(k+1)} + \sum_{i=k+2}^{n-k-1} x_{(i)} + (k+1)x_{(n-k)}\right)$$

where n is the number of observations and $x_{(i)}$ is the ith order statistic when the observations are arranged in increasing order:

$$x_{(1)} \leq x_{(2)} \leq \ldots \leq x_{(n)}$$

The Winsorized mean is computed as the ordinary mean after the k smallest observations are replaced by the $(k+1)$st smallest observation and the k largest observations are replaced by the $(k+1)$st largest observation.

For data from a symmetric distribution, the Winsorized mean is an unbiased estimate of the population mean. However, the Winsorized mean does not have a normal distribution even if the data are from a normal population.

The Winsorized sum of squared deviations is defined as

$$s_{wk}^2 = (k+1)(x_{(k+1)} - \bar{x}_{wk})^2 + \sum_{i=k+2}^{n-k-1}(x_{(i)} - \bar{x}_{wk})^2 + (k+1)(x_{(n-k)} - \bar{x}_{wk})^2$$

The Winsorized t statistic is given by

$$t_{wk} = \frac{\bar{x}_{wk} - \mu_0}{\text{SE}(\bar{x}_{wk})}$$

where μ_0 denotes the location under the null hypothesis and the standard error of the Winsorized mean is

$$\text{SE}(\bar{x}_{wk}) = \frac{n-1}{n-2k-1} \times \frac{s_{wk}}{\sqrt{n(n-1)}}$$

When the data are from a symmetric distribution, the distribution of t_{wk} is approximated by a Student's t distribution with $n - 2k - 1$ degrees of freedom (Tukey and McLaughlin 1963; Dixon and Tukey 1968).

The Winsorized $100(1 - \frac{\alpha}{2})\%$ confidence interval for the location parameter has upper and lower limits

$$\bar{x}_{wk} \pm t_{1-\frac{\alpha}{2};n-2k-1}\text{SE}(\bar{x}_{wk})$$

where $t_{1-\frac{\alpha}{2};n-2k-1}$ is the $(1 - \frac{\alpha}{2})100$th percentile of the Student's t distribution with $n - 2k - 1$ degrees of freedom.

Trimmed Means

Like the Winsorized mean, the trimmed mean is a robust estimator of the location that is relatively insensitive to outliers. The k-times trimmed mean is calculated as

$$\bar{x}_{tk} = \frac{1}{n-2k} \sum_{i=k+1}^{n-k} x_{(i)}$$

where n is the number of observations and $x_{(i)}$ is the ith order statistic when the observations are arranged in increasing order:

$$x_{(1)} \leq x_{(2)} \leq \cdots \leq x_{(n)}$$

The trimmed mean is computed after the k smallest and k largest observations are deleted from the sample. In other words, the observations are trimmed at each end.

For a symmetric distribution, the symmetrically trimmed mean is an unbiased estimate of the population mean. However, the trimmed mean does not have a normal distribution even if the data are from a normal population.

A robust estimate of the variance of the trimmed mean t_{tk} can be based on the Winsorized sum of squared deviations s_{wk}^2, which is defined in the section "Winsorized Means" on page 334; see Tukey and McLaughlin (1963). This can be used to compute a trimmed t test which is based on the test statistic

$$t_{tk} = \frac{(\bar{x}_{tk} - \mu_0)}{\text{SE}(\bar{x}_{tk})}$$

where the standard error of the trimmed mean is

$$\text{SE}(\bar{x}_{tk}) = \frac{s_{wk}}{\sqrt{(n-2k)(n-2k-1)}}$$

When the data are from a symmetric distribution, the distribution of t_{tk} is approximated by a Student's t distribution with $n - 2k - 1$ degrees of freedom (Tukey and McLaughlin 1963; Dixon and Tukey 1968).

The "trimmed" $100(1 - \alpha)\%$ confidence interval for the location parameter has upper and lower limits

$$\bar{x}_{tk} \pm t_{1-\frac{\alpha}{2}; n-2k-1} \text{SE}(\bar{x}_{tk})$$

where $t_{1-\frac{\alpha}{2}; n-2k-1}$ is the $(1 - \frac{\alpha}{2})100$th percentile of the Student's t distribution with $n - 2k - 1$ degrees of freedom.

Robust Estimates of Scale

The sample standard deviation, which is the most commonly used estimator of scale, is sensitive to outliers. Robust scale estimators, on the other hand, remain bounded when a single data value is replaced by an arbitrarily large or small value. The UNIVARIATE procedure computes several robust measures of scale, including the interquartile range, Gini's mean difference G, the median absolute deviation about the median (MAD), Q_n, and S_n. In addition, the procedure computes estimates of the normal standard deviation σ derived from each of these measures.

The interquartile range (IQR) is simply the difference between the upper and lower quartiles. For a normal population, σ can be estimated as IQR/1.34898.

Gini's mean difference is computed as

$$G = \frac{1}{\binom{n}{2}} \sum_{i<j} |x_i - x_j|$$

For a normal population, the expected value of G is $2\sigma/\sqrt{\pi}$. Thus $G\sqrt{\pi}/2$ is a robust estimator of σ when the data are from a normal sample. For the normal distribution, this estimator has high efficiency relative to the usual sample standard deviation, and it is also less sensitive to the presence of outliers.

A very robust scale estimator is the MAD, the median absolute deviation from the median (Hampel 1974), which is computed as

$$\text{MAD} = \text{med}_i(|x_i - \text{med}_j(x_j)|)$$

where the inner median, $\text{med}_j(x_j)$, is the median of the n observations, and the outer median (taken over i) is the median of the n absolute values of the deviations about the inner median. For a normal population, $1.4826 \times \text{MAD}$ is an estimator of σ.

The MAD has low efficiency for normal distributions, and it may not always be appropriate for symmetric distributions. Rousseeuw and Croux (1993) proposed two statistics as alternatives to the MAD. The first is

$$S_n = 1.1926 \times \text{med}_i(\text{med}_j(|x_i - x_j|))$$

where the outer median (taken over i) is the median of the n medians of $|x_i - x_j|$, $j = 1, 2, \ldots, n$. To reduce small-sample bias, $c_{sn} S_n$ is used to estimate σ, where c_{sn} is a correction factor; see Croux and Rousseeuw (1992).

The second statistic proposed by Rousseeuw and Croux (1993) is

$$Q_n = 2.2219\{|x_i - x_j|; i < j\}_{(k)}$$

where

$$k = \binom{\left[\frac{n}{2}\right]+1}{2}$$

In other words, Q_n is 2.2219 times the kth order statistic of the $\binom{n}{2}$ distances between the data points. The bias-corrected statistic $c_{qn} Q_n$ is used to estimate σ, where c_{qn} is a correction factor; see Croux and Rousseeuw (1992).

Creating Line Printer Plots

The PLOTS option in the PROC UNIVARIATE statement provides up to four diagnostic line printer plots to examine the data distribution. These plots are the stem-and-leaf plot or horizontal bar chart, the box plot, the normal probability plot, and the side-by-side box plots. If you specify the WEIGHT statement, PROC UNIVARIATE provides a weighted histogram, a weighted box plot based on the weighted quantiles, and a weighted normal probability plot.

Note that these plots are a legacy feature of the UNIVARIATE procedure in earlier versions of SAS. They predate the addition of the CDFPLOT, HISTOGRAM, PPPLOT, PROBPLOT, and QQPLOT statements, which provide high-resolution graphics displays. Also note that line printer plots requested with the PLOTS option are mainly intended for use with the ODS LISTING destination. See Example 4.5.

Stem-and-Leaf Plot

The first plot in the output is either a stem-and-leaf plot (Tukey 1977) or a horizontal bar chart. If any single interval contains more than 49 observations, the horizontal bar chart appears. Otherwise, the stem-and-leaf plot appears. The stem-and-leaf plot is like a horizontal bar chart in that both plots provide a method to visualize the overall distribution of the data. The stem-and-leaf plot provides more detail because each point in the plot represents an individual data value.

To change the number of stems that the plot displays, use PLOTSIZE= to increase or decrease the number of rows. Instructions that appear below the plot explain how to determine the values of the variable. If no instructions appear, you multiply *Stem.Leaf* by 1 to determine the values of the variable. For example, if the stem value is 10 and the leaf value is 1, then the variable value is approximately 10.1. For the stem-and-leaf plot, the procedure rounds a variable value to the nearest leaf. If the variable value is exactly halfway between two leaves, the value rounds to the nearest leaf with an even integer value. For example, a variable value of 3.15 has a stem value of 3 and a leaf value of 2.

Box Plot

The box plot, also known as a schematic box plot, appears beside the stem-and-leaf plot. Both plots use the same vertical scale. The box plot provides a visual summary of the data and identifies outliers. The bottom and top edges of the box correspond to the sample 25th (Q1) and 75th (Q3) percentiles. The box length is one *interquartile range* (Q3 − Q1). The center horizontal line with asterisk endpoints corresponds to the sample median. The central plus sign (+) corresponds to the sample mean. If the mean and median are equal, the plus sign falls on the line inside the box. The vertical lines that project out from the box, called *whiskers*, extend as far as the data extend, up to a distance of 1.5 interquartile ranges. Values farther away are potential outliers. The procedure identifies the extreme values with a zero or an asterisk (*). If zero appears, the value is between 1.5 and 3 interquartile ranges from the top or bottom edge of the box. If an asterisk appears, the value is more extreme.

NOTE: To produce box plots that use high-resolution graphics, use the BOXPLOT procedure in SAS/STAT software. See Chapter 24, "The BOXPLOT Procedure" (*SAS/STAT User's Guide*).

Normal Probability Plot

The normal probability plot plots the empirical quantiles against the quantiles of a standard normal distribution. Asterisks (*) indicate the data values. The plus signs (+) provide a straight reference line that is drawn by using the sample mean and standard deviation. If the data are from a normal distribution, the asterisks tend to fall along the reference line. The vertical coordinate is the data value, and the horizontal coordinate is $\Phi^{-1}(v_i)$ where

$$v_i = \frac{r_i - \frac{3}{8}}{n + \frac{1}{4}}$$

$\Phi^{-1}(\cdot)$ = inverse of the standard normal distribution function
r_i = rank of the ith data value when ordered from smallest to largest
n = number of nonmissing observations

For a weighted normal probability plot, the ith ordered observation is plotted against $\Phi^{-1}(v_i)$ where

$$v_i = \frac{(1 - \frac{3}{8i}) \sum_{j=1}^{i} w_{(j)}}{(1 + \frac{1}{4n}) \sum_{i=1}^{n} w_i}$$

$w_{(j)}$ = weight associated with the jth ordered observation

When each observation has an identical weight, $w_j = w$, the formula for v_i reduces to the expression for v_i in the unweighted normal probability plot:

$$v_i = \frac{i - \frac{3}{8}}{n + \frac{1}{4}}$$

When the value of VARDEF= is WDF or WEIGHT, a reference line with intercept $\hat{\mu}$ and slope $\hat{\sigma}$ is added to the plot. When the value of VARDEF= is DF or N, the slope is $\frac{\hat{sigma}}{\sqrt{\bar{w}}}$ where $\bar{w} = \frac{\sum_{i=1}^{n} w_i}{n}$ is the average weight.

When each observation has an identical weight and the value of VARDEF= is DF, N, or WEIGHT, the reference line reduces to the usual reference line with intercept $\hat{mu}$ and slope $\hat{\sigma}$ in the unweighted normal probability plot.

If the data are normally distributed with mean μ and standard deviation σ, and each observation has an identical weight w, then the points on the plot should lie approximately on a straight line. The intercept for this line is μ. The slope is σ when VARDEF= is WDF or WEIGHT, and the slope is $\frac{\sigma}{\sqrt{w}}$ when VARDEF= is DF or N.

NOTE: To produce high-resolution probability plots, use the PROBPLOT statement in PROC UNIVARIATE; see the section "PROBPLOT Statement" on page 294.

Side-by-Side Box Plots

When you use a BY statement with the PLOT option, PROC UNIVARIATE produces side-by-side box plots, one for each BY group. The box plots (also known as schematic plots) use a common scale that enables you to compare the data distribution across BY groups. This plot appears after the univariate analyses of all BY groups. Use the NOBYPLOT option to suppress this plot.

NOTE: To produce high-resolution side-by-side box plots, use the BOXPLOT procedure in SAS/STAT software. See Chapter 24, "The BOXPLOT Procedure" (*SAS/STAT User's Guide*).

Creating High-Resolution Graphics

If your site licenses SAS/GRAPH software, you can use the CDFPLOT, HISTOGRAM, PPPLOT, PROBPLOT, and QQPLOT statements to create high-resolution graphs.

The CDFPLOT statement plots the observed cumulative distribution function of a variable. You can optionally superimpose a fitted theoretical distribution on the plot.

The HISTOGRAM statement creates histograms that enable you to examine the data distribution. You can optionally fit families of density curves and superimpose kernel density estimates on the histograms. For additional information about the fitted distributions and kernel density estimates, see the sections "Formulas for Fitted Continuous Distributions" on page 348 and "Kernel Density Estimates" on page 360.

The PPPLOT statement creates a probability-probability (P-P) plot, which compares the empirical cumulative distribution function (ecdf) of a variable with a specified theoretical cumulative distribution function. You can use a P-P plot to determine how well a theoretical distribution models a set of measurements.

The PROBPLOT statement creates a probability plot, which compares ordered values of a variable with percentiles of a specified theoretical distribution. Probability plots are useful for graphical estimation of percentiles.

The QQPLOT statement creates a quantile-quantile plot, which compares ordered values of a variable with quantiles of a specified theoretical distribution. Q-Q plots are useful for graphical estimation of distribution parameters.

NOTE: You can use the CLASS statement with any of these plot statements to produce comparative versions of the plots.

Alternatives for Producing Graphics

The UNIVARIATE procedure supports two kinds of graphical output:

- traditional graphics
- ODS Statistical Graphics output, supported on an experimental basis for SAS 9.2

PROC UNIVARIATE produces traditional graphics by default. These graphs are saved in graphics catalogs. Their appearance is controlled by the SAS/GRAPH GOPTIONS, AXIS, and SYMBOL statements (as described in *SAS/GRAPH: Reference*) and numerous specialized plot statement options.

ODS Statistical Graphics (or ODS Graphics for short) is an extension to the Output Delivery System (ODS) that is invoked when you use the ODS GRAPHICS statement prior to your procedure statements. An ODS graph is produced in ODS output (not a graphics catalog), and the details of its appearance and layout are controlled by ODS styles and templates rather than by SAS/GRAPH statements and procedure options. See Chapter 21, "Statistical Graphics Using ODS" (*SAS/STAT User's Guide*), for a thorough discussion of ODS Graphics.

Prior to SAS 9.2, the plots produced by PROC UNIVARIATE were extremely basic by default. Producing attractive graphical output required the careful selection of colors, fonts, and other elements, which were specified via SAS/GRAPH statements and plot statement options. Beginning with SAS 9.2, the default appearance of traditional graphs is governed by the prevailing ODS style, which automatically produces attractive, consistent output. The SAS/GRAPH statements and procedure options for controlling graph appearance continue to be honored for traditional graphics. You can specify the NOGSTYLE system option to prevent the ODS style from affecting the appearance of traditional graphs. This enables existing PROC UNIVARIATE programs to produce customized graphs that appear as they did under previous SAS releases.

The appearance of ODS Graphics output is also controlled by the ODS style, but it is not affected by SAS/GRAPH statements or plot statement options that govern traditional graphics, For example, the CAXIS= option used to specify the color of graph axes in traditional graphics is ignored when producing ODS Graphics output. **NOTE:** Some features available with traditional graphics are not supported in ODS Graphics by the UNIVARIATE procedure for SAS 9.2.

The traditional graphics system enables you to control every detail of a graph through convenient procedure syntax. ODS Graphics provides the highest quality output with minimal syntax and full compatibility with graphics produced by SAS/STAT and SAS/ETS procedures.

The following code produces a histogram with a fitted lognormal distribution of the LoanToValueRatio data introduced in the section "Summarizing a Data Distribution" on page 225:

```
options nogstyle;
proc univariate data=HomeLoans noprint;
   histogram LoanToValueRatio / lognormal;
   inset lognormal(theta sigma zeta) / position=ne;
run;
```

The NOGSTYLE system option keeps the ODS style from influencing the output, and no SAS/GRAPH statements or procedure options affecting the appearance of the plot are specified. Figure 4.8 shows the resulting histogram, which is essentially identical to the default output produced under releases prior to SAS 9.2.

Figure 4.8 Traditional Graph with NOGSTYLE

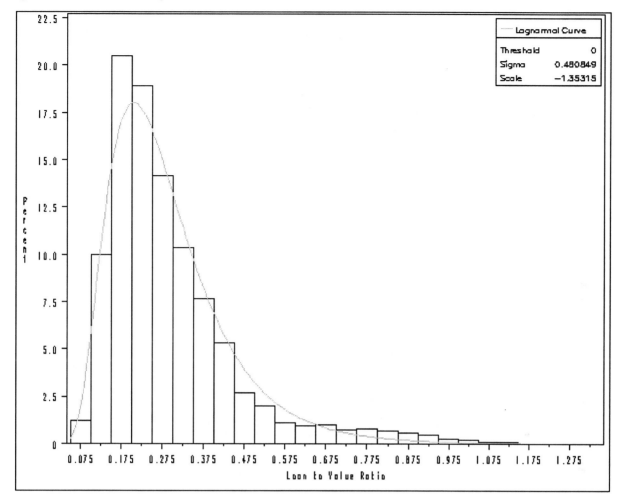

Figure 4.9 shows the result of executing the same code with the GSTYLE system option turned on (the default). Note the influence of the ODS style on the histogram's appearance. For example, the quality of the text is improved and histogram bars are filled by default.

Figure 4.9 Traditional Graph with GSTYLE

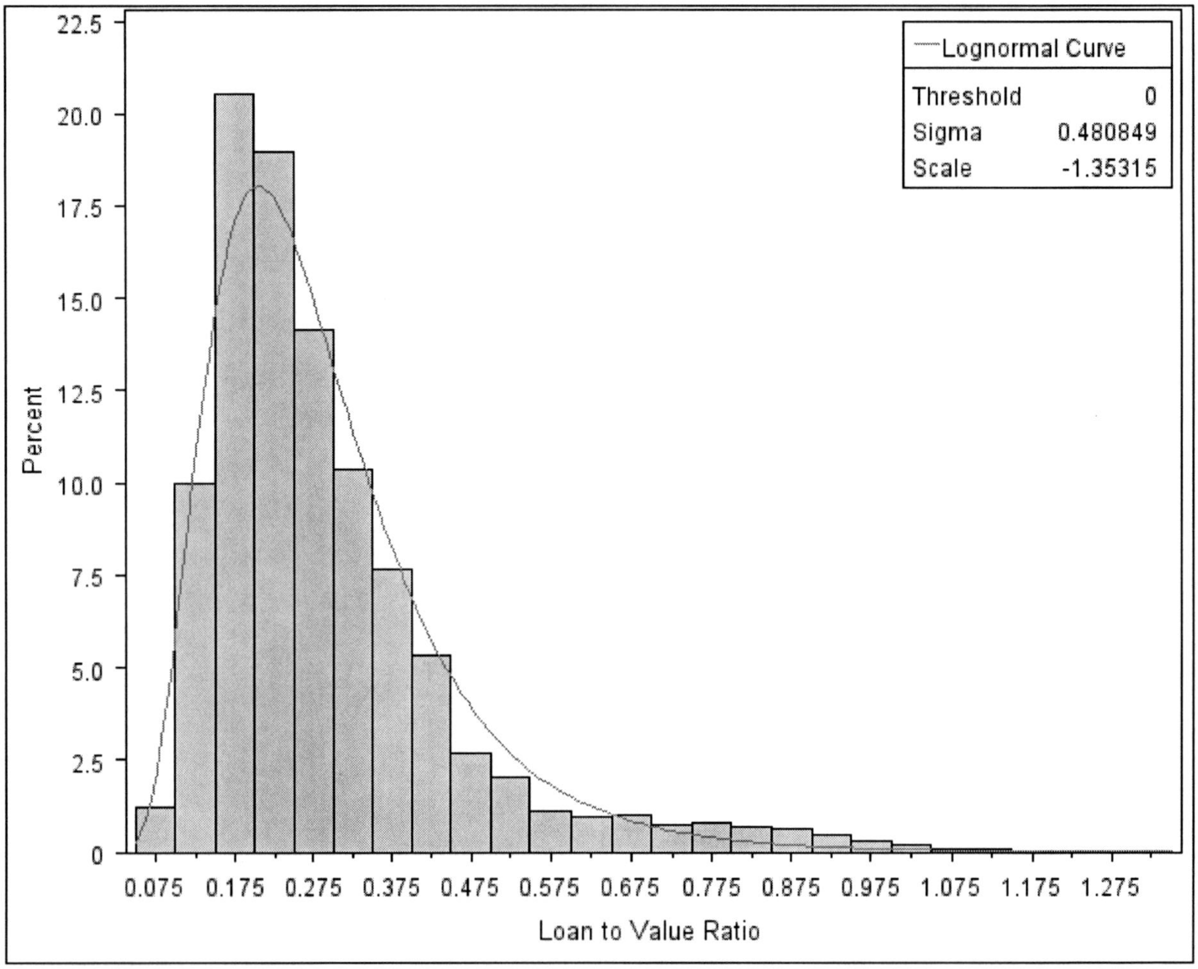

Figure 4.10 shows the same histogram produced using ODS Graphics. The histogram's appearance is governed by the same style elements as in Figure 4.9, but the plots are not identical. Note, for example, the title incorporated in the ODS Graphics output and the smoother appearance of the fitted curve.

Figure 4.10 ODS Graphics Output

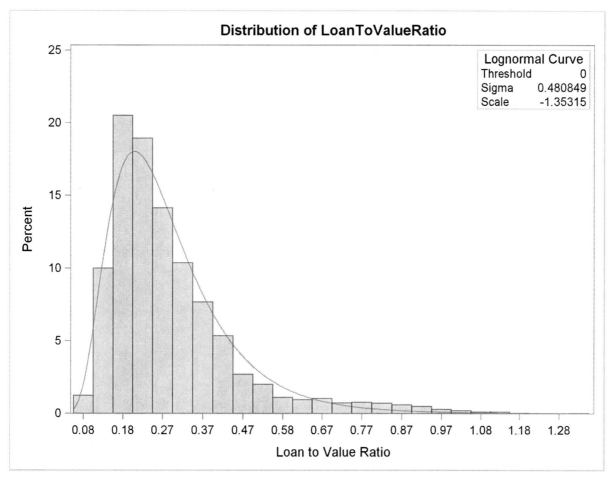

Using the CLASS Statement to Create Comparative Plots

When you use the CLASS statement with the CDFPLOT, HISTOGRAM, PPPLOT, PROBPLOT, or QQPLOT statements, PROC UNIVARIATE creates comparative versions of the plots. You can use these plot statements with the CLASS statement to create one-way and two-way comparative plots. When you use one CLASS variable, PROC UNIVARIATE displays an array of component plots (stacked or side-by-side), one for each level of the classification variable. When you use two CLASS variables, PROC UNIVARIATE displays a matrix of component plots, one for each combination of levels of the classification variables. The observations in a given level are referred to collectively as a *cell*.

When you create a one-way comparative plot, the observations in the input data set are sorted by the method specified in the ORDER= option. PROC UNIVARIATE creates a separate plot for the analysis variable values in each level and arranges these component plots in an array to form the comparative plot with uniform horizontal and vertical axes. See Example 4.15.

When you create a two-way comparative plot, the observations in the input data set are cross-classified according to the values (levels) of these variables. PROC UNIVARIATE creates a separate plot for the analysis variable values in each cell of the cross-classification and arranges these component plots in a matrix to form the comparative plot with uniform horizontal and vertical axes. The levels of the first CLASS variable are the labels for the rows of the matrix, and the levels of the second CLASS variable are the labels for the columns of the matrix. See Example 4.16.

PROC UNIVARIATE determines the layout of a two-way comparative plot by using the order for the first CLASS variable to obtain the order of the rows from top to bottom. Then it applies the order for the second CLASS variable to the observations that correspond to the first row to obtain the order of the columns from left to right. If any columns remain unordered (that is, the categories are unbalanced), PROC UNIVARIATE applies the order for the second CLASS variable to the observations in the second row, and so on, until all the columns have been ordered.

If you associate a label with a CLASS variable, PROC UNIVARIATE displays the variable label in the comparative plot and this label is parallel to the column (or row) labels.

Use the MISSING option to treat missing values as valid levels.

To reduce the number of classification levels, use a FORMAT statement to combine variable values.

Positioning Insets

Positioning an Inset Using Compass Point Values

To position an inset by using a compass point position, specify the value N, NE, E, SE, S, SW, W, or NW with the POSITION= option. The default position of the inset is NW. The following statements produce a histogram to show the position of the inset for the eight compass points:

```
data Score;
   input Student $ PreTest PostTest @@;
   label ScoreChange = 'Change in Test Scores';
   ScoreChange = PostTest - PreTest;
datalines;
Capalleti  94 91  Dubose      51 65
Engles     95 97  Grant       63 75
Krupski    80 75  Lundsford   92 55
Mcbane     75 78  Mullen      89 82
Nguyen     79 76  Patel       71 77
Si         75 70  Tanaka      87 73
;
run;
title 'Test Scores for a College Course';
proc univariate data=Score noprint;
   histogram PreTest / midpoints = 45 to 95 by 10;
   inset n        / cfill=blank
                    header='Position = NW' pos=nw;
   inset mean     / cfill=blank
                    header='Position = N ' pos=n ;
   inset sum      / cfill=blank
                    header='Position = NE' pos=ne;
   inset max      / cfill=blank
                    header='Position = E ' pos=e ;
```

```
        inset min    / cfill=blank
                       header='Position = SE' pos=se;
        inset nobs   / cfill=blank
                       header='Position = S ' pos=s ;
        inset range  / cfill=blank
                       header='Position = SW' pos=sw;
        inset mode   / cfill=blank
                       header='Position = W ' pos=w ;
        label PreTest = 'Pretest Score';
    run;
```

Figure 4.11 Compass Positions for Inset

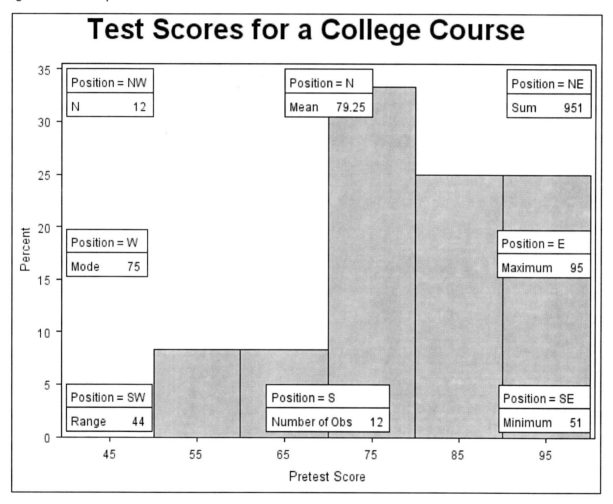

Positioning Insets in the Margins

To position an inset in one of the four margins that surround the plot area, specify the value LM, RM, TM, or BM with the POSITION= option. Margin positions are recommended if you list a large number of statistics in the INSET statement. If you attempt to display a lengthy inset in the interior of the plot, the inset is likely to collide with the data display.

Positioning an Inset Using Coordinates

To position an inset with coordinates, use POSITION=(x,y). You specify the coordinates in axis data units or in axis percentage units (the default). **NOTE:** You cannot position an inset with coordinates when producing ODS Graphics output.

If you specify the DATA option immediately following the coordinates, PROC UNIVARIATE positions the inset by using axis data units. For example, the following statements place the bottom left corner of the inset at 45 on the horizontal axis and 10 on the vertical axis:

```
title 'Test Scores for a College Course';
proc univariate data=Score noprint;
   histogram PreTest / midpoints = 45 to 95 by 10;
   inset n / header    = 'Position=(45,10)'
             position  = (45,10) data;
run;
```

Figure 4.12 Coordinate Position for Inset

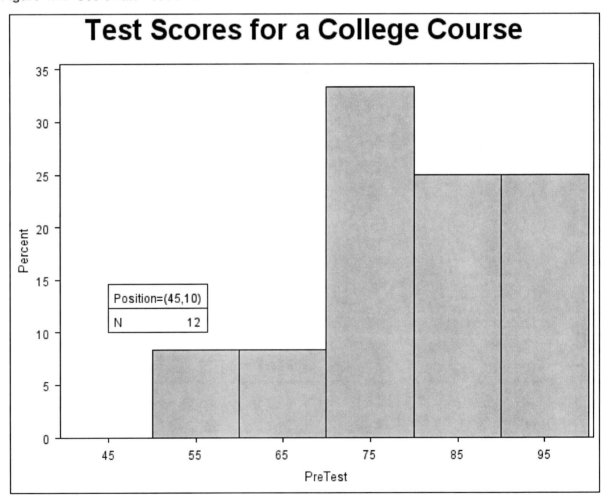

By default, the specified coordinates determine the position of the bottom left corner of the inset. To change this reference point, use the REFPOINT= option (see below).

If you omit the DATA option, PROC UNIVARIATE positions the inset by using axis percentage units. The coordinates in axis percentage units must be between 0 and 100. The coordinates of the bottom left corner of the display are (0,0), while the upper right corner is (100, 100). For example, the following statements create a histogram and use coordinates in axis percentage units to position the two insets:

```
title 'Test Scores for a College Course';
proc univariate data=Score noprint;
   histogram PreTest / midpoints = 45 to 95 by 10;
   inset min / position = (5,25)
               header   = 'Position=(5,25)'
               refpoint = tl;
   inset max / position = (95,95)
               header   = 'Position=(95,95)'
               refpoint = tr;
run;
```

The REFPOINT= option determines which corner of the inset to place at the coordinates that are specified with the POSITION= option. The first inset uses REFPOINT=TL, so that the top left corner of the inset is positioned 5% of the way across the horizontal axis and 25% of the way up the vertical axis. The second inset uses REFPOINT=TR, so that the top right corner of the inset is positioned 95% of the way across the horizontal axis and 95% of the way up the vertical axis.

Figure 4.13 Reference Point for Inset

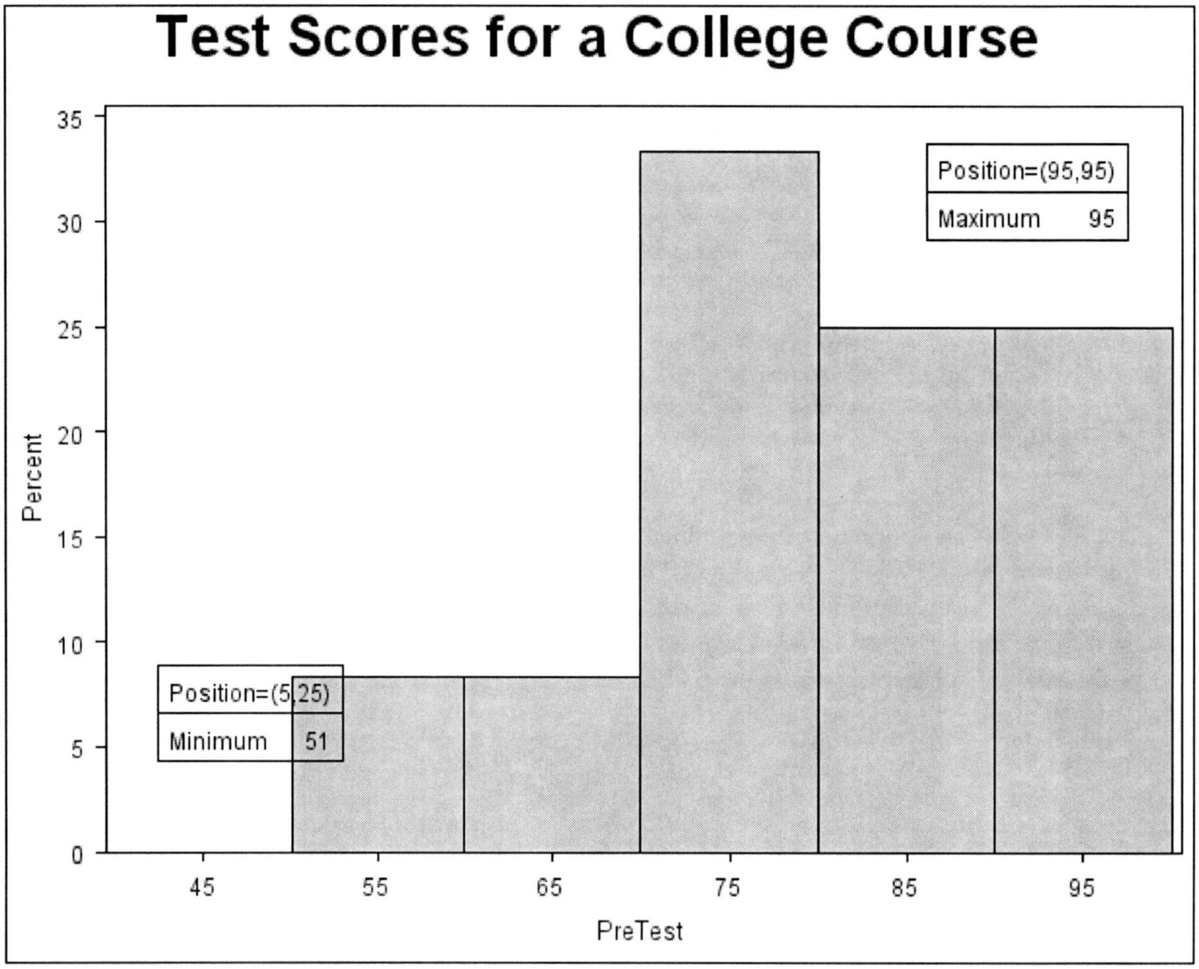

Test Scores for a College Course

(histogram of PreTest scores with insets: "Position=(95,95) Maximum 95" and "Position=(5,25) Minimum 51")

A sample program for these examples, *univar3.sas*, is available in the SAS Sample Library for Base SAS software.

Formulas for Fitted Continuous Distributions

The following sections provide information about the families of parametric distributions that you can fit with the HISTOGRAM statement. Properties of these distributions are discussed by Johnson, Kotz, and Balakrishnan (1994, 1995).

Beta Distribution

The fitted density function is

$$p(x) = \begin{cases} hv \frac{(x-\theta)^{\alpha-1}(\sigma+\theta-x)^{\beta-1}}{B(\alpha,\beta)\sigma^{(\alpha+\beta-1)}} & \text{for } \theta < x < \theta + \sigma \\ 0 & \text{for } x \leq \theta \text{ or } x \geq \theta + \sigma \end{cases}$$

where $B(\alpha, \beta) = \frac{\Gamma(\alpha)\Gamma(\beta)}{\Gamma(\alpha+\beta)}$ and

θ = lower threshold parameter (lower endpoint parameter)
σ = scale parameter ($\sigma > 0$)
α = shape parameter ($\alpha > 0$)
β = shape parameter ($\beta > 0$)
h = width of histogram interval
v = vertical scaling factor

and

$$v = \begin{cases} n & \text{the sample size, for VSCALE=COUNT} \\ 100 & \text{for VSCALE=PERCENT} \\ 1 & \text{for VSCALE=PROPPORTION} \end{cases}$$

NOTE: This notation is consistent with that of other distributions that you can fit with the HISTOGRAM statement. However, many texts, including Johnson, Kotz, and Balakrishnan (1995), write the beta density function as

$$p(x) = \begin{cases} \frac{(x-a)^{p-1}(b-x)^{q-1}}{B(p,q)(b-a)^{p+q-1}} & \text{for } a < x < b \\ 0 & \text{for } x \leq a \text{ or } x \geq b \end{cases}$$

The two parameterizations are related as follows:

$\sigma = b - a$
$\theta = a$
$\alpha = p$
$\beta = q$

The range of the beta distribution is bounded below by a threshold parameter $\theta = a$ and above by $\theta + \sigma = b$. If you specify a fitted beta curve by using the BETA option, θ must be less than the minimum data value and $\theta + \sigma$ must be greater than the maximum data value. You can specify θ and σ with the THETA= and SIGMA= *beta-options* in parentheses after the keyword BETA. By default, $\sigma = 1$ and $\theta = 0$. If you specify THETA=EST and SIGMA=EST, maximum likelihood estimates are computed for θ and σ. However, three- and four-parameter maximum likelihood estimation does not always converge.

In addition, you can specify α and β with the ALPHA= and BETA= *beta-options*, respectively. By default, the procedure calculates maximum likelihood estimates for α and β. For example, to fit a beta density curve to a set of data bounded below by 32 and above by 212 with maximum likelihood estimates for α and β, use the following statement:

```
histogram Length / beta(theta=32 sigma=180);
```

The beta distributions are also referred to as Pearson Type I or II distributions. These include the power function distribution ($\beta = 1$), the arc sine distribution ($\alpha = \beta = \frac{1}{2}$), and the generalized arc sine distributions ($\alpha + \beta = 1, \beta \neq \frac{1}{2}$).

You can use the DATA step function BETAINV to compute beta quantiles and the DATA step function PROBBETA to compute beta probabilities.

Exponential Distribution

The fitted density function is

$$p(x) = \begin{cases} \frac{hv}{\sigma} \exp(-(\frac{x-\theta}{\sigma})) & \text{for } x \geq \theta \\ 0 & \text{for } x < \theta \end{cases}$$

where

θ = threshold parameter
σ = scale parameter ($\sigma > 0$)
h = width of histogram interval
v = vertical scaling factor

and

$$v = \begin{cases} n & \text{the sample size, for VSCALE=COUNT} \\ 100 & \text{for VSCALE=PERCENT} \\ 1 & \text{for VSCALE=PROPPORTION} \end{cases}$$

The threshold parameter θ must be less than or equal to the minimum data value. You can specify θ with the THRESHOLD= *exponential-option*. By default, $\theta = 0$. If you specify THETA=EST, a maximum likelihood estimate is computed for θ. In addition, you can specify σ with the SCALE= *exponential-option*. By default, the procedure calculates a maximum likelihood estimate for σ. Note that some authors define the scale parameter as $\frac{1}{\sigma}$.

The exponential distribution is a special case of both the gamma distribution (with $\alpha = 1$) and the Weibull distribution (with $c = 1$). A related distribution is the extreme value distribution. If $Y = \exp(-X)$ has an exponential distribution, then X has an extreme value distribution.

Gamma Distribution

The fitted density function is

$$p(x) = \begin{cases} \frac{hv}{\Gamma(\alpha)\sigma} (\frac{x-\theta}{\sigma})^{\alpha-1} \exp(-(\frac{x-\theta}{\sigma})) & \text{for } x > \theta \\ 0 & \text{for } x \leq \theta \end{cases}$$

where

θ = threshold parameter
σ = scale parameter ($\sigma > 0$)
α = shape parameter ($\alpha > 0$)
h = width of histogram interval
v = vertical scaling factor

and

$$v = \begin{cases} n & \text{the sample size, for VSCALE=COUNT} \\ 100 & \text{for VSCALE=PERCENT} \\ 1 & \text{for VSCALE=PROPPORTION} \end{cases}$$

The threshold parameter θ must be less than the minimum data value. You can specify θ with the THRESHOLD= *gamma-option*. By default, $\theta = 0$. If you specify THETA=EST, a maximum likelihood estimate is computed for θ. In addition, you can specify σ and α with the SCALE= and ALPHA= *gamma-options*. By default, the procedure calculates maximum likelihood estimates for σ and α.

The gamma distributions are also referred to as Pearson Type III distributions, and they include the chi-square, exponential, and Erlang distributions. The probability density function for the chi-square distribution is

$$p(x) = \begin{cases} \frac{1}{2\Gamma(\frac{\nu}{2})} \left(\frac{x}{2}\right)^{\frac{\nu}{2}-1} \exp(-\frac{x}{2}) & \text{for } x > 0 \\ 0 & \text{for } x \leq 0 \end{cases}$$

Notice that this is a gamma distribution with $\alpha = \frac{\nu}{2}, \sigma = 2$, and $\theta = 0$. The exponential distribution is a gamma distribution with $\alpha = 1$, and the Erlang distribution is a gamma distribution with α being a positive integer. A related distribution is the Rayleigh distribution. If $R = \frac{\max(X_1,\ldots,X_n)}{\min(X_1,\ldots,X_n)}$ where the X_i's are independent χ^2_ν variables, then $\log R$ is distributed with a χ_ν distribution having a probability density function of

$$p(x) = \begin{cases} \left[2^{\frac{\nu}{2}-1}\Gamma(\frac{\nu}{2})\right]^{-1} x^{\nu-1} \exp(-\frac{x^2}{2}) & \text{for } x > 0 \\ 0 & \text{for } x \leq 0 \end{cases}$$

If $\nu = 2$, the preceding distribution is referred to as the Rayleigh distribution.

You can use the DATA step function GAMINV to compute gamma quantiles and the DATA step function PROBGAM to compute gamma probabilities.

Lognormal Distribution

The fitted density function is

$$p(x) = \begin{cases} \frac{h\upsilon}{\sigma\sqrt{2\pi}(x-\theta)} \exp\left(-\frac{(\log(x-\theta)-\zeta)^2}{2\sigma^2}\right) & \text{for } x > \theta \\ 0 & \text{for } x \leq \theta \end{cases}$$

where

θ = threshold parameter
ζ = scale parameter ($-\infty < \zeta < \infty$)
σ = shape parameter ($\sigma > 0$)
h = width of histogram interval
υ = vertical scaling factor

and

$$v = \begin{cases} n & \text{the sample size, for VSCALE=COUNT} \\ 100 & \text{for VSCALE=PERCENT} \\ 1 & \text{for VSCALE=PROPPORTION} \end{cases}$$

The threshold parameter θ must be less than the minimum data value. You can specify θ with the THRESHOLD= *lognormal-option*. By default, $\theta = 0$. If you specify THETA=EST, a maximum likelihood estimate is computed for θ. You can specify ζ and σ with the SCALE= and SHAPE= *lognormal-options*, respectively. By default, the procedure calculates maximum likelihood estimates for these parameters.

NOTE: The lognormal distribution is also referred to as the S_L distribution in the Johnson system of distributions.

NOTE: This book uses σ to denote the shape parameter of the lognormal distribution, whereas σ is used to denote the scale parameter of the beta, exponential, gamma, normal, and Weibull distributions. The use of σ to denote the lognormal shape parameter is based on the fact that $\frac{1}{\sigma}(\log(X - \theta) - \zeta)$ has a standard normal distribution if X is lognormally distributed. Based on this relationship, you can use the DATA step function PROBIT to compute lognormal quantiles and the DATA step function PROBNORM to compute probabilities.

Normal Distribution

The fitted density function is

$$p(x) = \frac{hv}{\sigma\sqrt{2\pi}} \exp\left(-\frac{1}{2}\left(\frac{x-\mu}{\sigma}\right)^2\right) \quad \text{for } -\infty < x < \infty$$

where

μ = mean
σ = standard deviation ($\sigma > 0$)
h = width of histogram interval
v = vertical scaling factor

and

$$v = \begin{cases} n & \text{the sample size, for VSCALE=COUNT} \\ 100 & \text{for VSCALE=PERCENT} \\ 1 & \text{for VSCALE=PROPPORTION} \end{cases}$$

You can specify μ and σ with the MU= and SIGMA= *normal-options*, respectively. By default, the procedure estimates μ with the sample mean and σ with the sample standard deviation.

You can use the DATA step function PROBIT to compute normal quantiles and the DATA step function PROBNORM to compute probabilities.

NOTE: The normal distribution is also referred to as the S_N distribution in the Johnson system of distributions.

Johnson S_B Distribution

The fitted density function is

$$p(x) = \begin{cases} \frac{\delta h v}{\sigma \sqrt{2\pi}} \left[\left(\frac{x-\theta}{\sigma} \right) \left(1 - \frac{x-\theta}{\sigma} \right) \right]^{-1} \times \\ \exp\left[-\frac{1}{2} \left(\gamma + \delta \log(\frac{x-\theta}{\theta+\sigma-x}) \right)^2 \right] & \text{for } \theta < x < \theta + \sigma \\ 0 & \text{for } x \leq \theta \text{ or } x \geq \theta + \sigma \end{cases}$$

where

θ = threshold parameter ($-\infty < \theta < \infty$)
σ = scale parameter ($\sigma > 0$)
δ = shape parameter ($\delta > 0$)
γ = shape parameter ($-\infty < \gamma < \infty$)
h = width of histogram interval
v = vertical scaling factor

and

$$v = \begin{cases} n & \text{the sample size, for VSCALE=COUNT} \\ 100 & \text{for VSCALE=PERCENT} \\ 1 & \text{for VSCALE=PROPPORTION} \end{cases}$$

The S_B distribution is bounded below by the parameter θ and above by the value $\theta + \sigma$. The parameter θ must be less than the minimum data value. You can specify θ with the THETA= S_B-*option*, or you can request that θ be estimated with the THETA = EST S_B-*option*. The default value for θ is zero. The sum $\theta + \sigma$ must be greater than the maximum data value. The default value for σ is one. You can specify σ with the SIGMA= S_B-*option*, or you can request that σ be estimated with the SIGMA = EST S_B-*option*.

By default, the method of percentiles given by Slifker and Shapiro (1980) is used to estimate the parameters. This method is based on four data percentiles, denoted by x_{-3z}, x_{-z}, x_z, and x_{3z}, which correspond to the four equally spaced percentiles of a standard normal distribution, denoted by $-3z$, $-z$, z, and $3z$, under the transformation

$$z = \gamma + \delta \log \left(\frac{x - \theta}{\theta + \sigma - x} \right)$$

The default value of z is 0.524. The results of the fit are dependent on the choice of z, and you can specify other values with the FITINTERVAL= option (specified in parentheses after the SB option). If you use the method of percentiles, you should select a value of z that corresponds to percentiles which are critical to your application.

The following values are computed from the data percentiles:

$m = x_{3z} - x_z$
$n = x_{-z} - x_{-3z}$
$p = x_z - x_{-z}$

It was demonstrated by Slifker and Shapiro (1980) that

$\frac{mn}{p^2} > 1$ for any S_U distribution
$\frac{mn}{p^2} < 1$ for any S_B distribution
$\frac{mn}{p^2} = 1$ for any S_L (lognormal) distribution

A tolerance interval around one is used to discriminate among the three families with this ratio criterion. You can specify the tolerance with the FITTOLERANCE= option (specified in parentheses after the SB option). The default tolerance is 0.01. Assuming that the criterion satisfies the inequality

$$\frac{mn}{p^2} < 1 - \text{tolerance}$$

the parameters of the S_B distribution are computed using the explicit formulas derived by Slifker and Shapiro (1980).

If you specify FITMETHOD = MOMENTS (in parentheses after the SB option), the method of moments is used to estimate the parameters. If you specify FITMETHOD = MLE (in parentheses after the SB option), the method of maximum likelihood is used to estimate the parameters. Note that maximum likelihood estimates may not always exist. Refer to Bowman and Shenton (1983) for discussion of methods for fitting Johnson distributions.

Johnson S_U Distribution

The fitted density function is

$$p(x) = \begin{cases} \frac{\delta h v}{\sigma \sqrt{2\pi}} \frac{1}{\sqrt{1+((x-\theta)/\sigma)^2}} \times \\ \exp\left[-\frac{1}{2}\left(\gamma + \delta \sinh^{-1}\left(\frac{x-\theta}{\sigma}\right)\right)^2\right] & \text{for } x > \theta \\ 0 & \text{for } x \leq \theta \end{cases}$$

where

θ = location parameter ($-\infty < \theta < \infty$)
σ = scale parameter ($\sigma > 0$)
δ = shape parameter ($\delta > 0$)
γ = shape parameter ($-\infty < \gamma < \infty$)
h = width of histogram interval
v = vertical scaling factor

and

$$v = \begin{cases} n & \text{the sample size, for VSCALE=COUNT} \\ 100 & \text{for VSCALE=PERCENT} \\ 1 & \text{for VSCALE=PROPPORTION} \end{cases}$$

You can specify the parameters with the THETA=, SIGMA=, DELTA=, and GAMMA= S_U-*options*, which are enclosed in parentheses after the SU option. If you do not specify these parameters, they are estimated.

By default, the method of percentiles given by Slifker and Shapiro (1980) is used to estimate the parameters. This method is based on four data percentiles, denoted by x_{-3z}, x_{-z}, x_z, and x_{3z}, which correspond to the four equally spaced percentiles of a standard normal distribution, denoted by $-3z, -z, z$, and $3z$, under the transformation

$$z = \gamma + \delta \sinh^{-1}\left(\frac{x - \theta}{\sigma}\right)$$

The default value of z is 0.524. The results of the fit are dependent on the choice of z, and you can specify other values with the FITINTERVAL= option (specified in parentheses after the SB option). If you use the method of percentiles, you should select a value of z that corresponds to percentiles that are critical to your application. You can specify the value of z with the FITINTERVAL= option (specified in parentheses after the SU option).

The following values are computed from the data percentiles:

$$\begin{aligned} m &= x_{3z} - x_z \\ n &= x_{-z} - x_{-3z} \\ p &= x_z - x_{-z} \end{aligned}$$

It was demonstrated by Slifker and Shapiro (1980) that

$\frac{mn}{p^2} > 1$ for any S_U distribution

$\frac{mn}{p^2} < 1$ for any S_B distribution

$\frac{mn}{p^2} = 1$ for any S_L (lognormal) distribution

A tolerance interval around one is used to discriminate among the three families with this ratio criterion. You can specify the tolerance with the FITTOLERANCE= option (specified in parentheses after the SU option). The default tolerance is 0.01. Assuming that the criterion satisfies the inequality

$$\frac{mn}{p^2} > 1 + \text{tolerance}$$

the parameters of the S_U distribution are computed using the explicit formulas derived by Slifker and Shapiro (1980).

If you specify FITMETHOD = MOMENTS (in parentheses after the SU option), the method of moments is used to estimate the parameters. If you specify FITMETHOD = MLE (in parentheses after the SU option), the method of maximum likelihood is used to estimate the parameters. Note that maximum likelihood estimates do not always exist. Refer to Bowman and Shenton (1983) for discussion of methods for fitting Johnson distributions.

Weibull Distribution

The fitted density function is

$$p(x) = \begin{cases} hv\frac{c}{\sigma}(\frac{x-\theta}{\sigma})^{c-1} \exp(-(\frac{x-\theta}{\sigma})^c) & \text{for } x > \theta \\ 0 & \text{for } x \leq \theta \end{cases}$$

where

θ = threshold parameter
σ = scale parameter ($\sigma > 0$)
c = shape parameter ($c > 0$)
h = width of histogram interval
v = vertical scaling factor

and

$$v = \begin{cases} n & \text{the sample size, for VSCALE=COUNT} \\ 100 & \text{for VSCALE=PERCENT} \\ 1 & \text{for VSCALE=PROPPORTION} \end{cases}$$

The threshold parameter θ must be less than the minimum data value. You can specify θ with the THRESHOLD= *Weibull-option*. By default, $\theta = 0$. If you specify THETA=EST, a maximum likelihood estimate is computed for θ. You can specify σ and c with the SCALE= and SHAPE= *Weibull-options*, respectively. By default, the procedure calculates maximum likelihood estimates for σ and c.

The exponential distribution is a special case of the Weibull distribution where $c = 1$.

Goodness-of-Fit Tests

When you specify the NORMAL option in the PROC UNIVARIATE statement or you request a fitted parametric distribution in the HISTOGRAM statement, the procedure computes goodness-of-fit tests for the null hypothesis that the values of the analysis variable are a random sample from the specified theoretical distribution. See Example 4.22.

When you specify the NORMAL option, these tests, which are summarized in the output table labeled "Tests for Normality," include the following:

- Shapiro-Wilk test
- Kolmogorov-Smirnov test
- Anderson-Darling test
- Cramér-von Mises test

The Kolmogorov-Smirnov D statistic, the Anderson-Darling statistic, and the Cramér-von Mises statistic are based on the empirical distribution function (EDF). However, some EDF tests are not supported when certain combinations of the parameters of a specified distribution are estimated. See Table 4.84 for a list of the EDF tests available. You determine whether to reject the null hypothesis by examining the p-value that is associated with a goodness-of-fit statistic. When the p-value is less than the predetermined critical value (α), you reject the null hypothesis and conclude that the data did not come from the specified distribution.

If you want to test the normality assumptions for analysis of variance methods, beware of using a statistical test for normality alone. A test's ability to reject the null hypothesis (known as the *power* of the test) increases with the sample size. As the sample size becomes larger, increasingly smaller departures from normality can be detected. Because small deviations from normality do not severely affect the validity of analysis of variance tests, it is important to examine other statistics and plots to make a final assessment of normality. The skewness and kurtosis measures and the plots that are provided by the PLOTS option, the HISTOGRAM statement, the PROBPLOT statement, and the QQPLOT statement can be very helpful. For small sample sizes, power is low for detecting larger departures from normality that may be important. To increase the test's ability to detect such deviations, you may want to declare significance at higher levels, such as 0.15 or 0.20, rather than the often-used 0.05 level. Again, consulting plots and additional statistics can help you assess the severity of the deviations from normality.

Shapiro-Wilk Statistic

If the sample size is less than or equal to 2000 and you specify the NORMAL option, PROC UNIVARIATE computes the Shapiro-Wilk statistic, W (also denoted as W_n to emphasize its dependence on the sample size n). The W statistic is the ratio of the best estimator of the variance (based on the square of a linear combination of the order statistics) to the usual corrected sum of squares estimator of the variance (Shapiro and Wilk 1965). When n is greater than three, the coefficients to compute the linear combination of the order statistics are approximated by the method of Royston (1992). The statistic W is always greater than zero and less than or equal to one ($0 < W \leq 1$).

Small values of W lead to the rejection of the null hypothesis of normality. The distribution of W is highly skewed. Seemingly large values of W (such as 0.90) may be considered small and lead you to reject the null hypothesis. The method for computing the p-value (the probability of obtaining a W statistic less than or equal to the observed value) depends on n. For $n = 3$, the probability distribution of W is known and is used to determine the p-value. For $n > 4$, a normalizing transformation is computed:

$$Z_n = \begin{cases} (-\log(\gamma - \log(1 - W_n)) - \mu)/\sigma & \text{if } 4 \leq n \leq 11 \\ (\log(1 - W_n) - \mu)/\sigma & \text{if } 12 \leq n \leq 2000 \end{cases}$$

The values of σ, γ, and μ are functions of n obtained from simulation results. Large values of Z_n indicate departure from normality, and because the statistic Z_n has an approximately standard normal distribution, this distribution is used to determine the p-values for $n > 4$.

EDF Goodness-of-Fit Tests

When you fit a parametric distribution, PROC UNIVARIATE provides a series of goodness-of-fit tests based on the empirical distribution function (EDF). The EDF tests offer advantages over traditional chi-square goodness-of-fit test, including improved power and invariance with respect to the histogram midpoints. For a thorough discussion, refer to D'Agostino and Stephens (1986).

The empirical distribution function is defined for a set of n independent observations $X_1, \ldots, X_n$ with a common distribution function $F(x)$. Denote the observations ordered from smallest to largest as $X_{(1)}, \ldots, X_{(n)}$. The empirical distribution function, $F_n(x)$, is defined as

$$F_n(x) = 0, \quad x < X_{(1)}$$
$$F_n(x) = \frac{i}{n}, \quad X_{(i)} \leq x < X_{(i+1)} \quad i = 1, \ldots, n-1$$
$$F_n(x) = 1, \quad X_{(n)} \leq x$$

Note that $F_n(x)$ is a step function that takes a step of height $\frac{1}{n}$ at each observation. This function estimates the distribution function $F(x)$. At any value x, $F_n(x)$ is the proportion of observations less than or equal to x, while $F(x)$ is the probability of an observation less than or equal to x. EDF statistics measure the discrepancy between $F_n(x)$ and $F(x)$.

The computational formulas for the EDF statistics make use of the probability integral transformation $U = F(X)$. If $F(X)$ is the distribution function of X, the random variable U is uniformly distributed between 0 and 1.

Given n observations $X_{(1)}, \ldots, X_{(n)}$, the values $U_{(i)} = F(X_{(i)})$ are computed by applying the transformation, as discussed in the next three sections.

PROC UNIVARIATE provides three EDF tests:

- Kolmogorov-Smirnov
- Anderson-Darling
- Cramér-von Mises

The following sections provide formal definitions of these EDF statistics.

Kolmogorov D Statistic

The Kolmogorov-Smirnov statistic (D) is defined as

$$D = \sup_x |F_n(x) - F(x)|$$

The Kolmogorov-Smirnov statistic belongs to the supremum class of EDF statistics. This class of statistics is based on the largest vertical difference between $F(x)$ and $F_n(x)$.

The Kolmogorov-Smirnov statistic is computed as the maximum of D^+ and D^-, where D^+ is the largest vertical distance between the EDF and the distribution function when the EDF is greater than the distribution function, and D^- is the largest vertical distance when the EDF is less than the distribution function.

$$D^+ = \max_i \left(\frac{i}{n} - U_{(i)} \right)$$
$$D^- = \max_i \left(U_{(i)} - \frac{i-1}{n} \right)$$
$$D = \max(D^+, D^-)$$

PROC UNIVARIATE uses a modified Kolmogorov D statistic to test the data against a normal distribution with mean and variance equal to the sample mean and variance.

Anderson-Darling Statistic

The Anderson-Darling statistic and the Cramér-von Mises statistic belong to the quadratic class of EDF statistics. This class of statistics is based on the squared difference $(F_n(x) - F(x))^2$. Quadratic statistics have the following general form:

$$Q = n \int_{-\infty}^{+\infty} (F_n(x) - F(x))^2 \psi(x) dF(x)$$

The function $\psi(x)$ weights the squared difference $(F_n(x) - F(x))^2$.

The Anderson-Darling statistic (A^2) is defined as

$$A^2 = n \int_{-\infty}^{+\infty} (F_n(x) - F(x))^2 [F(x)(1 - F(x))]^{-1} dF(x)$$

Here the weight function is $\psi(x) = [F(x)(1 - F(x))]^{-1}$.

The Anderson-Darling statistic is computed as

$$A^2 = -n - \frac{1}{n} \sum_{i=1}^{n} \left[(2i - 1) \log U_{(i)} + (2n + 1 - 2i) \log(1 - U_{(i)}) \right]$$

Cramér-von Mises Statistic

The Cramér-von Mises statistic (W^2) is defined as

$$W^2 = n \int_{-\infty}^{+\infty} (F_n(x) - F(x))^2 dF(x)$$

Here the weight function is $\psi(x) = 1$.

The Cramér-von Mises statistic is computed as

$$W^2 = \sum_{i=1}^{n} \left(U_{(i)} - \frac{2i-1}{2n} \right)^2 + \frac{1}{12n}$$

Probability Values of EDF Tests

Once the EDF test statistics are computed, PROC UNIVARIATE computes the associated probability values (p-values). The UNIVARIATE procedure uses internal tables of probability levels similar to those given by D'Agostino and Stephens (1986). If the value is between two probability levels, then linear interpolation is used to estimate the probability value.

The probability value depends upon the parameters that are known and the parameters that are estimated for the distribution. Table 4.84 summarizes different combinations fitted for which EDF tests are available.

Table 4.84 Availability of EDF Tests

Distribution	Parameters			Tests Available
	Threshold	Scale	Shape	
beta	θ known	σ known	α, β known	all
	θ known	σ known	$\alpha, \beta < 5$ unknown	all
exponential	θ known,	σ known		all
	θ known	σ unknown		all
	θ unknown	σ known		all
	θ unknown	σ unknown		all
gamma	θ known	σ known	α known	all
	θ known	σ unknown	α known	all
	θ known	σ known	α unknown	all
	θ known	σ unknown	α unknown	all
	θ unknown	σ known	$\alpha > 1$ known	all
	θ unknown	σ unknown	$\alpha > 1$ known	all
	θ unknown	σ known	$\alpha > 1$ unknown	all
	θ unknown	σ unknown	$\alpha > 1$ unknown	all
lognormal	θ known	ζ known	σ known	all
	θ known	ζ known	σ unknown	A^2 and W^2
	θ known	ζ unknown	σ known	A^2 and W^2
	θ known	ζ unknown	σ unknown	all
	θ unknown	ζ known	$\sigma < 3$ known	all
	θ unknown	ζ known	$\sigma < 3$ unknown	all
	θ unknown	ζ unknown	$\sigma < 3$ known	all
	θ unknown	ζ unknown	$\sigma < 3$ unknown	all
normal	θ known	σ known		all
	θ known	σ unknown		A^2 and W^2
	θ unknown	σ known		A^2 and W^2
	θ unknown	σ unknown		all
Weibull	θ known	σ known	c known	all
	θ known	σ unknown	c known	A^2 and W^2
	θ known	σ known	c unknown	A^2 and W^2
	θ known	σ unknown	c unknown	A^2 and W^2
	θ unknown	σ known	$c > 2$ known	all
	θ unknown	σ unknown	$c > 2$ known	all
	θ unknown	σ known	$c > 2$ unknown	all
	θ unknown	σ unknown	$c > 2$ unknown	all

Kernel Density Estimates

You can use the KERNEL option to superimpose kernel density estimates on histograms. Smoothing the data distribution with a kernel density estimate can be more effective than using a histogram to identify features that might be obscured by the choice of histogram bins or sampling variation.

A kernel density estimate can also be more effective than a parametric curve fit when the process distribution is multi-modal. See Example 4.23.

The general form of the kernel density estimator is

$$\hat{f}_\lambda(x) = \frac{hv}{n\lambda} \sum_{i=1}^{n} K_0\left(\frac{x - x_i}{\lambda}\right)$$

where

$K_0(\cdot)$ is the kernel function
λ is the bandwidth
n is the sample size
x_i is the ith observation
$v = $ vertical scaling factor

and

$$v = \begin{cases} n & \text{for VSCALE=COUNT} \\ 100 & \text{for VSCALE=PERCENT} \\ 1 & \text{for VSCALE=PROPPORTION} \end{cases}$$

The KERNEL option provides three kernel functions (K_0): normal, quadratic, and triangular. You can specify the function with the K= *kernel-option* in parentheses after the KERNEL option. Values for the K= option are NORMAL, QUADRATIC, and TRIANGULAR (with aliases of N, Q, and T, respectively). By default, a normal kernel is used. The formulas for the kernel functions are

Normal $\quad K_0(t) = \frac{1}{\sqrt{2\pi}} \exp(-\frac{1}{2}t^2) \quad$ for $-\infty < t < \infty$
Quadratic $\quad K_0(t) = \frac{3}{4}(1 - t^2) \quad$ for $|t| \leq 1$
Triangular $\quad K_0(t) = 1 - |t| \quad$ for $|t| \leq 1$

The value of λ, referred to as the bandwidth parameter, determines the degree of smoothness in the estimated density function. You specify λ indirectly by specifying a standardized bandwidth c with the C= *kernel-option*. If Q is the interquartile range and n is the sample size, then c is related to λ by the formula

$$\lambda = cQn^{-\frac{1}{5}}$$

For a specific kernel function, the discrepancy between the density estimator $\hat{f}_\lambda(x)$ and the true density $f(x)$ is measured by the mean integrated square error (MISE):

$$\text{MISE}(\lambda) = \int_x \{E(\hat{f}_\lambda(x)) - f(x)\}^2 dx + \int_x var(\hat{f}_\lambda(x)) dx$$

The MISE is the sum of the integrated squared bias and the variance. An approximate mean integrated square error (AMISE) is:

$$\text{AMISE}(\lambda) = \frac{1}{4}\lambda^4 \left(\int_t t^2 K(t) dt\right)^2 \int_x (f''(x))^2 dx + \frac{1}{n\lambda} \int_t K(t)^2 dt$$

A bandwidth that minimizes AMISE can be derived by treating $f(x)$ as the normal density that has parameters μ and σ estimated by the sample mean and standard deviation. If you do not specify a bandwidth parameter or if you specify C=MISE, the bandwidth that minimizes AMISE is used. The value of AMISE can be used to compare different density estimates. You can also specify C=SJPI to select the bandwidth by using a plug-in formula of Sheather and Jones (Jones, Marron, and Sheather 1996). For each estimate, the bandwidth parameter c, the kernel function type, and the value of AMISE are reported in the SAS log.

The general kernel density estimates assume that the domain of the density to estimate can take on all values on a real line. However, sometimes the domain of a density is an interval bounded on one or both sides. For example, if a variable Y is a measurement of only positive values, then the kernel density curve should be bounded so that is zero for negative Y values. You can use the LOWER= and UPPER= *kernel-options* to specify the bounds.

The UNIVARIATE procedure uses a reflection technique to create the bounded kernel density curve, as described in Silverman (1986, pp. 30-31). It adds the reflections of the kernel density that are outside the boundary to the bounded kernel estimates. The general form of the bounded kernel density estimator is computed by replacing $K_0\left(\frac{x-x_i}{\lambda}\right)$ in the original equation with

$$\left\{ K_0\left(\frac{x-x_i}{\lambda}\right) + K_0\left(\frac{(x-x_l)+(x_i-x_l)}{\lambda}\right) + K_0\left(\frac{(x_u-x)+(x_u-x_i)}{\lambda}\right) \right\}$$

where x_l is the lower bound and x_u is the upper bound.

Without a lower bound, $x_l = -\infty$ and $K_0\left(\frac{(x-x_l)+(x_i-x_l)}{\lambda}\right) = 0$. Similarly, without an upper bound, $x_u = \infty$ and $K_0\left(\frac{(x_u-x)+(x_u-x_i)}{\lambda}\right) = 0$.

When C=MISE is used with a bounded kernel density, the UNIVARIATE procedure uses a bandwidth that minimizes the AMISE for its corresponding unbounded kernel.

Construction of Quantile-Quantile and Probability Plots

Figure 4.14 illustrates how a Q-Q plot is constructed for a specified theoretical distribution. First, the n nonmissing values of the variable are ordered from smallest to largest:

$$x_{(1)} \leq x_{(2)} \leq \cdots \leq x_{(n)}$$

Then the ith ordered value $x_{(i)}$ is plotted as a point whose y-coordinate is $x_{(i)}$ and whose x-coordinate is $F^{-1}\left(\frac{i-0.375}{n+0.25}\right)$, where $F(\cdot)$ is the specified distribution with zero location parameter and unit scale parameter.

You can modify the adjustment constants -0.375 and 0.25 with the RANKADJ= and NADJ= options. This default combination is recommended by Blom (1958). For additional information, see Chambers et al. (1983). Because $x_{(i)}$ is a quantile of the empirical cumulative distribution function (ecdf), a Q-Q plot compares quantiles of the ecdf with quantiles of a theoretical distribution. Probability plots (see the section "PROBPLOT Statement" on page 294) are constructed the same way, except that the x-axis is scaled nonlinearly in percentiles.

Figure 4.14 Construction of a Q-Q Plot

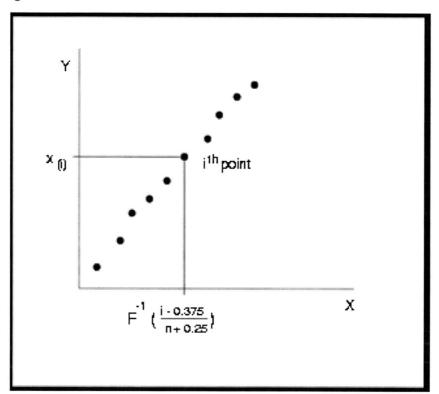

Interpretation of Quantile-Quantile and Probability Plots

The following properties of Q-Q plots and probability plots make them useful diagnostics of how well a specified theoretical distribution fits a set of measurements:

- If the quantiles of the theoretical and data distributions agree, the plotted points fall on or near the line $y = x$.

- If the theoretical and data distributions differ only in their location or scale, the points on the plot fall on or near the line $y = ax + b$. The slope a and intercept b are visual estimates of the scale and location parameters of the theoretical distribution.

Q-Q plots are more convenient than probability plots for graphical estimation of the location and scale parameters because the x-axis of a Q-Q plot is scaled linearly. On the other hand, probability plots are more convenient for estimating percentiles or probabilities.

There are many reasons why the point pattern in a Q-Q plot may not be linear. Chambers et al. (1983) and Fowlkes (1987) discuss the interpretations of commonly encountered departures from linearity, and these are summarized in Table 4.85.

In some applications, a nonlinear pattern may be more revealing than a linear pattern. However, Chambers et al. (1983) note that departures from linearity can also be due to chance variation.

Table 4.85 Quantile-Quantile Plot Diagnostics

Description of Point Pattern	Possible Interpretation
all but a few points fall on a line	outliers in the data
left end of pattern is below the line; right end of pattern is above the line	long tails at both ends of the data distribution
left end of pattern is above the line; right end of pattern is below the line	short tails at both ends of the data distribution
curved pattern with slope increasing from left to right	data distribution is skewed to the right
curved pattern with slope decreasing from left to right	data distribution is skewed to the left
staircase pattern (plateaus and gaps)	data have been rounded or are discrete

When the pattern is linear, you can use Q-Q plots to estimate shape, location, and scale parameters and to estimate percentiles. See Example 4.26 through Example 4.34.

Distributions for Probability and Q-Q Plots

You can use the PROBPLOT and QQPLOT statements to request probability and Q-Q plots that are based on the theoretical distributions summarized in Table 4.86.

Table 4.86 Distributions and Parameters

Distribution	Density Function $p(x)$	Range	Location	Scale	Shape
beta	$\frac{(x-\theta)^{\alpha-1}(\theta+\sigma-x)^{\beta-1}}{B(\alpha,\beta)\sigma^{(\alpha+\beta-1)}}$	$\theta < x < \theta + \sigma$	θ	σ	α, β
exponential	$\frac{1}{\sigma} \exp\left(-\frac{x-\theta}{\sigma}\right)$	$x \geq \theta$	θ	σ	
gamma	$\frac{1}{\sigma \Gamma(\alpha)} \left(\frac{x-\theta}{\sigma}\right)^{\alpha-1} \exp\left(-\frac{x-\theta}{\sigma}\right)$	$x > \theta$	θ	σ	α
lognormal (3-parameter)	$\frac{1}{\sigma\sqrt{2\pi}(x-\theta)} \exp\left(-\frac{(\log(x-\theta)-\zeta)^2}{2\sigma^2}\right)$	$x > \theta$	θ	ζ	σ
normal	$\frac{1}{\sigma\sqrt{2\pi}} \exp\left(-\frac{(x-\mu)^2}{2\sigma^2}\right)$	all x	μ	σ	
Weibull (3-parameter)	$\frac{c}{\sigma}\left(\frac{x-\theta}{\sigma}\right)^{c-1} \exp\left(-\left(\frac{x-\theta}{\sigma}\right)^c\right)$	$x > \theta$	θ	σ	c
Weibull (2-parameter)	$\frac{c}{\sigma}\left(\frac{x-\theta_0}{\sigma}\right)^{c-1} \exp\left(-\left(\frac{x-\theta_0}{\sigma}\right)^c\right)$	$x > \theta_0$	θ_0 (known)	σ	c

You can request these distributions with the BETA, EXPONENTIAL, GAMMA, LOGNORMAL, NORMAL, WEIBULL, and WEIBULL2 options, respectively. If you do not specify a distribution option, a normal probability plot or a normal Q-Q plot is created.

The following sections provide details for constructing Q-Q plots that are based on these distributions. Probability plots are constructed similarly except that the horizontal axis is scaled in percentile units.

Beta Distribution

To create the plot, the observations are ordered from smallest to largest, and the ith ordered observation is plotted against the quantile $B^{-1}_{\alpha\beta}\left(\frac{i-0.375}{n+0.25}\right)$, where $B^{-1}_{\alpha\beta}(\cdot)$ is the inverse normalized incomplete beta function, n is the number of nonmissing observations, and α and β are the shape parameters of the beta distribution. In a probability plot, the horizontal axis is scaled in percentile units.

The pattern on the plot for ALPHA=α and BETA=β tends to be linear with intercept θ and slope σ if the data are beta distributed with the specific density function

$$p(x) = \begin{cases} \frac{(x-\theta)^{\alpha-1}(\theta+\sigma-x)^{\beta-1}}{B(\alpha,\beta)\sigma^{(\alpha+\beta-1)}} & \text{for } \theta < x < \theta + \sigma \\ 0 & \text{for } x \leq \theta \text{ or } x \geq \theta + \sigma \end{cases}$$

where $B(\alpha, \beta) = \frac{\Gamma(\alpha)\Gamma(\beta)}{\Gamma(\alpha+\beta)}$ and

θ = lower threshold parameter
σ = scale parameter ($\sigma > 0$)
α = first shape parameter ($\alpha > 0$)
β = second shape parameter ($\beta > 0$)

Exponential Distribution

To create the plot, the observations are ordered from smallest to largest, and the ith ordered observation is plotted against the quantile $-\log\left(1 - \frac{i-0.375}{n+0.25}\right)$, where n is the number of nonmissing observations. In a probability plot, the horizontal axis is scaled in percentile units.

The pattern on the plot tends to be linear with intercept θ and slope σ if the data are exponentially distributed with the specific density function

$$p(x) = \begin{cases} \frac{1}{\sigma}\exp\left(-\frac{x-\theta}{\sigma}\right) & \text{for } x \geq \theta \\ 0 & \text{for } x < \theta \end{cases}$$

where θ is a threshold parameter, and σ is a positive scale parameter.

Gamma Distribution

To create the plot, the observations are ordered from smallest to largest, and the ith ordered observation is plotted against the quantile $G^{-1}_{\alpha}\left(\frac{i-0.375}{n+0.25}\right)$, where $G^{-1}_{\alpha}(\cdot)$ is the inverse normalized incomplete gamma function, n is the number of nonmissing observations, and α is the shape parameter of the gamma distribution. In a probability plot, the horizontal axis is scaled in percentile units.

The pattern on the plot for ALPHA=α tends to be linear with intercept θ and slope σ if the data are gamma distributed with the specific density function

$$p(x) = \begin{cases} \frac{1}{\sigma \Gamma(\alpha)} \left(\frac{x-\theta}{\sigma}\right)^{\alpha-1} \exp\left(-\frac{x-\theta}{\sigma}\right) & \text{for } x > \theta \\ 0 & \text{for } x \leq \theta \end{cases}$$

where

θ = threshold parameter
σ = scale parameter ($\sigma > 0$)
α = shape parameter ($\alpha > 0$)

Lognormal Distribution

To create the plot, the observations are ordered from smallest to largest, and the ith ordered observation is plotted against the quantile $\exp\left(\sigma \Phi^{-1}\left(\frac{i-0.375}{n+0.25}\right)\right)$, where $\Phi^{-1}(\cdot)$ is the inverse cumulative standard normal distribution, n is the number of nonmissing observations, and σ is the shape parameter of the lognormal distribution. In a probability plot, the horizontal axis is scaled in percentile units.

The pattern on the plot for SIGMA=σ tends to be linear with intercept θ and slope $\exp(\zeta)$ if the data are lognormally distributed with the specific density function

$$p(x) = \begin{cases} \frac{1}{\sigma \sqrt{2\pi}(x-\theta)} \exp\left(-\frac{(\log(x-\theta)-\zeta)^2}{2\sigma^2}\right) & \text{for } x > \theta \\ 0 & \text{for } x \leq \theta \end{cases}$$

where

θ = threshold parameter
ζ = scale parameter
σ = shape parameter ($\sigma > 0$)

See Example 4.26 and Example 4.33.

Normal Distribution

To create the plot, the observations are ordered from smallest to largest, and the ith ordered observation is plotted against the quantile $\Phi^{-1}\left(\frac{i-0.375}{n+0.25}\right)$, where $\Phi^{-1}(\cdot)$ is the inverse cumulative standard normal distribution and n is the number of nonmissing observations. In a probability plot, the horizontal axis is scaled in percentile units.

The point pattern on the plot tends to be linear with intercept μ and slope σ if the data are normally distributed with the specific density function

$$p(x) = \frac{1}{\sigma\sqrt{2\pi}} \exp\left(-\frac{(x-\mu)^2}{2\sigma^2}\right) \quad \text{for all } x$$

where μ is the mean and σ is the standard deviation ($\sigma > 0$).

Three-Parameter Weibull Distribution

To create the plot, the observations are ordered from smallest to largest, and the ith ordered observation is plotted against the quantile $\left(-\log\left(1 - \frac{i-0.375}{n+0.25}\right)\right)^{\frac{1}{c}}$, where n is the number of nonmissing observations, and c is the Weibull distribution shape parameter. In a probability plot, the horizontal axis is scaled in percentile units.

The pattern on the plot for C=c tends to be linear with intercept θ and slope σ if the data are Weibull distributed with the specific density function

$$p(x) = \begin{cases} \frac{c}{\sigma}\left(\frac{x-\theta}{\sigma}\right)^{c-1} \exp\left(-\left(\frac{x-\theta}{\sigma}\right)^c\right) & \text{for } x > \theta \\ 0 & \text{for } x \leq \theta \end{cases}$$

where

θ = threshold parameter
σ = scale parameter ($\sigma > 0$)
c = shape parameter ($c > 0$)

See Example 4.34.

Two-Parameter Weibull Distribution

To create the plot, the observations are ordered from smallest to largest, and the log of the shifted ith ordered observation $x_{(i)}$, denoted by $\log(x_{(i)} - \theta_0)$, is plotted against the quantile $\log\left(-\log\left(1 - \frac{i-0.375}{n+0.25}\right)\right)$, where n is the number of nonmissing observations. In a probability plot, the horizontal axis is scaled in percentile units.

Unlike the three-parameter Weibull quantile, the preceding expression is free of distribution parameters. Consequently, the C= shape parameter is not mandatory with the WEIBULL2 distribution option.

The pattern on the plot for THETA=θ_0 tends to be linear with intercept $\log(\sigma)$ and slope $\frac{1}{c}$ if the data are Weibull distributed with the specific density function

$$p(x) = \begin{cases} \frac{c}{\sigma}\left(\frac{x-\theta_0}{\sigma}\right)^{c-1} \exp\left(-\left(\frac{x-\theta_0}{\sigma}\right)^c\right) & \text{for } x > \theta_0 \\ 0 & \text{for } x \leq \theta_0 \end{cases}$$

where

θ_0 = known lower threshold
σ = scale parameter ($\sigma > 0$)
c = shape parameter ($c > 0$)

See Example 4.34.

Estimating Shape Parameters Using Q-Q Plots

Some of the distribution options in the PROBPLOT or QQPLOT statements require you to specify one or two shape parameters in parentheses after the distribution keyword. These are summarized in Table 4.87.

You can visually estimate the value of a shape parameter by specifying a list of values for the shape parameter option. A separate plot is produced for each value, and you can then select the value of the shape parameter that produces the most nearly linear point pattern. Alternatively, you can request that the plot be created using an estimated shape parameter. See the entries for the distribution options in the section "Dictionary of Options" on page 300 (for the PROBPLOT statement) and in the section "Dictionary of Options" on page 310 (for the QQPLOT statement).

NOTE: For Q-Q plots created with the WEIBULL2 option, you can estimate the shape parameter c from a linear pattern by using the fact that the slope of the pattern is $\frac{1}{c}$.

Table 4.87 Shape Parameter Options

Distribution Keyword	Mandatory Shape Parameter Option	Range
BETA	ALPHA=α, BETA=β	$\alpha > 0, \beta > 0$
EXPONENTIAL	none	
GAMMA	ALPHA=α	$\alpha > 0$
LOGNORMAL	SIGMA=σ	$\sigma > 0$
NORMAL	none	
WEIBULL	C=c	$c > 0$
WEIBULL2	none	

Estimating Location and Scale Parameters Using Q-Q Plots

If you specify location and scale parameters for a distribution in a PROBPLOT or QQPLOT statement (or if you request estimates for these parameters), a diagonal distribution reference line is displayed on the plot. (An exception is the two-parameter Weibull distribution, for which a line is displayed when you specify or estimate the scale and shape parameters.) Agreement between this line and the point pattern indicates that the distribution with these parameters is a good fit.

When the point pattern on a Q-Q plot is linear, its intercept and slope provide estimates of the location and scale parameters. (An exception to this rule is the two-parameter Weibull distribution, for which the intercept and slope are related to the scale and shape parameters.)

Table 4.88 shows how the specified parameters determine the intercept and slope of the line. The intercept and slope are based on the quantile scale for the horizontal axis, which is used in Q-Q plots.

Table 4.88 Intercept and Slope of Distribution Reference Line

Distribution	Parameters			Linear Pattern	
	Location	Scale	Shape	Intercept	Slope
Beta	θ	σ	α, β	θ	σ
Exponential	θ	σ		θ	σ
Gamma	θ	σ	α	θ	σ
Lognormal	θ	ζ	σ	θ	$\exp(\zeta)$
Normal	μ	σ		μ	σ
Weibull (3-parameter)	θ	σ	c	θ	σ
Weibull (2-parameter)	θ_0 (known)	σ	c	$\log(\sigma)$	$\frac{1}{c}$

For instance, specifying MU=3 and SIGMA=2 with the NORMAL option requests a line with intercept 3 and slope 2. Specifying SIGMA=1 and C=2 with the WEIBULL2 option requests a line with intercept $\log(1) = 0$ and slope $\frac{1}{2}$. On a probability plot with the LOGNORMAL and WEIBULL2 options, you can specify the slope directly with the SLOPE= option. That is, for the LOGNORMAL option, specifying THETA= θ_0 and SLOPE=$\exp(\zeta_0)$ displays the same line as specifying THETA= θ_0 and ZETA= ζ_0. For the WEIBULL2 option, specifying SIGMA= σ_0 and SLOPE= $\frac{1}{c_0}$ displays the same line as specifying SIGMA= σ_0 and C= c_0.

Estimating Percentiles Using Q-Q Plots

There are two ways to estimate percentiles from a Q-Q plot:

- Specify the PCTLAXIS option, which adds a percentile axis opposite the theoretical quantile axis. The scale for the percentile axis ranges between 0 and 100 with tick marks at percentile values such as 1, 5, 10, 25, 50, 75, 90, 95, and 99.

- Specify the PCTLSCALE option, which relabels the horizontal axis tick marks with their percentile equivalents but does not alter their spacing. For example, on a normal Q-Q plot, the tick mark labeled "0" is relabeled as "50" because the 50th percentile corresponds to the zero quantile.

You can also estimate percentiles by using probability plots created with the PROBPLOT statement. See Example 4.32.

Input Data Sets

DATA= Data Set

The DATA= data set provides the set of variables that are analyzed. The UNIVARIATE procedure must have a DATA= data set. If you do not specify one with the DATA= option in the PROC UNIVARIATE statement, the procedure uses the last data set created.

ANNOTATE= Data Sets

You can add features to plots by specifying ANNOTATE= data sets either in the PROC UNIVARIATE statement or in individual plot statements.

Information contained in an ANNOTATE= data set specified in the PROC UNIVARIATE statement is used for all plots produced in a given PROC step; this is a "global" ANNOTATE= data set. By using this global data set, you can keep information common to all high-resolution plots in one data set.

Information contained in the ANNOTATE= data set specified in a plot statement is used only for plots produced by that statement; this is a "local" ANNOTATE= data set. By using this data set, you can add statement-specific features to plots. For example, you can add different features to plots produced by the HISTOGRAM and QQPLOT statements by specifying an ANNOTATE= data set in each plot statement.

You can specify an ANNOTATE= data set in the PROC UNIVARIATE statement and in plot statements. This enables you to add some features to all plots and also add statement-specific features to plots. See Example 4.25.

OUT= Output Data Set in the OUTPUT Statement

PROC UNIVARIATE creates an OUT= data set for each OUTPUT statement. This data set contains an observation for each combination of levels of the variables in the BY statement, or a single observation if you do not specify a BY statement. Thus the number of observations in the new data set corresponds to the number of groups for which statistics are calculated. Without a BY statement, the procedure computes statistics and percentiles by using all the observations in the input data set. With a BY statement, the procedure computes statistics and percentiles by using the observations within each BY group.

The variables in the OUT= data set are as follows:

- BY statement variables. The values of these variables match the values in the corresponding BY group in the DATA= data set and indicate which BY group each observation summarizes.

- variables created by selecting statistics in the OUTPUT statement. The statistics are computed using all the nonmissing data, or they are computed for each BY group if you use a BY statement.

- variables created by requesting new percentiles with the PCTLPTS= option. The names of these new variables depend on the values of the PCTLPRE= and PCTLNAME= options.

If the output data set contains a percentile variable or a quartile variable, the percentile definition assigned with the PCTLDEF= option in the PROC UNIVARIATE statement is recorded in the output data set label. See Example 4.8.

The following table lists variables available in the OUT= data set.

Table 4.89 Variables Available in the OUT= Data Set

Variable Name	Description
Descriptive Statistics	
CSS	sum of squares corrected for the mean
CV	percent coefficient of variation
KURTOSIS	measurement of the heaviness of tails
MAX	largest (maximum) value
MEAN	arithmetic mean
MIN	smallest (minimum) value
MODE	most frequent value (if not unique, the smallest mode)
N	number of observations on which calculations are based
NMISS	number of missing observations
NOBS	total number of observations
RANGE	difference between the maximum and minimum values
SKEWNESS	measurement of the tendency of the deviations to be larger in one direction than in the other
STD	standard deviation
STDMEAN	standard error of the mean
SUM	sum
SUMWGT	sum of the weights
USS	uncorrected sum of squares
VAR	variance
Quantile Statistics	
MEDIAN | P50	middle value (50th percentile)
P1	1st percentile
P5	5th percentile
P10	10th percentile
P90	90th percentile
P95	95th percentile
P99	99th percentile
Q1 | P25	lower quartile (25th percentile)
Q3 | P75	upper quartile (75th percentile)
QRANGE	difference between the upper and lower quartiles (also known as the inner quartile range)

Table 4.89 (continued)

Variable Name	Description
Robust Statistics	
GINI	Gini's mean difference
MAD	median absolute difference
QN	2nd variation of median absolute difference
SN	1st variation of median absolute difference
STD_GINI	standard deviation for Gini's mean difference
STD_MAD	standard deviation for median absolute difference
STD_QN	standard deviation for the second variation of the median absolute difference
STD_QRANGE	estimate of the standard deviation, based on interquartile range
STD_SN	standard deviation for the first variation of the median absolute difference
Hypothesis Test Statistics	
MSIGN	sign statistic
NORMAL	test statistic for normality. If the sample size is less than or equal to 2000, this is the Shapiro-Wilk W statistic. Otherwise, it is the Kolmogorov D statistic.
PROBM	probability of a greater absolute value for the sign statistic
PROBN	probability that the data came from a normal distribution
PROBS	probability of a greater absolute value for the signed rank statistic
PROBT	two-tailed p-value for Student's t statistic with $n-1$ degrees of freedom
SIGNRANK	signed rank statistic
T	Student's t statistic to test the null hypothesis that the population mean is equal to μ_0

OUTHISTOGRAM= Output Data Set

You can create an OUTHISTOGRAM= data set with the HISTOGRAM statement. This data set contains information about histogram intervals. Because you can specify multiple HISTOGRAM statements with the UNIVARIATE procedure, you can create multiple OUTHISTOGRAM= data sets.

An OUTHISTOGRAM= data set contains a group of observations for each variable in the HISTOGRAM statement. The group contains an observation for each interval of the histogram, beginning with the leftmost interval that contains a value of the variable and ending with the rightmost interval that contains a value of the variable. These intervals do not necessarily coincide with the intervals displayed in the histogram because the histogram might be padded with empty intervals at either end. If you superimpose one or more fitted curves on the histogram, the OUTHISTOGRAM= data set contains multiple groups of observations for each variable (one group for each curve). If you use a BY statement, the OUTHISTOGRAM= data set contains groups of observations for each BY group. ID variables are not saved in an OUTHISTOGRAM= data set.

By default, an OUTHISTOGRAM= data set contains the _MIDPT_ variable, whose values identify histogram intervals by their midpoints. When the ENDPOINTS= or NENDPOINTS option is specified, intervals are identified by endpoint values instead. If the RTINCLUDE option is specified, the _MAXPT_ variable contains upper endpoint values. Otherwise, the _MINPT_ variable contains lower endpoint values. See Example 4.18.

Table 4.90 Variables in the OUTHISTOGRAM= Data Set

Variable	Description
CURVE	name of fitted distribution (if requested in HISTOGRAM statement)
EXPPCT	estimated percent of population in histogram interval determined from optional fitted distribution
MAXPT	upper endpoint of histogram interval
MIDPT	midpoint of histogram interval
MINPT	lower endpoint of histogram interval
OBSPCT	percent of variable values in histogram interval
VAR	variable name

OUTKERNEL= Output Data Set

You can create an OUTKERNEL= data set with the HISTOGRAM statement. This data set contains information about histogram intervals. Because you can specify multiple HISTOGRAM statements with the UNIVARIATE procedure, you can create multiple OUTKERNEL= data sets.

An OUTKERNEL= data set contains a group of observations for each kernel density estimate requested with the HISTOGRAM statement. These observations span a range of analysis variable values recorded in the _VALUE_ variable. The procedure determines the increment between values, and therefore the number of observations in the group. The variable _DENSITY_ contains the kernel density calculated for the corresponding analysis variable value.

When a density curve is overlaid on a histogram, the curve is scaled so that the area under the curve equals the total area of the histogram bars. The scaled density values are saved in the variable _COUNT_, _PERCENT_, or _PROPORTION_, depending on the histogram's vertical axis scale, determined by the VSCALE= option. Only one of these variables appears in a given OUTKERNEL= data set.

Table 4.91 lists the variables in an OUTKERNEL= data set.

Table 4.91 Variables in the OUTKERNEL= Data Set

Variable	Description
_C_	standardized bandwidth parameter
COUNT	kernel density scaled for VSCALE=COUNT
DENSITY	kernel density
PERCENT	kernel density scaled for VSCALE=PERCENT (default)
PROPORTION	kernel density scaled for VSCALE=PROPORTION
TYPE	kernel function
VALUE	variable value at which kernel function is calculated
VAR	variable name

OUTTABLE= Output Data Set

The OUTTABLE= data set saves univariate statistics in a data set that contains one observation per analysis variable. The following variables are saved:

Table 4.92 Variables in the OUTTABLE= Data Set

Variable	Description
KURT	kurtosis
MAX	maximum
MEAN	mean
MEDIAN	median
MIN	minimum
MODE	mode
NMISS	number of missing observations
_N_	number of nonmissing observations
_P1_	1st percentile
_P5_	5th percentile
_P10_	10th percentile
_P90_	90th percentile
_P95_	95th percentile
_P99_	99th percentile
_Q1_	25th percentile (lower quartile)
_Q3_	75th percentile (upper quartile)
QRANGE	interquartile range (upper quartile minus lower quartile)
RANGE	range
SGNRNK	centered sign rank
SKEW	skewness
STD	standard deviation
SUMWGT	sum of the weights
SUM	sum
VAR	variance
VAR	variable name

The OUTTABLE= data set and the OUT= data set (see the section "OUT= Output Data Set in the OUTPUT Statement" on page 370) contain essentially the same information. However, the structure of the OUTTABLE= data set may be more appropriate when you are computing summary statistics for more than one analysis variable in the same invocation of the UNIVARIATE procedure. Each observation in the OUTTABLE= data set corresponds to a different analysis variable, and the variables in the data set correspond to summary statistics and indices.

For example, suppose you have 10 analysis variables (P1-P10). The following statements create an OUTTABLE= data set named Table, which contains summary statistics for each of these variables:

```
data Analysis;
   input A1-A10;
   datalines;
72   223  332  138  110  145   23  293  353  458
97    54   61  196  275  171  117   72   81  141
56   170  140  400  371   72   60   20  484  138
124    6  332  493  214   43  125   55  372   30
152  236  222   76  187  126  192  334  109  546
  5  260  194  277  176   96  109  184  240  261
161  253  153  300   37  156  282  293  451  299
128  121  254  297  363  132  209  257  429  295
116  152  331   27  442  103   80  393  383   94
 43  178  278  159   25  180  253  333   51  225
 34  128  182  415  524  112   13  186  145  131
142  236  234  255  211   80  281  135  179   11
108  215  335   66  254  196  190  363  226  379
 62  232  219  474   31  139   15   56  429  298
177  218  275  171  457  146  163   18  155  129
  0  235   83  239  398   99  226  389  498   18
147  199  324  258  504    2  218  295  422  287
 39  161  156  198  214   58  238   19  231  548
120   42  372  420  232  112  157   79  197  166
178   83  238  492  463   68   46  386   45   81
161  267  372  296  501   96   11  288  330   74
 14    2   52   81  169   63  194  161  173   54
 22  181   92  272  417   94  188  180  367  342
 55  248  214  422  133  193  144  318  271  479
 56   83  169   30  379    5  296  320  396  597
;
run;
proc univariate data=Analysis outtable=Table noprint;
   var A1-A10;
run;
```

The following statements create the table shown in Figure 4.15, which contains the mean, standard deviation, and so on, for each analysis variable:

```
proc print data=Table label noobs;
   var _VAR_ _MIN_ _MEAN_ _MAX_ _STD_;
   label _VAR_='Analysis';
run;
```

Figure 4.15 Tabulating Results for Multiple Process Variables

```
              Test Scores for a College Course

                                                 Standard
     Analysis    Minimum      Mean     Maximum   Deviation

        A1          0         90.76      178      57.024
        A2          2        167.32      267      81.628
        A3         52        224.56      372      96.525
        A4         27        258.08      493     145.218
        A5         25        283.48      524     157.033
        A6          2        107.48      196      52.437
        A7         11        153.20      296      90.031
        A8         18        217.08      393     130.031
        A9         45        280.68      498     140.943
        A10        11        243.24      597     178.799
```

Tables for Summary Statistics

By default, PROC UNIVARIATE produces ODS tables of moments, basic statistical measures, tests for location, quantiles, and extreme observations. You must specify options in the PROC UNIVARIATE statement to request other statistics and tables. The CIBASIC option produces a table that displays confidence limits for the mean, standard deviation, and variance. The CIPCTLDF and CIPCTLNORMAL options request tables of confidence limits for the quantiles. The LOCCOUNT option requests a table that shows the number of values greater than, not equal to, and less than the value of MU0=. The FREQ option requests a table of frequencies counts. The NEXTRVAL= option requests a table of extreme values. The NORMAL option requests a table with tests for normality.

The TRIMMED=, WINSORIZED=, and ROBUSTSCALE options request tables with robust estimators. The table of trimmed or Winsorized means includes the percentage and the number of observations that are trimmed or Winsorized at each end, the mean and standard error, confidence limits, and the Student's t test. The table with robust measures of scale includes interquartile range, Gini's mean difference G, MAD, Q_n, and S_n, with their corresponding estimates of σ.

See the section "ODS Table Names" on page 376 for the names of ODS tables created by PROC UNIVARIATE.

ODS Table Names

PROC UNIVARIATE assigns a name to each table that it creates. You can use these names to reference the table when you use the Output Delivery System (ODS) to select tables and create output data sets.

Table 4.93 ODS Tables Produced with the PROC UNIVARIATE Statement

ODS Table Name	Description	Option
BasicIntervals	confidence intervals for mean, standard deviation, variance	CIBASIC
BasicMeasures	measures of location and variability	default
ExtremeObs	extreme observations	default
ExtremeValues	extreme values	NEXTRVAL=
Frequencies	frequencies	FREQ
LocationCounts	counts used for sign test and signed rank test	LOCCOUNT
MissingValues	missing values	default, if missing values exist
Modes	modes	MODES
Moments	sample moments	default
Plots	line printer plots	PLOTS
Quantiles	quantiles	default
RobustScale	robust measures of scale	ROBUSTSCALE
SSPlots	line printer side-by-side box plots	PLOTS (with BY statement)
TestsForLocation	tests for location	default
TestsForNormality	tests for normality	NORMALTEST
TrimmedMeans	trimmed means	TRIMMED=
WinsorizedMeans	Winsorized means	WINSORIZED=

Table 4.94 ODS Tables Produced with the HISTOGRAM Statement

ODS Table Name	Description	Option
Bins	histogram bins	MIDPERCENTS secondary option
FitQuantiles	quantiles of fitted distribution	any distribution option
GoodnessOfFit	goodness-of-fit tests for fitted distribution	any distribution option
HistogramBins	histogram bins	MIDPERCENTS option
ParameterEstimates	parameter estimates for fitted distribution	any distribution option

ODS Tables for Fitted Distributions

If you request a fitted parametric distribution with a HISTOGRAM statement, PROC UNIVARIATE creates a summary that is organized into the ODS tables described in this section.

Parameters

The ParameterEstimates table lists the estimated (or specified) parameters for the fitted curve as well as the estimated mean and estimated standard deviation. See "Formulas for Fitted Continuous Distributions" on page 348.

EDF Goodness-of-Fit Tests

When you fit a parametric distribution, the HISTOGRAM statement provides a series of goodness-of-fit tests based on the empirical distribution function (EDF). See "EDF Goodness-of-Fit Tests" on page 357. These are displayed in the GoodnessOfFit table.

Histogram Intervals

The Bins table is included in the summary only if you specify the MIDPERCENTS option in parentheses after the distribution option. This table lists the midpoints for the histogram bins along with the observed and estimated percentages of the observations that lie in each bin. The estimated percentages are based on the fitted distribution.

If you specify the MIDPERCENTS option without requesting a fitted distribution, the Histogram-Bins table is included in the summary. This table lists the interval midpoints with the observed percent of observations that lie in the interval. See the entry for the MIDPERCENTS option on page 267.

Quantiles

The FitQuantiles table lists observed and estimated quantiles. You can use the PERCENTS= option to specify the list of quantiles in this table. See the entry for the PERCENTS= option on page 268. By default, the table lists observed and estimated quantiles for the 1, 5, 10, 25, 50, 75, 90, 95, and 99 percent of a fitted parametric distribution.

ODS Graphics (Experimental)

The UNIVARIATE procedure supports ODS Graphics on an experimental basis in SAS 9.2. To use ODS Graphics, you must specify the ODS GRAPHICS statement prior to the PROC UNIVARIATE statement. For more information about ODS Graphics, see Chapter 21, "Statistical Graphics Using ODS" (*SAS/STAT User's Guide*).

PROC UNIVARIATE assigns a name to each graph it creates by using ODS Graphics. You can use these names to reference the graphs when you use ODS. The names are listed in Table 4.95.

Table 4.95 ODS Graphics Produced by PROC UNIVARIATE

ODS Graph Name	Plot Description	Statement
CDFPlot	cdf plot	CDFPLOT
Histogram	histogram	HISTOGRAM
PPPlot	P-P plot	PPPLOT
ProbPlot	probability plot	PROBPLOT
QQPlot	Q-Q plot	QQPLOT

Computational Resources

Because the UNIVARIATE procedure computes quantile statistics, it requires additional memory to store a copy of the data in memory. By default, the MEANS, SUMMARY, and TABULATE procedures require less memory because they do not automatically compute quantiles. These procedures also provide an option to use a new fixed-memory quantiles estimation method that is usually less memory-intensive.

In the UNIVARIATE procedure, the only factor that limits the number of variables that you can analyze is the computer resources that are available. The amount of temporary storage and CPU time required depends on the statements and the options that you specify. To calculate the computer resources the procedure needs, let

- N be the number of observations in the data set
- V be the number of variables in the VAR statement
- U_i be the number of unique values for the ith variable

Then the minimum memory requirement in bytes to process all variables is $M = 24 \sum_i U_i$. If M bytes are not available, PROC UNIVARIATE must process the data multiple times to compute all the statistics. This reduces the minimum memory requirement to $M = 24 \max(U_i)$.

Using the ROUND= option reduces the number of unique values (U_i), thereby reducing memory requirements. The ROBUSTSCALE option requires $40U_i$ bytes of temporary storage.

Several factors affect the CPU time:

- The time to create V tree structures to internally store the observations is proportional to $NV \log(N)$.

- The time to compute moments and quantiles for the ith variable is proportional to U_i.

- The time to compute the NORMAL option test statistics is proportional to N.

- The time to compute the ROBUSTSCALE option test statistics is proportional to $U_i \log(U_i)$.

- The time to compute the exact significance level of the sign rank statistic can increase when the number of nonzero values is less than or equal to 20.

Each of these factors has a different constant of proportionality. For additional information about optimizing CPU performance and memory usage, see the SAS documentation for your operating environment.

Examples: UNIVARIATE Procedure

Example 4.1: Computing Descriptive Statistics for Multiple Variables

This example computes univariate statistics for two variables. The following statements create the data set BPressure, which contains the systolic (Systolic) and diastolic (Diastolic) blood pressure readings for 22 patients:

```
data BPressure;
   length PatientID $2;
   input PatientID $ Systolic Diastolic @@;
   datalines;
CK 120 50   SS 96  60  FR 100 70
CP 120 75   BL 140 90  ES 120 70
CP 165 110  JI 110 40  MC 119 66
FC 125 76   RW 133 60  KD 108 54
DS 110 50   JW 130 80  BH 120 65
JW 134 80   SB 118 76  NS 122 78
GS 122 70   AB 122 78  EC 112 62
HH 122 82
;
run;
```

The following statements produce descriptive statistics and quantiles for the variables Systolic and Diastolic:

```
title 'Systolic and Diastolic Blood Pressure';
ods select BasicMeasures Quantiles;
proc univariate data=BPressure;
   var Systolic Diastolic;
run;
```

The ODS SELECT statement restricts the output, which is shown in Output 4.1.1, to the "BasicMeasures" and "Quantiles" tables; see the section "ODS Table Names" on page 376. You use the PROC UNIVARIATE statement to request univariate statistics for the variables listed in the VAR statement, which specifies the analysis variables and their order in the output. Formulas for computing the statistics in the "BasicMeasures" table are provided in the section "Descriptive Statistics" on page 325. The quantiles are calculated using *Definition 5*, which is the default definition; see the section "Calculating Percentiles" on page 328.

A sample program for this example, *uniex01.sas*, is available in the SAS Sample Library for Base SAS software.

Output 4.1.1 Display Basic Measures and Quantiles

```
                 Systolic and Diastolic Blood Pressure

                          The UNIVARIATE Procedure
                             Variable:  Systolic

                          Basic Statistical Measures

             Location                      Variability

         Mean     121.2727     Std Deviation           14.28346
         Median   120.0000     Variance               204.01732
         Mode     120.0000     Range                   69.00000
                               Interquartile Range     13.00000

     NOTE: The mode displayed is the smallest of 2 modes with a count of 4.

                             Quantiles (Definition 5)

                         Quantile          Estimate

                         100% Max             165
                         99%                  165
                         95%                  140
                         90%                  134
                         75% Q3               125
                         50% Median           120
                         25% Q1               112
                         10%                  108
                         5%                   100
                         1%                    96
                         0% Min                96


                 Systolic and Diastolic Blood Pressure

                          The UNIVARIATE Procedure
                             Variable:  Diastolic

                          Basic Statistical Measures

             Location                      Variability

         Mean     70.09091     Std Deviation           15.16547
         Median   70.00000     Variance               229.99134
         Mode     70.00000     Range                   70.00000
                               Interquartile Range     18.00000
```

Output 4.1.1 *continued*

```
              Quantiles (Definition 5)

              Quantile        Estimate

              100% Max           110
              99%                110
              95%                 90
              90%                 82
              75% Q3              78
              50% Median          70
              25% Q1              60
              10%                 50
              5%                  50
              1%                  40
              0% Min              40
```

Example 4.2: Calculating Modes

An instructor is interested in calculating all the modes of the scores on a recent exam. The following statements create a data set named Exam, which contains the exam scores in the variable Score:

```
data Exam;
   label Score = 'Exam Score';
   input Score @@;
   datalines;
81 97 78 99 77 81 84 86 86 97
85 86 94 76 75 42 91 90 88 86
97 97 89 69 72 82 83 81 80 81
;
run;
```

The following statements use the MODES option to request a table of all possible modes:

```
title 'Table of Modes for Exam Scores';
ods select Modes;
proc univariate data=Exam modes;
   var Score;
run;
```

The ODS SELECT statement restricts the output to the "Modes" table; see the section "ODS Table Names" on page 376.

Output 4.2.1 Table of Modes Display

```
                Table of Modes for Exam Scores

                    The UNIVARIATE Procedure
              Variable:  Score   (Exam Score)

                              Modes

                      Mode       Count

                       81          4
                       86          4
                       97          4
```

By default, when the MODES option is used and there is more than one mode, the lowest mode is displayed in the "BasicMeasures" table. The following statements illustrate the default behavior:

```
title 'Default Output';
ods select BasicMeasures;
proc univariate data=Exam;
   var Score;
run;
```

Output 4.2.2 Default Output (Without MODES Option)

```
                            Default Output

                      The UNIVARIATE Procedure
                Variable:  Score   (Exam Score)

                     Basic Statistical Measures

          Location                       Variability

      Mean     83.66667     Std Deviation          11.08069
      Median   84.50000     Variance              122.78161
      Mode     81.00000     Range                  57.00000
                            Interquartile Range    10.00000

    NOTE: The mode displayed is the smallest of 3 modes with a count of 4.
```

The default output displays a mode of 81 and includes a note regarding the number of modes; the modes 86 and 97 are not displayed. The ODS SELECT statement restricts the output to the "BasicMeasures" table; see the section "ODS Table Names" on page 376.

A sample program for this example, *uniex02.sas*, is available in the SAS Sample Library for Base SAS software.

Example 4.3: Identifying Extreme Observations and Extreme Values

This example, which uses the data set **BPressure** introduced in Example 4.1, illustrates how to produce a table of the extreme observations and a table of the extreme values in a data set. The following statements generate the "Extreme Observations" tables for Systolic and Diastolic, which enable you to identify the extreme observations for each variable:

```
title 'Extreme Blood Pressure Observations';
ods select ExtremeObs;
proc univariate data=BPressure;
   var Systolic Diastolic;
   id PatientID;
run;
```

The ODS SELECT statement restricts the output to the "ExtremeObs" table; see the section "ODS Table Names" on page 376. The ID statement requests that the extreme observations are to be identified using the value of PatientID as well as the observation number. By default, the five lowest and five highest observations are displayed. You can use the NEXTROBS= option to request a different number of extreme observations.

Output 4.3.1 shows that the patient identified as 'CP' (Observation 7) has the highest values for both Systolic and Diastolic. To visualize extreme observations, you can create histograms; see Example 4.14.

Output 4.3.1 Blood Pressure Extreme Observations

```
              Extreme Blood Pressure Observations

                       The UNIVARIATE Procedure
                         Variable:  Systolic

                         Extreme Observations

       ---------Lowest---------        ---------Highest--------

                   Patient                         Patient
         Value     ID         Obs         Value    ID         Obs

            96    SS            2          130    JW           14
           100    FR            3          133    RW           11
           108    KD           12          134    JW           16
           110    DS           13          140    BL            5
           110    JI            8          165    CP            7
```

Output 4.3.1 *continued*

```
                Extreme Blood Pressure Observations

                        The UNIVARIATE Procedure
                          Variable:  Diastolic

                           Extreme Observations

        ---------Lowest---------         ---------Highest--------

                    Patient                       Patient
           Value    ID        Obs         Value    ID        Obs

              40    JI          8            80    JW         14
              50    DS         13            80    JW         16
              50    CK          1            82    HH         22
              54    KD         12            90    BL          5
              60    RW         11           110    CP          7
```

The following statements generate the "Extreme Values" tables for Systolic and Diastolic, which tabulate the tails of the distributions:

```
title 'Extreme Blood Pressure Values';
ods select ExtremeValues;
proc univariate data=BPressure nextrval=5;
   var Systolic Diastolic;
run;
```

The ODS SELECT statement restricts the output to the "ExtremeValues" table; see the section "ODS Table Names" on page 376. The NEXTRVAL= option specifies the number of extreme values at each end of the distribution to be shown in the tables in Output 4.3.2.

Output 4.3.2 shows that the values 78 and 80 occurred twice for Diastolic and the maximum of Diastolic is 110. Note that Output 4.3.1 displays the value of 80 twice for Diastolic because there are two observations with that value. In Output 4.3.2, the value 80 is only displayed once.

Output 4.3.2 Blood Pressure Extreme Values

```
                  Extreme Blood Pressure Values

                     The UNIVARIATE Procedure
                        Variable:  Systolic

                           Extreme Values

        ---------Lowest--------         --------Highest--------

         Order    Value    Freq         Order    Value    Freq

             1       96       1            11      130       1
             2      100       1            12      133       1
             3      108       1            13      134       1
             4      110       2            14      140       1
             5      112       1            15      165       1
```

Output 4.3.2 *continued*

```
                     Extreme Blood Pressure Values

                        The UNIVARIATE Procedure
                          Variable:  Diastolic

                              Extreme Values

             ---------Lowest--------      --------Highest--------

            Order     Value     Freq     Order     Value     Freq

              1         40        1        11        78        2
              2         50        2        12        80        2
              3         54        1        13        82        1
              4         60        2        14        90        1
              5         62        1        15       110        1
```

A sample program for this example, *uniex01.sas*, is available in the SAS Sample Library for Base SAS software.

Example 4.4: Creating a Frequency Table

An instructor is interested in creating a frequency table of score changes between a pair of tests given in one of his college courses. The data set Score contains test scores for his students who took a pretest and a posttest on the same material. The variable ScoreChange contains the difference between the two test scores. The following statements create the data set:

```
data Score;
   input Student $ PreTest PostTest @@;
   label ScoreChange = 'Change in Test Scores';
   ScoreChange = PostTest - PreTest;
   datalines;
Capalleti  94 91  Dubose      51 65
Engles     95 97  Grant       63 75
Krupski    80 75  Lundsford   92 55
Mcbane     75 78  Mullen      89 82
Nguyen     79 76  Patel       71 77
Si         75 70  Tanaka      87 73
;
run;
```

The following statements produce a frequency table for the variable ScoreChange:

```
title 'Analysis of Score Changes';
ods select Frequencies;
proc univariate data=Score freq;
   var ScoreChange;
run;
```

The ODS SELECT statement restricts the output to the "Frequencies" table; see the section "ODS Table Names" on page 376. The FREQ option on the PROC UNIVARIATE statement requests the table of frequencies shown in Output 4.4.1.

Output 4.4.1 Table of Frequencies

```
                         Analysis of Score Changes

                           The UNIVARIATE Procedure
                 Variable:  ScoreChange   (Change in Test Scores)

                              Frequency Counts

                 Percents                     Percents                     Percents
Value Count   Cell    Cum     Value Count   Cell    Cum     Value Count   Cell    Cum

 -37    1     8.3     8.3      -3     2    16.7    58.3       6     1     8.3    83.3
 -14    1     8.3    16.7       2     1     8.3    66.7      12     1     8.3    91.7
  -7    1     8.3    25.0       3     1     8.3    75.0      14     1     8.3   100.0
  -5    2    16.7    41.7
```

From Output 4.4.1, the instructor sees that only score changes of −3 and −5 occurred more than once.

A sample program for this example, *uniex03.sas*, is available in the SAS Sample Library for Base SAS software.

Example 4.5: Creating Plots for Line Printer Output

The PLOT option in the PROC UNIVARIATE statement requests several basic plots for display in line printer output. For more information about plots created by the PLOT option, see the section "Creating Line Printer Plots" on page 337. This example illustrates the use of the PLOT option as well as BY processing in PROC UNIVARIATE.

A researcher is analyzing a data set consisting of air pollution data from three different measurement sites. The data set AirPoll, created by the following statements, contains the variables Site and Ozone, which are the site number and ozone level, respectively.

```
data AirPoll (keep = Site Ozone);
   label Site  = 'Site Number'
         Ozone = 'Ozone level (in ppb)';
   do i = 1 to 3;
      input Site @@;
      do j = 1 to 15;
         input Ozone @@;
         output;
      end;
   end;
   datalines;
102 4 6 3 4 7 8 2 3 4 1 3 8 9 5 6
```

388 ♦ Chapter 4: The UNIVARIATE Procedure

```
134 5 3 6 2 1 2 4 3 2 4 6 4 6 3 1
137 8 9 7 8 6 7 6 7 9 8 9 8 7 8 5
;
run;
```

The following statements produce stem-and-leaf plots, box plots, and normal probability plots for each site in the AirPoll data set:

```
ods select Plots SSPlots;
proc univariate data=AirPoll plot;
   by Site;
   var Ozone;
run;
```

The PLOT option produces a stem-and-leaf plot, a box plot, and a normal probability plot for the Ozone variable at each site. Because the BY statement is used, a side-by-side box plot is also created to compare the ozone levels across sites. Note that AirPoll is sorted by Site; in general, the data set should be sorted by the BY variable by using the SORT procedure. The ODS SELECT statement restricts the output to the "Plots" and "SSPlots" tables; see the section "ODS Table Names" on page 376. Optionally, you can specify the PLOTSIZE=n option to control the approximate number of rows (between 8 and the page size) that the plots occupy.

Output 4.5.1 through Output 4.5.3 show the plots produced for each BY group. Output 4.5.4 shows the side-by-side box plot for comparing Ozone values across sites.

Output 4.5.1 Ozone Plots for BY Group Site = 102

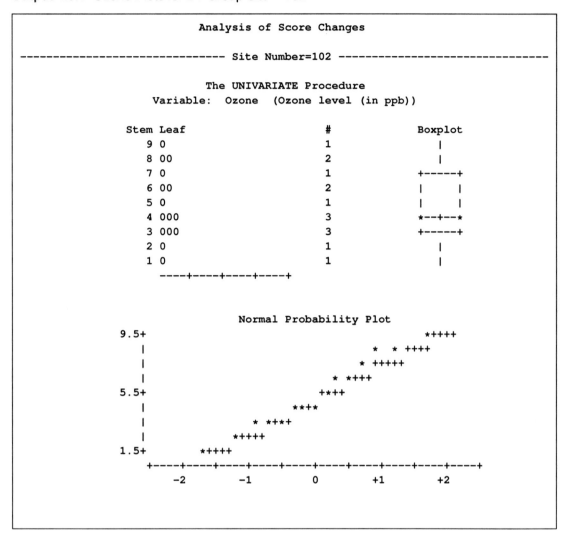

Output 4.5.2 Ozone Plots for BY Group Site = 134

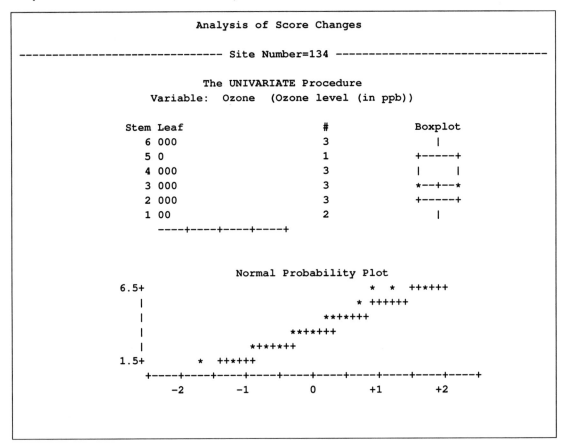

Output 4.5.3 Ozone Plots for BY Group Site = 137

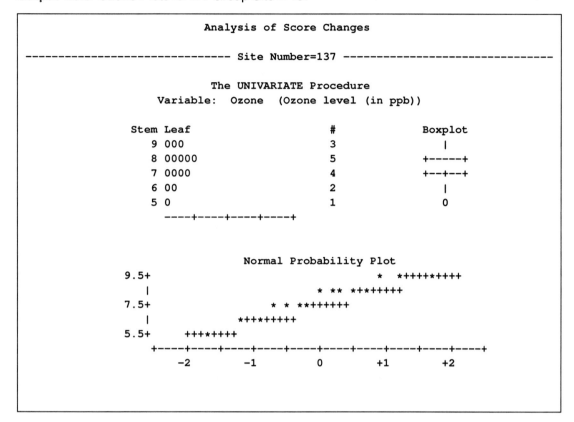

Output 4.5.4 Ozone Side-by-Side Boxplot for All BY Groups

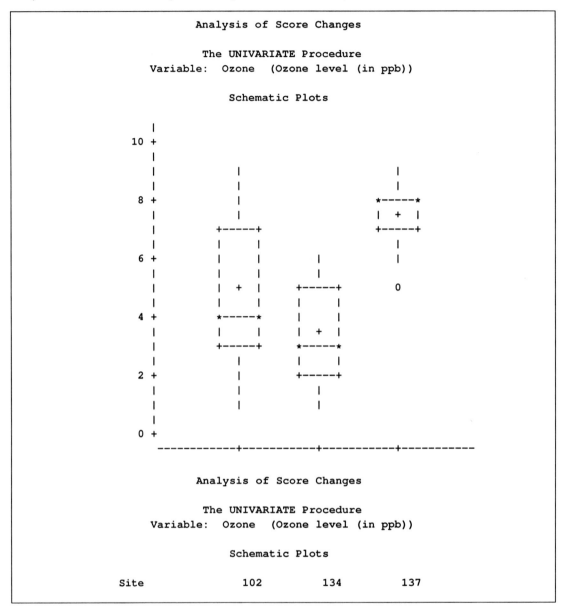

Note that you can use the PROBPLOT statement with the NORMAL option to produce high-resolution normal probability plots; see the section "Modeling a Data Distribution" on page 229.

Note that you can use the BOXPLOT procedure to produce box plots that use high-resolution graphics. See Chapter 24, "The BOXPLOT Procedure" (*SAS/STAT User's Guide*).

A sample program for this example, *uniex04.sas*, is available in the SAS Sample Library for Base SAS software.

Example 4.6: Analyzing a Data Set With a FREQ Variable

This example illustrates how to use PROC UNIVARIATE to analyze a data set with a variable that contains the frequency of each observation. The data set Speeding contains data on the number of cars pulled over for speeding on a stretch of highway with a 65 mile per hour speed limit. Speed is the speed at which the cars were traveling, and Number is the number of cars at each speed. The following statements create the data set:

```
data Speeding;
   label Speed = 'Speed (in miles per hour)';
   do Speed = 66 to 85;
      input Number @@;
      output;
   end;
   datalines;
 2  3  2  1  3  6  8  9 10 13
12 14  6  2  0  0  1  1  0  1
;
run;
```

The following statements create a table of moments for the variable Speed:

```
title 'Analysis of Speeding Data';
ods select Moments;
proc univariate data=Speeding;
   freq Number;
   var Speed;
run;
```

The ODS SELECT statement restricts the output, which is shown in Output 4.6.1, to the "Moments" table; see the section "ODS Table Names" on page 376. The FREQ statement specifies that the value of the variable Number represents the frequency of each observation.

For the formulas used to compute these moments, see the section "Descriptive Statistics" on page 325. A sample program for this example, *uniex05.sas*, is available in the SAS Sample Library for Base SAS software.

Output 4.6.1 Table of Moments

```
                       Analysis of Speeding Data

                         The UNIVARIATE Procedure
             Variable:  Speed   (Speed (in miles per hour))

                              Freq:  Number

                                Moments

    N                           94    Sum Weights              94
    Mean                  74.3404255  Sum Observations       6988
    Std Deviation         3.44403237  Variance           11.861359
    Skewness             -0.1275543   Kurtosis           0.92002287
    Uncorrected SS           520594   Corrected SS       1103.10638
    Coeff Variation       4.63278538  Std Error Mean     0.35522482
```

Example 4.7: Saving Summary Statistics in an OUT= Output Data Set

This example illustrates how to save summary statistics in an output data set. The following statements create a data set named Belts, which contains the breaking strengths (Strength) and widths (Width) of a sample of 50 automotive seat belts:

```
data Belts;
   label Strength = 'Breaking Strength (lb/in)'
         Width    = 'Width in Inches';
   input Strength Width @@;
   datalines;
1243.51  3.036  1221.95  2.995  1131.67  2.983  1129.70  3.019
1198.08  3.106  1273.31  2.947  1250.24  3.018  1225.47  2.980
1126.78  2.965  1174.62  3.033  1250.79  2.941  1216.75  3.037
1285.30  2.893  1214.14  3.035  1270.24  2.957  1249.55  2.958
1166.02  3.067  1278.85  3.037  1280.74  2.984  1201.96  3.002
1101.73  2.961  1165.79  3.075  1186.19  3.058  1124.46  2.929
1213.62  2.984  1213.93  3.029  1289.59  2.956  1208.27  3.029
1247.48  3.027  1284.34  3.073  1209.09  3.004  1146.78  3.061
1224.03  2.915  1200.43  2.974  1183.42  3.033  1195.66  2.995
1258.31  2.958  1136.05  3.022  1177.44  3.090  1246.13  3.022
1183.67  3.045  1206.50  3.024  1195.69  3.005  1223.49  2.971
1147.47  2.944  1171.76  3.005  1207.28  3.065  1131.33  2.984
1215.92  3.003  1202.17  3.058
;
run;
```

The following statements produce two output data sets containing summary statistics:

```
proc univariate data=Belts noprint;
   var Strength Width;
   output out=Means          mean=StrengthMean WidthMean;
   output out=StrengthStats mean=StrengthMean std=StrengthSD
                            min=StrengthMin   max=StrengthMax;
run;
```

When you specify an OUTPUT statement, you must also specify a VAR statement. You can use multiple OUTPUT statements with a single procedure statement. Each OUTPUT statement creates a new data set with the name specified by the OUT= option. In this example, two data sets, Means and StrengthStats, are created. See Output 4.7.1 for a listing of Means and Output 4.7.2 for a listing of StrengthStats.

Output 4.7.1 Listing of Output Data Set Means

	Analysis of Speeding Data	
Obs	Strength Mean	Width Mean
1	1205.75	3.00584

Output 4.7.2 Listing of Output Data Set StrengthStats

	Analysis of Speeding Data			
Obs	Strength Mean	Strength SD	Strength Max	Strength Min
1	1205.75	48.3290	1289.59	1101.73

Summary statistics are saved in an output data set by specifying *keyword=names* after the OUT= option. In the preceding statements, the first OUTPUT statement specifies the *keyword* MEAN followed by the *names* StrengthMean and WidthMean. The second OUTPUT statement specifies the *keywords* MEAN, STD, MAX, and MIN, for which the *names* StrengthMean, StrengthSD, StrengthMax, and StrengthMin are given.

The *keyword* specifies the statistic to be saved in the output data set, and the *names* determine the names for the new variables. The first *name* listed after a keyword contains that statistic for the first variable listed in the VAR statement; the second *name* contains that statistic for the second variable in the VAR statement, and so on.

The data set Means contains the mean of Strength in a variable named StrengthMean and the mean of Width in a variable named WidthMean. The data set StrengthStats contains the mean, standard deviation, maximum value, and minimum value of Strength in the variables StrengthMean, StrengthSD, StrengthMax, and StrengthMin, respectively.

See the section "OUT= Output Data Set in the OUTPUT Statement" on page 370 for more information about OUT= output data sets.

A sample program for this example, *uniex06.sas*, is available in the SAS Sample Library for Base SAS software.

Example 4.8: Saving Percentiles in an Output Data Set

This example, which uses the Belts data set from the previous example, illustrates how to save percentiles in an output data set. The UNIVARIATE procedure automatically computes the 1st, 5th, 10th, 25th, 75th, 90th, 95th, and 99th percentiles for each variable. You can save these percentiles in an output data set by specifying the appropriate keywords. For example, the following statements create an output data set named PctlStrength, which contains the 5th and 95th percentiles of the variable Strength:

```
proc univariate data=Belts noprint;
   var Strength Width;
   output out=PctlStrength p5=p5str p95=p95str;
run;
```

The output data set PctlStrength is listed in Output 4.8.1.

Output 4.8.1 Listing of Output Data Set PctlStrength

	Analysis of Speeding Data	
Obs	p95str	p5str
1	1284.34	1126.78

You can use the PCTLPTS=, PCTLPRE=, and PCTLNAME= options to save percentiles not automatically computed by the UNIVARIATE procedure. For example, the following statements create an output data set named Pctls, which contains the 20th and 40th percentiles of the variables Strength and Width:

```
proc univariate data=Belts noprint;
   var Strength Width;
   output out=Pctls pctlpts  = 20 40
                    pctlpre  = Strength Width
                    pctlname = pct20 pct40;
run;
```

The PCTLPTS= option specifies the percentiles to compute (in this case, the 20th and 40th percentiles). The PCTLPRE= and PCTLNAME= options build the names for the variables containing the percentiles. The PCTLPRE= option gives prefixes for the new variables, and the PCTLNAME= option gives a suffix to add to the prefix. When you use the PCTLPTS= specification, you must also use the PCTLPRE= specification.

The OUTPUT statement saves the 20th and 40th percentiles of Strength and Width in the variables Strengthpct20, Widthpct20, Strengthpct40, and Weightpct40. The output data set Pctls is listed in Output 4.8.2.

Output 4.8.2 Listing of Output Data Set Pctls

		Analysis of Speeding Data		
Obs	Strengthpct20	Strengthpct40	Widthpct20	Widthpct40
1	1165.91	1199.26	2.9595	2.995

A sample program for this example, *uniex06.sas*, is available in the SAS Sample Library for Base SAS software.

Example 4.9: Computing Confidence Limits for the Mean, Standard Deviation, and Variance

This example illustrates how to compute confidence limits for the mean, standard deviation, and variance of a population. A researcher is studying the heights of a certain population of adult females. She has collected a random sample of heights of 75 females, which are saved in the data set Heights:

```
data Heights;
   label Height = 'Height (in)';
   input Height @@;
   datalines;
64.1 60.9 64.1 64.7 66.7 65.0 63.7 67.4 64.9 63.7
64.0 67.5 62.8 63.9 65.9 62.3 64.1 60.6 68.6 68.6
63.7 63.0 64.7 68.2 66.7 62.8 64.0 64.1 62.1 62.9
62.7 60.9 61.6 64.6 65.7 66.6 66.7 66.0 68.5 64.4
60.5 63.0 60.0 61.6 64.3 60.2 63.5 64.7 66.0 65.1
63.6 62.0 63.6 65.8 66.0 65.4 63.5 66.3 66.2 67.5
65.8 63.1 65.8 64.4 64.0 64.9 65.7 61.0 64.1 65.5
68.6 66.6 65.7 65.1 70.0
;
run;
```

The following statements produce confidence limits for the mean, standard deviation, and variance of the population of heights:

```
title 'Analysis of Female Heights';
ods select BasicIntervals;
proc univariate data=Heights cibasic;
   var Height;
run;
```

The CIBASIC option requests confidence limits for the mean, standard deviation, and variance. For example, Output 4.9.1 shows that the 95% confidence interval for the population mean is (64.06, 65.07). The ODS SELECT statement restricts the output to the "BasicIntervals" table; see the section "ODS Table Names" on page 376.

The confidence limits in Output 4.9.1 assume that the heights are normally distributed, so you should check this assumption before using these confidence limits. See the section "Shapiro-Wilk Statistic" on page 357 for information about the Shapiro-Wilk test for normality in PROC UNIVARIATE. See Example 4.19 for an example that uses the test for normality.

Output 4.9.1 Default 95% Confidence Limits

```
                    Analysis of Female Heights

                       The UNIVARIATE Procedure
                    Variable:  Height   (Height (in))

              Basic Confidence Limits Assuming Normality

       Parameter            Estimate      95% Confidence Limits

       Mean                  64.56667     64.06302     65.07031
       Std Deviation          2.18900      1.88608      2.60874
       Variance               4.79171      3.55731      6.80552
```

By default, the confidence limits produced by the CIBASIC option produce 95% confidence intervals. You can request different level confidence limits by using the ALPHA= option in parentheses after the CIBASIC option. The following statements produce 90% confidence limits:

```
title 'Analysis of Female Heights';
ods select BasicIntervals;
proc univariate data=Heights cibasic(alpha=.1);
   var Height;
run;
```

The 90% confidence limits are displayed in Output 4.9.2.

Output 4.9.2 90% Confidence Limits

```
                    Analysis of Female Heights

                       The UNIVARIATE Procedure
                    Variable:  Height   (Height (in))

              Basic Confidence Limits Assuming Normality

       Parameter            Estimate      90% Confidence Limits

       Mean                  64.56667     64.14564     64.98770
       Std Deviation          2.18900      1.93114      2.53474
       Variance               4.79171      3.72929      6.42492
```

For the formulas used to compute these limits, see the section "Confidence Limits for Parameters of the Normal Distribution" on page 333.

A sample program for this example, *uniex07.sas*, is available in the SAS Sample Library for Base SAS software.

Example 4.10: Computing Confidence Limits for Quantiles and Percentiles

This example, which is a continuation of Example 4.9, illustrates how to compute confidence limits for quantiles and percentiles. A second researcher is more interested in summarizing the heights with quantiles than the mean and standard deviation. He is also interested in computing 90% confidence intervals for the quantiles. The following statements produce estimated quantiles and confidence limits for the population quantiles:

```
title 'Analysis of Female Heights';
ods select Quantiles;
proc univariate data=Heights ciquantnormal(alpha=.1);
   var Height;
run;
```

The ODS SELECT statement restricts the output to the "Quantiles" table; see the section "ODS Table Names" on page 376. The CIQUANTNORMAL option produces confidence limits for the quantiles. As noted in Output 4.10.1, these limits assume that the data are normally distributed. You should check this assumption before using these confidence limits. See the section "Shapiro-Wilk Statistic" on page 357 for information about the Shapiro-Wilk test for normality in PROC UNIVARIATE; see Example 4.19 for an example that uses the test for normality.

Output 4.10.1 Normal-Based Quantile Confidence Limits

```
               Analysis of Female Heights

                  The UNIVARIATE Procedure
            Variable:  Height   (Height (in))

                 Quantiles (Definition 5)

                                90% Confidence Limits
        Quantile      Estimate     Assuming Normality

        100% Max        70.0
        99%             70.0      68.94553     70.58228
        95%             68.6      67.59184     68.89311
        90%             67.5      66.85981     68.00273
        75% Q3          66.0      65.60757     66.54262
        50% Median      64.4      64.14564     64.98770
        25% Q1          63.1      62.59071     63.52576
        10%             61.6      61.13060     62.27352
        5%              60.6      60.24022     61.54149
        1%              60.0      58.55106     60.18781
        0% Min          60.0
```

It is also possible to use PROC UNIVARIATE to compute confidence limits for quantiles without assuming normality. The following statements use the CIQUANTDF option to request distribution-free confidence limits for the quantiles of the population of heights:

```
title 'Analysis of Female Heights';
ods select Quantiles;
proc univariate data=Heights ciquantdf(alpha=.1);
   var Height;
run;
```

The distribution-free confidence limits are shown in Output 4.10.2.

Output 4.10.2 Distribution-Free Quantile Confidence Limits

```
                   Analysis of Female Heights

                     The UNIVARIATE Procedure
              Variable:  Height  (Height (in))

                     Quantiles (Definition 5)

                        Quantile        Estimate

                        100% Max           70.0
                        99%                70.0
                        95%                68.6
                        90%                67.5
                        75% Q3             66.0
                        50% Median         64.4
                        25% Q1             63.1
                        10%                61.6
                        5%                 60.6
                        1%                 60.0
                        0% Min             60.0

                     Quantiles (Definition 5)

                 90% Confidence Limits      -------Order Statistics-------
   Quantile         Distribution Free       LCL Rank   UCL Rank    Coverage

   100% Max
   99%                 68.6       70.0         73         75         48.97
   95%                 67.5       70.0         68         75         94.50
   90%                 66.6       68.6         63         72         91.53
   75% Q3              65.7       66.6         50         63         91.77
   50% Median          64.1       65.1         31         46         91.54
   25% Q1              62.7       63.7         13         26         91.77
   10%                 60.6       62.7          4         13         91.53
   5%                  60.0       61.6          1          8         94.50
   1%                  60.0       60.5          1          3         48.97
   0% Min
```

The table in Output 4.10.2 includes the ranks from which the confidence limits are computed. For more information about how these confidence limits are calculated, see the section "Confidence Limits for Percentiles" on page 329. Note that confidence limits for quantiles are not produced when the WEIGHT statement is used.

A sample program for this example, *uniex07.sas*, is available in the SAS Sample Library for Base SAS software.

Example 4.11: Computing Robust Estimates

This example illustrates how you can use the UNIVARIATE procedure to compute robust estimates of location and scale. The following statements compute these estimates for the variable Systolic in the data set BPressure, which was introduced in Example 4.1:

```
title 'Robust Estimates for Blood Pressure Data';
ods select TrimmedMeans WinsorizedMeans RobustScale;
proc univariate data=BPressure trimmed=1 .1
                winsorized=.1  robustscale;
   var Systolic;
run;
```

The ODS SELECT statement restricts the output to the "TrimmedMeans," "WinsorizedMeans," and "RobustScale" tables; see the section "ODS Table Names" on page 376. The TRIMMED= option computes two trimmed means, the first after removing one observation and the second after removing 10% of the observations. If the value of TRIMMED= is greater than or equal to one, it is interpreted as the number of observations to be trimmed. The WINSORIZED= option computes a Winsorized mean that replaces three observations from the tails with the next closest observations. (Three observations are replaced because $np = (22)(.1) = 2.2$, and three is the smallest integer greater than 2.2.) The trimmed and Winsorized means for Systolic are displayed in Output 4.11.1.

Output 4.11.1 Computation of Trimmed and Winsorized Means

```
              Robust Estimates for Blood Pressure Data

                         The UNIVARIATE Procedure
                           Variable:  Systolic

                              Trimmed Means

 Percent     Number                   Std Error
 Trimmed    Trimmed     Trimmed       Trimmed       95% Confidence
 in Tail    in Tail      Mean          Mean              Limits         DF

    4.55        1      120.3500      2.573536      114.9635  125.7365    19
   13.64        3      120.3125      2.395387      115.2069  125.4181    15

                              Trimmed Means

                    Percent
                    Trimmed     t for H0:
                    in Tail     Mu0=0.00     Pr > |t|

                      4.55      46.76446      <.0001
                     13.64      50.22675      <.0001
```

Output 4.11.1 *continued*

```
                              Winsorized Means

   Percent     Number                 Std Error
 Winsorized  Winsorized  Winsorized  Winsorized      95% Confidence
   in Tail    in Tail       Mean        Mean            Limits          DF

    13.64         3        120.6364    2.417065    115.4845  125.7882    15

                              Winsorized Means

                  Percent
                Winsorized    t for H0:
                 in Tail      Mu0=0.00     Pr > |t|

                  13.64       49.91027     <.0001
```

Output 4.11.1 shows the trimmed mean for Systolic is 120.35 after one observation has been trimmed, and 120.31 after 3 observations are trimmed. The Winsorized mean for Systolic is 120.64. For details on trimmed and Winsorized means, see the section "Robust Estimators" on page 333. The trimmed means can be compared with the means shown in Output 4.1.1 (from Example 4.1), which displays the mean for Systolic as 121.273.

The ROBUSTSCALE option requests a table, displayed in Output 4.11.2, which includes the interquartile range, Gini's mean difference, the median absolute deviation about the median, Q_n, and S_n.

Output 4.11.2 shows the robust estimates of scale for Systolic. For instance, the interquartile range is 13. The estimates of σ range from 9.54 to 13.32. See the section "Robust Estimators" on page 333.

A sample program for this example, *uniex01.sas*, is available in the SAS Sample Library for Base SAS software.

Output 4.11.2 Computation of Robust Estimates of Scale

```
                Robust Measures of Scale

                                           Estimate
        Measure                   Value    of Sigma

        Interquartile Range     13.00000    9.63691
        Gini's Mean Difference  15.03030   13.32026
        MAD                      6.50000    9.63690
        Sn                       9.54080    9.54080
        Qn                      13.33140   11.36786
```

Example 4.12: Testing for Location

This example, which is a continuation of Example 4.9, illustrates how to carry out three tests for location: the Student's t test, the sign test, and the Wilcoxon signed rank test. These tests are discussed in the section "Tests for Location" on page 331.

The following statements demonstrate the tests for location by using the Heights data set introduced in Example 4.9. Because the data consists of adult female heights, the researchers are not interested in testing whether the mean of the population is equal to zero inches, which is the default μ_0 value. Instead, they are interested in testing whether the mean is equal to 66 inches. The following statements test the null hypothesis $H_0: \mu_0 = 66$:

```
title 'Analysis of Female Height Data';
ods select TestsForLocation LocationCounts;
proc univariate data=Heights mu0=66 loccount;
   var Height;
run;
```

The ODS SELECT statement restricts the output to the "TestsForLocation" and "LocationCounts" tables; see the section "ODS Table Names" on page 376. The MU0= option specifies the null hypothesis value of μ_0 for the tests for location; by default, $\mu_0 = 0$. The LOCCOUNT option produces the table of the number of observations greater than, not equal to, and less than 66 inches.

Output 4.12.1 contains the results of the tests for location. All three tests are highly significant, causing the researchers to reject the hypothesis that the mean is 66 inches.

A sample program for this example, *uniex07.sas*, is available in the SAS Sample Library for Base SAS software.

Output 4.12.1 Tests for Location with MU0=66 and LOCCOUNT

```
                Analysis of Female Height Data

                    The UNIVARIATE Procedure
             Variable:  Height   (Height (in))

                 Tests for Location: Mu0=66

         Test            -Statistic-      -----p Value------

         Student's t    t   -5.67065      Pr > |t|    <.0001
         Sign           M        -20      Pr >= |M|   <.0001
         Signed Rank    S       -849      Pr >= |S|   <.0001


                 Location Counts: Mu0=66.00

                    Count               Value

                    Num Obs >  Mu0         16
                    Num Obs ^= Mu0         72
                    Num Obs <  Mu0         56
```

Example 4.13: Performing a Sign Test Using Paired Data

This example demonstrates a sign test for paired data, which is a specific application of the tests for location discussed in Example 4.12.

The instructor from Example 4.4 is now interested in performing a sign test for the pairs of test scores in his college course. The following statements request basic statistical measures and tests for location:

```
title 'Test Scores for a College Course';
ods select BasicMeasures TestsForLocation;
proc univariate data=Score;
   var ScoreChange;
run;
```

The ODS SELECT statement restricts the output to the "BasicMeasures" and "TestsForLocation" tables; see the section "ODS Table Names" on page 376. The instructor is not willing to assume that the ScoreChange variable is normal or even symmetric, so he decides to examine the sign test. The large *p*-value (0.7744) of the sign test provides insufficient evidence of a difference in test score medians.

Output 4.13.1 Sign Test for ScoreChange

```
                 Test Scores for a College Course

                      The UNIVARIATE Procedure
           Variable:  ScoreChange   (Change in Test Scores)

                     Basic Statistical Measures

           Location                    Variability

       Mean     -3.08333     Std Deviation           13.33797
       Median   -3.00000     Variance               177.90152
       Mode     -5.00000     Range                   51.00000
                             Interquartile Range     10.50000

   NOTE: The mode displayed is the smallest of 2 modes with a count of 2.

                       Tests for Location: Mu0=0

           Test            -Statistic-      -----p Value------

           Student's t    t  -0.80079       Pr > |t|    0.4402
           Sign           M        -1       Pr >= |M|   0.7744
           Signed Rank    S      -8.5       Pr >= |S|   0.5278
```

A sample program for this example, *uniex03.sas*, is available in the SAS Sample Library for Base SAS software.

Example 4.14: Creating a Histogram

This example illustrates how to create a histogram. A semiconductor manufacturer produces printed circuit boards that are sampled to determine the thickness of their copper plating. The following statements create a data set named Trans, which contains the plating thicknesses (Thick) of 100 boards:

```
data Trans;
   input Thick @@;
   label Thick = 'Plating Thickness (mils)';
   datalines;
3.468 3.428 3.509 3.516 3.461 3.492 3.478 3.556 3.482 3.512
3.490 3.467 3.498 3.519 3.504 3.469 3.497 3.495 3.518 3.523
3.458 3.478 3.443 3.500 3.449 3.525 3.461 3.489 3.514 3.470
3.561 3.506 3.444 3.479 3.524 3.531 3.501 3.495 3.443 3.458
3.481 3.497 3.461 3.513 3.528 3.496 3.533 3.450 3.516 3.476
3.512 3.550 3.441 3.541 3.569 3.531 3.468 3.564 3.522 3.520
3.505 3.523 3.475 3.470 3.457 3.536 3.528 3.477 3.536 3.491
3.510 3.461 3.431 3.502 3.491 3.506 3.439 3.513 3.496 3.539
3.469 3.481 3.515 3.535 3.460 3.575 3.488 3.515 3.484 3.482
3.517 3.483 3.467 3.467 3.502 3.471 3.516 3.474 3.500 3.466
;
run;
```

The following statements create the histogram shown in Output 4.14.1.

```
title 'Analysis of Plating Thickness';
proc univariate data=Trans noprint;
   histogram Thick;
run;
```

The NOPRINT option in the PROC UNIVARIATE statement suppresses tables of summary statistics for the variable Thick that would be displayed by default. A histogram is created for each variable listed in the HISTOGRAM statement.

Output 4.14.1 Histogram for Plating Thickness

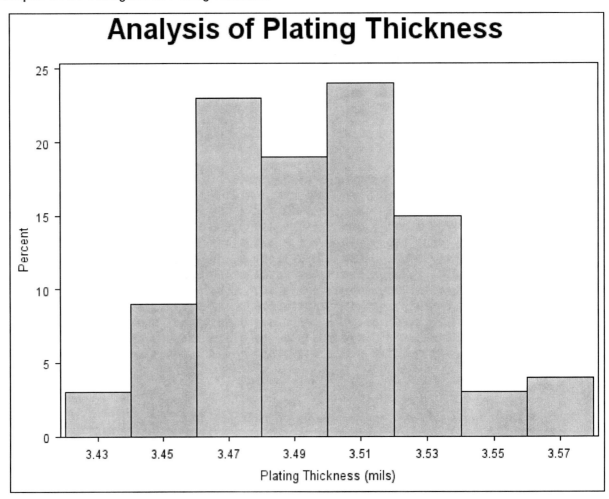

A sample program for this example, *uniex08.sas*, is available in the SAS Sample Library for Base SAS software.

Example 4.15: Creating a One-Way Comparative Histogram

This example illustrates how to create a comparative histogram. The effective channel length (in microns) is measured for 1225 field effect transistors. The channel lengths (Length) are stored in a data set named Channel, which is partially listed in Output 4.15.1:

Output 4.15.1 Partial Listing of Data Set Channel

```
            The Data Set Channel

         Lot        Length

         Lot 1        0.91
           .            .
         Lot 1        1.17
         Lot 2        1.47
           .            .
         Lot 2        1.39
         Lot 3        2.04
           .            .
         Lot 3        1.91
```

The following statements request a histogram of Length ignoring the lot source:

```
title 'Histogram of Length Ignoring Lot Source';
proc univariate data=Channel noprint;
   histogram Length;
run;
```

The resulting histogram is shown in Output 4.15.2.

Output 4.15.2 Histogram for Length Ignoring Lot Source

Histogram of Length Ignoring Lot Source

To investigate whether the peaks (modes) in Output 4.15.2 are related to the lot source, you can create a comparative histogram by using Lot as a classification variable. The following statements create the histogram shown in Output 4.15.3:

```
title 'Comparative Analysis of Lot Source';
proc univariate data=Channel noprint;
   class Lot;
   histogram Length / nrows = 3;
run;
```

The CLASS statement requests comparisons for each level (distinct value) of the classification variable Lot. The HISTOGRAM statement requests a comparative histogram for the variable Length. The NROWS= option specifies the number of rows per panel in the comparative histogram. By default, comparative histograms are displayed in two rows per panel.

Output 4.15.3 Comparison by Lot Source

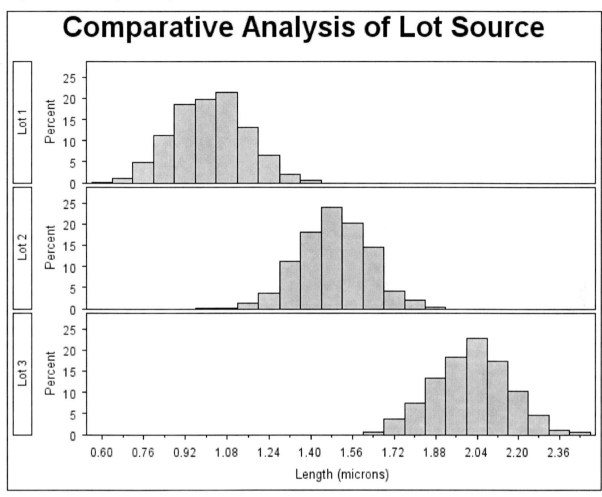

Output 4.15.3 reveals that the distributions of Length are similarly distributed except for shifts in mean.

A sample program for this example, *uniex09.sas*, is available in the SAS Sample Library for Base SAS software.

Example 4.16: Creating a Two-Way Comparative Histogram

This example illustrates how to create a two-way comparative histogram. Two suppliers (A and B) provide disk drives for a computer manufacturer. The manufacturer measures the disk drive opening width to determine whether there has been a change in variability from 2002 to 2003 for each supplier.

The following statements save the measurements in a data set named Disk. There are two classification variables, Supplier and Year, and a user-defined format is associated with Year.

```
proc format ;
   value mytime  1 = '2002' 2 = '2003';

data Disk;
   input @1 Supplier $10. Year Width;
   label Width = 'Opening Width (inches)';
   format Year mytime.;
datalines;
Supplier A   1    1.8932
    .        .      .
Supplier B   1    1.8986
Supplier A   2    1.8978
    .        .      .
Supplier B   2    1.8997
;
```

The following statements create the comparative histogram in Output 4.16.1:

```
title 'Results of Supplier Training Program';
proc univariate data=Disk noprint;
   class Supplier Year / keylevel = ('Supplier A' '2003');
   histogram Width / intertile = 1.0
                     vaxis     = 0 10 20 30
                     ncols     = 2
                     nrows     = 2;
run;
```

The KEYLEVEL= option specifies the key cell as the cell for which Supplier is equal to 'SUPPLIER A' and Year is equal to '2003.' This cell determines the binning for the other cells, and the columns are arranged so that this cell is displayed in the upper left corner. Without the KEYLEVEL= option, the default key cell would be the cell for which Supplier is equal to 'SUPPLIER A' and Year is equal to '2002'; the column labeled '2002' would be displayed to the left of the column labeled '2003.'

The VAXIS= option specifies the tick mark labels for the vertical axis. The NROWS=2 and NCOLS=2 options specify a 2 × 2 arrangement for the tiles. Output 4.16.1 provides evidence that both suppliers have reduced variability from 2002 to 2003.

Output 4.16.1 Two-Way Comparative Histogram

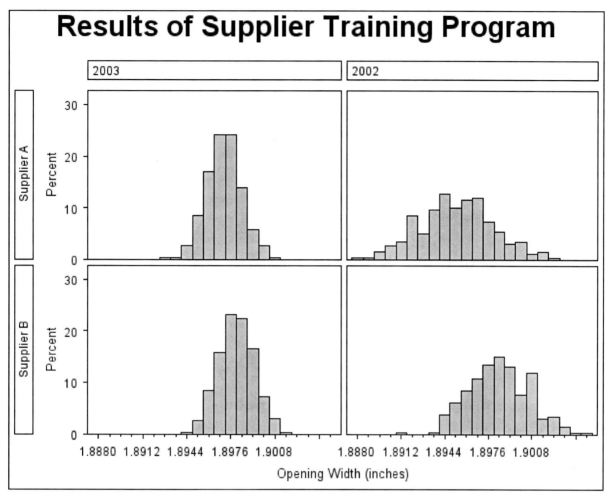

A sample program for this example, *uniex10.sas*, is available in the SAS Sample Library for Base SAS software.

Example 4.17: Adding Insets with Descriptive Statistics

This example illustrates how to add insets with descriptive statistics to a comparative histogram; see Output 4.17.1. Three similar machines are used to attach a part to an assembly. One hundred assemblies are sampled from the output of each machine, and a part position is measured in millimeters. The following statements create the data set Machines, which contains the measurements in a variable named Position:

```
data Machines;
   input Position @@;
   label Position = 'Position in Millimeters';
   if       (_n_ <= 100) then Machine = 'Machine 1';
   else if (_n_ <= 200) then Machine = 'Machine 2';
   else                      Machine = 'Machine 3';
   datalines;
-0.17  -0.19  -0.24  -0.24  -0.12   0.07  -0.61   0.22   1.91  -0.08
-0.59   0.05  -0.38   0.82  -0.14   0.32   0.12  -0.02   0.26   0.19

... more lines ...

 0.79   0.66   0.22   0.71   0.53   0.57   0.90   0.48   1.17   1.03
;
run;
```

The following statements create the comparative histogram in Output 4.17.1:

```
title 'Machine Comparison Study';
proc univariate data=Machines noprint;
   class Machine;
   histogram Position / nrows     = 3
                        intertile = 1
                        midpoints = -1.2 to 2.2 by 0.1
                        vaxis     =    0 to 16 by 4;
   inset mean std="Std Dev" / pos = ne format = 6.3;
run;
```

The INSET statement requests insets that contain the sample mean and standard deviation for each machine in the corresponding tile. The MIDPOINTS= option specifies the midpoints of the histogram bins.

Output 4.17.1 Comparative Histograms

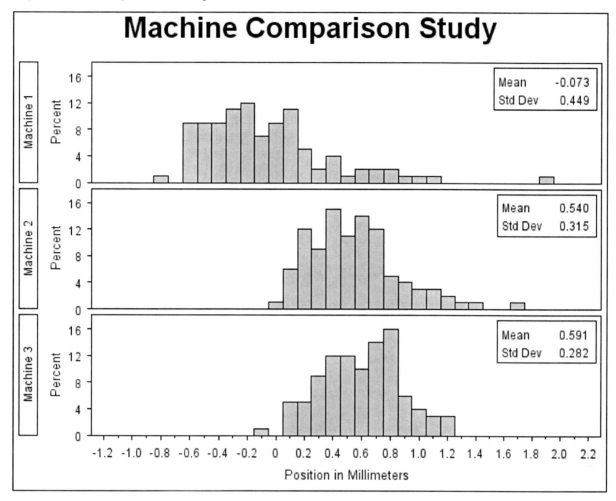

Output 4.17.1 shows that the average position for Machines 2 and 3 are similar and that the spread for Machine 1 is much larger than for Machines 2 and 3.

A sample program for this example, *uniex11.sas*, is available in the SAS Sample Library for Base SAS software.

Example 4.18: Binning a Histogram

This example, which is a continuation of Example 4.14, demonstrates various methods for binning a histogram. This example also illustrates how to save bin percentages in an OUTHISTOGRAM= data set.

The manufacturer from Example 4.14 now wants to enhance the histogram by using the ENDPOINTS= option to change the endpoints of the bins. The following statements create a histogram with bins that have end points 3.425 and 3.6 and width 0.025:

```
title 'Enhancing a Histogram';
ods select HistogramBins MyHist;
proc univariate data=Trans;
   histogram Thick / midpercents name='MyHist'
                    endpoints = 3.425 to 3.6 by .025;
run;
```

The ODS SELECT statement restricts the output to the "HistogramBins" table and the "MyHist" histogram; see the section "ODS Table Names" on page 376. The ENDPOINTS= option specifies the endpoints for the histogram bins. By default, if the ENDPOINTS= option is not specified, the automatic binning algorithm computes values for the midpoints of the bins. The MIDPERCENTS option requests a table of the midpoints of each histogram bin and the percent of the observations that fall in each bin. This table is displayed in Output 4.18.1; the histogram is displayed in Output 4.18.2. The NAME= option specifies a name for the histogram that can be used in the ODS SELECT statement.

Output 4.18.1 Table of Bin Percentages Requested with MIDPERCENTS Option

```
                   Enhancing a Histogram

                   The UNIVARIATE Procedure

                  Histogram Bins for Thick

                       Bin
                    Minimum      Observed
                     Point        Percent

                     3.425         8.000
                     3.450        21.000
                     3.475        25.000
                     3.500        29.000
                     3.525        11.000
                     3.550         5.000
                     3.575         1.000
```

Output 4.18.2 Histogram with ENDPOINTS= Option

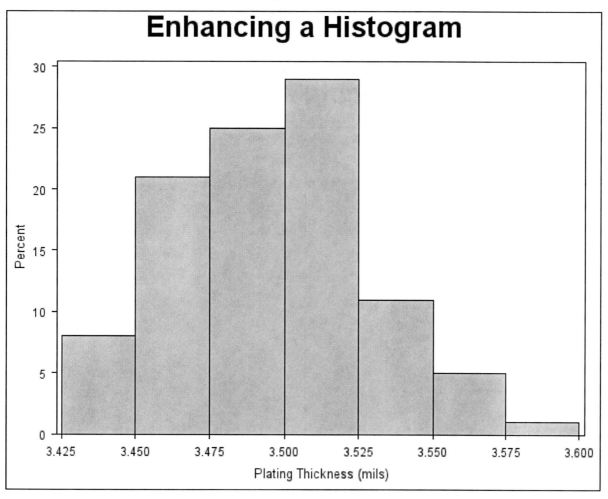

The MIDPOINTS= option is an alternative to the ENDPOINTS= option for specifying histogram bins. The following statements create a histogram, shown in Output 4.18.3, which is similar to the one in Output 4.18.2:

```
title 'Enhancing a Histogram';
proc univariate data=Trans noprint;
   histogram Thick / midpoints   = 3.4375 to 3.5875 by .025
                    rtinclude
                    outhistogram = OutMdpts;
run;
```

Output 4.18.3 differs from Output 4.18.2 in two ways:

- The MIDPOINTS= option specifies the bins for the histogram by specifying the midpoints of the bins instead of specifying the endpoints. Note that the histogram displays midpoints instead of endpoints.

- The RTINCLUDE option requests that the right endpoint of each bin be included in the histogram interval instead of the default, which is to include the left endpoint in the interval. This changes the histogram slightly from Output 4.18.2. Six observations have a thickness equal to an endpoint of an interval. For instance, there is one observation with a thickness of 3.45 mils. In Output 4.18.3, this observation is included in the bin from 3.425 to 3.45.

Output 4.18.3 Histogram with MIDPOINTS= and RTINCLUDE Options

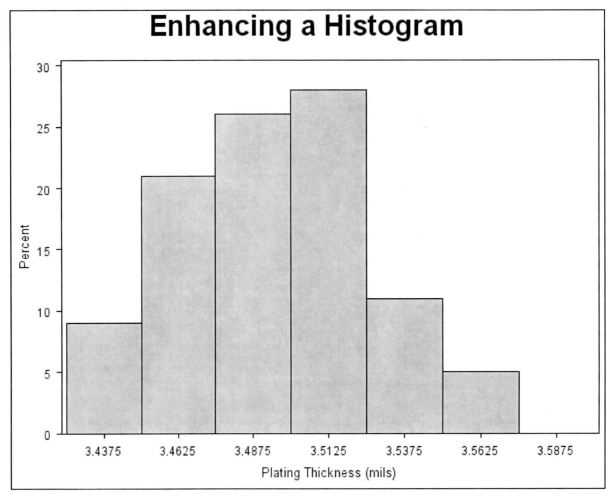

The OUTHISTOGRAM= option produces an output data set named OutMdpts, displayed in Output 4.18.4. This data set provides information about the bins of the histogram. For more information, see the section "OUTHISTOGRAM= Output Data Set" on page 372.

Output 4.18.4 The OUTHISTOGRAM= Data Set OutMdpts

```
                    Enhancing a Histogram

          Obs    _VAR_    _MIDPT_    _OBSPCT_    _COUNT_

           1     Thick     3.4375        9           9
           2     Thick     3.4625       21          21
           3     Thick     3.4875       26          26
           4     Thick     3.5125       28          28
           5     Thick     3.5375       11          11
           6     Thick     3.5625        5           5
```

A sample program for this example, *uniex08.sas*, is available in the SAS Sample Library for Base SAS software.

Example 4.19: Adding a Normal Curve to a Histogram

This example is a continuation of Example 4.14. The following statements fit a normal distribution to the thickness measurements in the Trans data set and superimpose the fitted density curve on the histogram:

```
title 'Analysis of Plating Thickness';
ods select ParameterEstimates GoodnessOfFit FitQuantiles Bins MyPlot;
proc univariate data=Trans;
   histogram Thick / normal(percents=20 40 60 80 midpercents)
                    name='MyPlot';
   inset n normal(ksdpval) / pos = ne format = 6.3;
run;
```

The ODS SELECT statement restricts the output to the "ParameterEstimates," "GoodnessOfFit," "FitQuantiles," and "Bins" tables; see the section "ODS Table Names" on page 376. The NORMAL option specifies that the normal curve be displayed on the histogram shown in Output 4.19.2. It also requests a summary of the fitted distribution, which is shown in Output 4.19.1. This summary includes goodness-of-fit tests, parameter estimates, and quantiles of the fitted distribution. (If you specify the NORMALTEST option in the PROC UNIVARIATE statement, the Shapiro-Wilk test for normality is included in the tables of statistical output.)

Two secondary options are specified in parentheses after the NORMAL primary option. The PERCENTS= option specifies quantiles, which are to be displayed in the "FitQuantiles" table. The MIDPERCENTS option requests a table that lists the midpoints, the observed percentage of observations, and the estimated percentage of the population in each interval (estimated from the fitted normal distribution). See Table 4.12 and Table 4.17 for the secondary options that can be specified with after the NORMAL primary option.

Output 4.19.1 Summary of Fitted Normal Distribution

```
            Analysis of Plating Thickness

               The UNIVARIATE Procedure
         Fitted Normal Distribution for Thick

         Parameters for Normal Distribution

            Parameter    Symbol    Estimate

            Mean         Mu         3.49533
            Std Dev      Sigma      0.032117
```

Output 4.19.1 *continued*

```
              Goodness-of-Fit Tests for Normal Distribution

Test                     ----Statistic-----      ------p Value------

Kolmogorov-Smirnov       D      0.05563823       Pr > D       >0.150
Cramer-von Mises         W-Sq   0.04307548       Pr > W-Sq    >0.250
Anderson-Darling         A-Sq   0.27840748       Pr > A-Sq    >0.250


           Histogram Bin Percents for Normal Distribution

               Bin       -------Percent------
            Midpoint    Observed     Estimated

              3.43        3.000         3.296
              3.45        9.000         9.319
              3.47       23.000        18.091
              3.49       19.000        24.124
              3.51       24.000        22.099
              3.53       15.000        13.907
              3.55        3.000         6.011
              3.57        4.000         1.784


                Quantiles for Normal Distribution

                         ------Quantile------
             Percent    Observed     Estimated

              20.0       3.46700      3.46830
              40.0       3.48350      3.48719
              60.0       3.50450      3.50347
              80.0       3.52250      3.52236
```

Output 4.19.2 Histogram Superimposed with Normal Curve

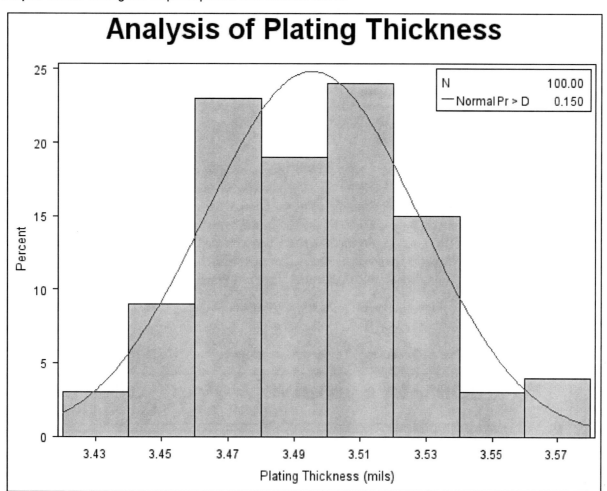

The histogram of the variable Thick with a superimposed normal curve is shown in Output 4.19.2.

The estimated parameters for the normal curve ($\hat{mu} = 3.50$ and $\hat{\sigma} = 0.03$) are shown in Output 4.19.1. By default, the parameters are estimated unless you specify values with the MU= and SIGMA= secondary options after the NORMAL primary option. The results of three goodness-of-fit tests based on the empirical distribution function (EDF) are displayed in Output 4.19.1. Because the *p*-values are all greater than 0.15, the hypothesis of normality is not rejected.

A sample program for this example, *uniex08.sas*, is available in the SAS Sample Library for Base SAS software.

Example 4.20: Adding Fitted Normal Curves to a Comparative Histogram

This example is a continuation of Example 4.15, which introduced the data set Channel on page 407. In Output 4.15.3, it appears that the channel lengths in each lot are normally distributed. The following statements use the NORMAL option to fit a normal distribution for each lot:

```
title 'Comparative Analysis of Lot Source';
proc univariate data=Channel noprint;
   class Lot;
   histogram Length / nrows     = 3
                      intertile = 1
                      cprop
                      normal(noprint);
   inset n = "N" / pos = nw;
run;
```

The NOPRINT option in the PROC UNIVARIATE statement suppresses the tables of statistical output produced by default; the NOPRINT option in parentheses after the NORMAL option suppresses the tables of statistical output related to the fit of the normal distribution. The normal parameters are estimated from the data for each lot, and the curves are superimposed on each component histogram. The INTERTILE= option specifies the space between the framed areas, which are referred to as tiles. The CPROP= option requests the shaded bars above each tile, which represent the relative frequencies of observations in each lot. The comparative histogram is displayed in Output 4.20.1.

A sample program for this example, *uniex09.sas*, is available in the SAS Sample Library for Base SAS software.

Output 4.20.1 Fitting Normal Curves to a Comparative Histogram

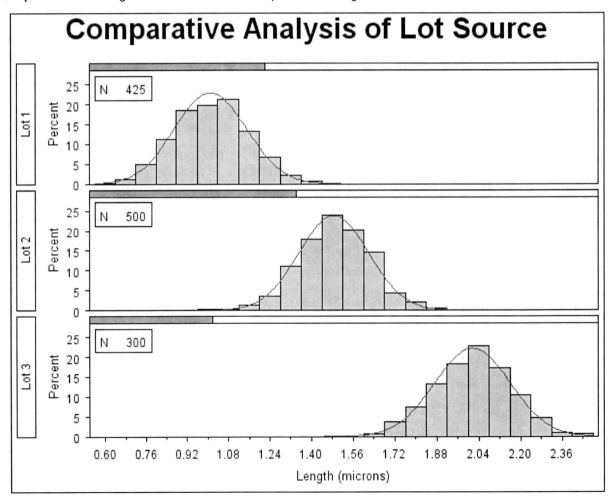

Example 4.21: Fitting a Beta Curve

You can use a beta distribution to model the distribution of a variable that is known to vary between lower and upper bounds. In this example, a manufacturing company uses a robotic arm to attach hinges on metal sheets. The attachment point should be offset 10.1 mm from the left edge of the sheet. The actual offset varies between 10.0 and 10.5 mm due to variation in the arm. The following statements save the offsets for 50 attachment points as the values of the variable Length in the data set Robots:

```
data Robots;
   input Length @@;
   label Length = 'Attachment Point Offset (in mm)';
   datalines;
10.147 10.070 10.032 10.042 10.102
10.034 10.143 10.278 10.114 10.127
10.122 10.018 10.271 10.293 10.136
10.240 10.205 10.186 10.186 10.080
10.158 10.114 10.018 10.201 10.065
10.061 10.133 10.153 10.201 10.109
10.122 10.139 10.090 10.136 10.066
10.074 10.175 10.052 10.059 10.077
10.211 10.122 10.031 10.322 10.187
10.094 10.067 10.094 10.051 10.174
;
run;
```

The following statements create a histogram with a fitted beta density curve, shown in Output 4.21.1:

```
title 'Fitted Beta Distribution of Offsets';
ods select ParameterEstimates FitQuantiles MyHist;
proc univariate data=Robots;
   histogram Length /
      beta(theta=10 scale=0.5 color=red fill)
      href      = 10
      hreflabel = 'Lower Bound'
      lhref     = 2
      vaxis     = axis1
      name      = 'MyHist';
   axis1 label=(a=90 r=0);
   inset n = 'Sample Size'
         beta / pos=ne  cfill=blank;
run;
```

The ODS SELECT statement restricts the output to the "ParameterEstimates" and "FitQuantiles" tables; see the section "ODS Table Names" on page 376. The BETA primary option requests a fitted beta distribution. The THETA= secondary option specifies the lower threshold. The SCALE= secondary option specifies the range between the lower threshold and the upper threshold. Note that the default THETA= and SCALE= values are zero and one, respectively.

Output 4.21.1 Superimposing a Histogram with a Fitted Beta Curve

Fitted Beta Distribution of Offsets

[Histogram with fitted beta curve; x-axis: Attachment Point Offset (in mm) from 10.02 to 10.32; y-axis: Percent from 0 to 35; Lower Bound reference line shown; inset shows Sample Size 50, Beta]

The FILL secondary option specifies that the area under the curve is to be filled with the CFILL= color. (If FILL were omitted, the CFILL= color would be used to fill the histogram bars instead.)

The HREF= option draws a reference line at the lower bound, and the HREFLABEL= option adds the label *Lower Bound*. The LHREF= option specifies a dashed line type for the reference line. The INSET statement adds an inset with the sample size positioned in the northeast corner of the plot.

In addition to displaying the beta curve, the BETA option requests a summary of the curve fit. This summary, which includes parameters for the curve and the observed and estimated quantiles, is shown in Output 4.21.2.

A sample program for this example, *uniex12.sas*, is available in the SAS Sample Library for Base SAS software.

Output 4.21.2 Summary of Fitted Beta Distribution

```
                  Fitted Beta Distribution of Offsets

                         The UNIVARIATE Procedure
                    Fitted Beta Distribution for Length

                      Parameters for Beta Distribution

                   Parameter      Symbol      Estimate

                   Threshold      Theta             10
                   Scale          Sigma            0.5
                   Shape          Alpha        2.06832
                   Shape          Beta        6.022479
                   Mean                       10.12782
                   Std Dev                    0.072339


                     Quantiles for Beta Distribution
                                  ------Quantile------
                      Percent     Observed    Estimated

                          1.0      10.0180      10.0124
                          5.0      10.0310      10.0285
                         10.0      10.0380      10.0416
                         25.0      10.0670      10.0718
                         50.0      10.1220      10.1174
                         75.0      10.1750      10.1735
                         90.0      10.2255      10.2292
                         95.0      10.2780      10.2630
                         99.0      10.3220      10.3237
```

Example 4.22: Fitting Lognormal, Weibull, and Gamma Curves

To determine an appropriate model for a data distribution, you should consider curves from several distribution families. As shown in this example, you can use the HISTOGRAM statement to fit more than one distribution and display the density curves on a histogram. The gap between two plates is measured (in cm) for each of 50 welded assemblies selected at random from the output of a welding process. The following statements save the measurements (Gap) in a data set named Plates:

```
data Plates;
   label Gap = 'Plate Gap in cm';
   input Gap @@;
   datalines;
0.746  0.357  0.376  0.327  0.485 1.741  0.241  0.777  0.768  0.409
0.252  0.512  0.534  1.656  0.742 0.378  0.714  1.121  0.597  0.231
0.541  0.805  0.682  0.418  0.506 0.501  0.247  0.922  0.880  0.344
0.519  1.302  0.275  0.601  0.388 0.450  0.845  0.319  0.486  0.529
1.547  0.690  0.676  0.314  0.736 0.643  0.483  0.352  0.636  1.080
;
run;
```

The following statements fit three distributions (lognormal, Weibull, and gamma) and display their density curves on a single histogram:

```
title 'Distribution of Plate Gaps';
ods select ParameterEstimates GoodnessOfFit FitQuantiles MyHist;
proc univariate data=Plates;
   var Gap;
   histogram / midpoints=0.2 to 1.8 by 0.2
               lognormal
               weibull
               gamma
               vaxis  = axis1
               name   = 'MyHist';
   inset n mean(5.3) std='Std Dev'(5.3) skewness(5.3)
           / pos = ne  header = 'Summary Statistics';
   axis1 label=(a=90 r=0);
run;
```

The ODS SELECT statement restricts the output to the "ParameterEstimates," "GoodnessOfFit," and "FitQuantiles" tables; see the section "ODS Table Names" on page 376. The LOGNORMAL, WEIBULL, and GAMMA primary options request superimposed fitted curves on the histogram in Output 4.22.1. Note that a threshold parameter $\theta = 0$ is assumed for each curve. In applications where the threshold is not zero, you can specify θ with the THETA= secondary option.

The LOGNORMAL, WEIBULL, and GAMMA options also produce the summaries for the fitted distributions shown in Output 4.22.2 through Output 4.22.4.

Output 4.22.2 provides three EDF goodness-of-fit tests for the lognormal distribution: the Anderson-Darling, the Cramér-von Mises, and the Kolmogorov-Smirnov tests. At the $\alpha = 0.10$ significance level, all tests support the conclusion that the two-parameter lognormal distribution with scale parameter $\hat{\zeta} = -0.58$ and shape parameter $\hat{\sigma} = 0.50$ provides a good model for the distribution of plate gaps.

Output 4.22.1 Superimposing a Histogram with Fitted Curves

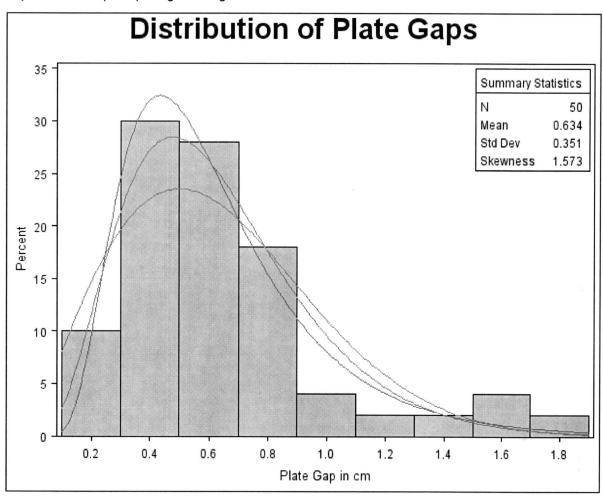

Output 4.22.2 Summary of Fitted Lognormal Distribution

```
                    Distribution of Plate Gaps

                      The UNIVARIATE Procedure
                Fitted Lognormal Distribution for Gap

                Parameters for Lognormal Distribution

              Parameter     Symbol    Estimate

              Threshold     Theta            0
              Scale         Zeta      -0.58375
              Shape         Sigma     0.499546
              Mean                    0.631932
              Std Dev                 0.336436
```

Output 4.22.2 *continued*

```
          Goodness-of-Fit Tests for Lognormal Distribution

Test                     ----Statistic-----      ------p Value------

Kolmogorov-Smirnov    D        0.06441431        Pr > D      >0.150
Cramer-von Mises      W-Sq     0.02823022        Pr > W-Sq   >0.500
Anderson-Darling      A-Sq     0.24308402        Pr > A-Sq   >0.500


             Quantiles for Lognormal Distribution

                      ------Quantile------
          Percent     Observed    Estimated

             1.0       0.23100      0.17449
             5.0       0.24700      0.24526
            10.0       0.29450      0.29407
            25.0       0.37800      0.39825
            50.0       0.53150      0.55780
            75.0       0.74600      0.78129
            90.0       1.10050      1.05807
            95.0       1.54700      1.26862
            99.0       1.74100      1.78313
```

Output 4.22.3 Summary of Fitted Weibull Distribution

```
                    Distribution of Plate Gaps

                     The UNIVARIATE Procedure
                Fitted Weibull Distribution for Gap

              Parameters for Weibull Distribution

             Parameter    Symbol     Estimate

             Threshold    Theta             0
             Scale        Sigma      0.719208
             Shape        C          1.961159
             Mean                    0.637641
             Std Dev                 0.339248

           Goodness-of-Fit Tests for Weibull Distribution

Test                     ----Statistic-----      ------p Value------

Cramer-von Mises      W-Sq     0.15937281        Pr > W-Sq    0.016
Anderson-Darling      A-Sq     1.15693542        Pr > A-Sq   <0.010
```

Output 4.22.3 *continued*

```
        Quantiles for Weibull Distribution

                 ------Quantile------
      Percent    Observed    Estimated

         1.0      0.23100      0.06889
         5.0      0.24700      0.15817
        10.0      0.29450      0.22831
        25.0      0.37800      0.38102
        50.0      0.53150      0.59661
        75.0      0.74600      0.84955
        90.0      1.10050      1.10040
        95.0      1.54700      1.25842
        99.0      1.74100      1.56691
```

Output 4.22.3 provides two EDF goodness-of-fit tests for the Weibull distribution: the Anderson-Darling and the Cramér-von Mises tests. The *p*-values for the EDF tests are all less than 0.10, indicating that the data do not support a Weibull model.

Output 4.22.4 Summary of Fitted Gamma Distribution

```
                 Distribution of Plate Gaps

                    The UNIVARIATE Procedure
              Fitted Gamma Distribution for Gap

             Parameters for Gamma Distribution

            Parameter    Symbol    Estimate

            Threshold    Theta            0
            Scale        Sigma     0.155198
            Shape        Alpha     4.082646
            Mean                   0.63362
            Std Dev                0.313587

         Goodness-of-Fit Tests for Gamma Distribution

   Test                 ----Statistic-----    ------p Value------

   Kolmogorov-Smirnov    D     0.09695325    Pr > D        >0.250
   Cramer-von Mises      W-Sq  0.07398467    Pr > W-Sq     >0.250
   Anderson-Darling      A-Sq  0.58106613    Pr > A-Sq      0.137
```

Output 4.22.4 *continued*

```
            Quantiles for Gamma Distribution

                        ------Quantile------
            Percent    Observed    Estimated

                1.0     0.23100      0.13326
                5.0     0.24700      0.21951
               10.0     0.29450      0.27938
               25.0     0.37800      0.40404
               50.0     0.53150      0.58271
               75.0     0.74600      0.80804
               90.0     1.10050      1.05392
               95.0     1.54700      1.22160
               99.0     1.74100      1.57939
```

Output 4.22.4 provides three EDF goodness-of-fit tests for the gamma distribution: the Anderson-Darling, the Cramér-von Mises, and the Kolmogorov-Smirnov tests. At the $\alpha = 0.10$ significance level, all tests support the conclusion that the gamma distribution with scale parameter $\sigma = 0.16$ and shape parameter $\alpha = 4.08$ provides a good model for the distribution of plate gaps.

Based on this analysis, the fitted lognormal distribution and the fitted gamma distribution are both good models for the distribution of plate gaps.

A sample program for this example, *uniex13.sas*, is available in the SAS Sample Library for Base SAS software.

Example 4.23: Computing Kernel Density Estimates

This example illustrates the use of kernel density estimates to visualize a nonnormal data distribution. This example uses the data set Channel, which is introduced in Example 4.15.

When you compute kernel density estimates, you should try several choices for the bandwidth parameter c because this determines the smoothness and closeness of the fit. You can specify a list of up to five C= values with the KERNEL option to request multiple density estimates, as shown in the following statements:

```
   title 'FET Channel Length Analysis';
proc univariate data=Channel noprint;
   histogram Length / kernel(c = 0.25 0.50 0.75 1.00
                     l = 1 20 2 34
                     noprint);
run;
```

The L= secondary option specifies distinct line types for the curves (the L= values are paired with the C= values in the order listed). Output 4.23.1 demonstrates the effect of c. In general, larger values of c yield smoother density estimates, and smaller values yield estimates that more closely fit the data distribution.

Output 4.23.1 Multiple Kernel Density Estimates

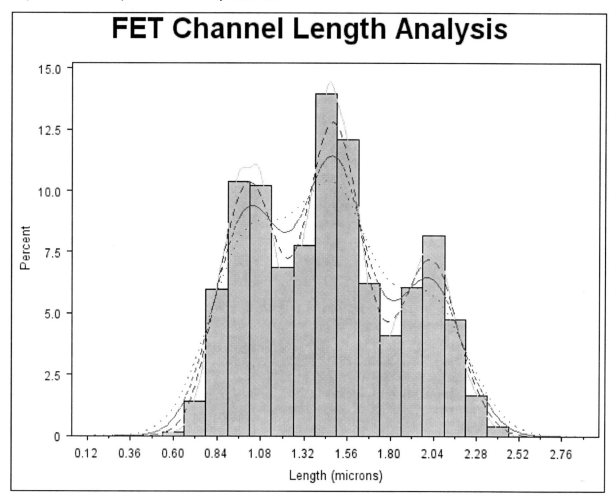

Output 4.23.1 reveals strong trimodality in the data, which is displayed with comparative histograms in Example 4.15.

A sample program for this example, *uniex09.sas*, is available in the SAS Sample Library for Base SAS software.

Example 4.24: Fitting a Three-Parameter Lognormal Curve

If you request a lognormal fit with the LOGNORMAL primary option, a two-parameter lognormal distribution is assumed. This means that the shape parameter σ and the scale parameter ζ are unknown (unless specified) and that the threshold θ is known (it is either specified with the THETA= option or assumed to be zero).

If it is necessary to estimate θ in addition to ζ and σ, the distribution is referred to as a three-parameter lognormal distribution. This example shows how you can request a three-parameter lognormal distribution.

A manufacturing process produces a plastic laminate whose strength must exceed a minimum of 25 pounds per square inch (psi). Samples are tested, and a lognormal distribution is observed for the strengths. It is important to estimate θ to determine whether the process meets the strength requirement. The following statements save the strengths for 49 samples in the data set Plastic:

```
data Plastic;
   label Strength = 'Strength in psi';
   input Strength @@;
   datalines;
30.26 31.23 71.96 47.39 33.93 76.15 42.21
81.37 78.48 72.65 61.63 34.90 24.83 68.93
43.27 41.76 57.24 23.80 34.03 33.38 21.87
31.29 32.48 51.54 44.06 42.66 47.98 33.73
25.80 29.95 60.89 55.33 39.44 34.50 73.51
43.41 54.67 99.43 50.76 48.81 31.86 33.88
35.57 60.41 54.92 35.66 59.30 41.96 45.32
;
run;
```

The following statements use the LOGNORMAL primary option in the HISTOGRAM statement to display the fitted three-parameter lognormal curve shown in Output 4.24.1:

```
title 'Three-Parameter Lognormal Fit';
proc univariate data=Plastic noprint;
   histogram Strength / lognormal(fill theta = est noprint);
   inset lognormal    / format=6.2 pos=ne;
run;
```

The NOPRINT option suppresses the tables of statistical output produced by default. Specifying THETA=EST requests a local maximum likelihood estimate (LMLE) for θ, as described by Cohen (1951). This estimate is then used to compute maximum likelihood estimates for σ and ζ.

NOTE: You can also specify THETA=EST with the WEIBULL primary option to fit a three-parameter Weibull distribution.

A sample program for this example, *uniex14.sas*, is available in the SAS Sample Library for Base SAS software.

Output 4.24.1 Three-Parameter Lognormal Fit

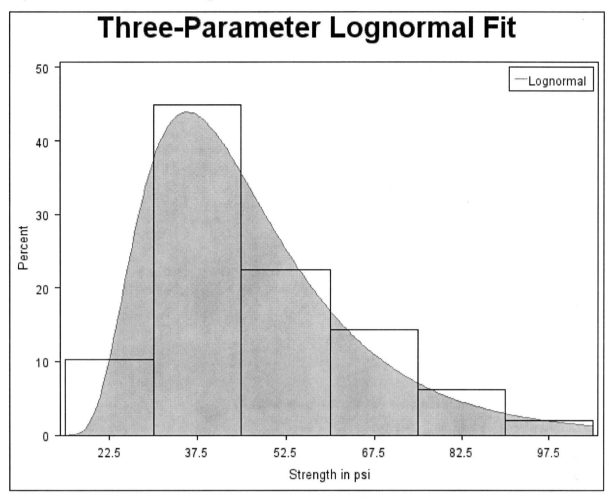

Example 4.25: Annotating a Folded Normal Curve

This example shows how to display a fitted curve that is not supported by the HISTOGRAM statement. The offset of an attachment point is measured (in mm) for a number of manufactured assemblies, and the measurements (Offset) are saved in a data set named Assembly. The following statements create the data set Assembly:

```
data Assembly;
   label Offset = 'Offset (in mm)';
   input Offset @@;
   datalines;
11.11 13.07 11.42  3.92 11.08  5.40 11.22 14.69  6.27  9.76
 9.18  5.07  3.51 16.65 14.10  9.69 16.61  5.67  2.89  8.13
 9.97  3.28 13.03 13.78  3.13  9.53  4.58  7.94 13.51 11.43
11.98  3.90  7.67  4.32 12.69  6.17 11.48  2.82 20.42  1.01
 3.18  6.02  6.63  1.72  2.42 11.32 16.49  1.22  9.13  3.34
 1.29  1.70  0.65  2.62  2.04 11.08 18.85 11.94  8.34  2.07
 0.31  8.91 13.62 14.94  4.83 16.84  7.09  3.37  0.49 15.19
 5.16  4.14  1.92 12.70  1.97  2.10  9.38  3.18  4.18  7.22
```

```
15.84 10.85  2.35  1.93  9.19  1.39 11.40 12.20 16.07  9.23
 0.05  2.15  1.95  4.39  0.48 10.16  4.81  8.28  5.68 22.81
 0.23  0.38 12.71  0.06 10.11 18.38  5.53  9.36  9.32  3.63
12.93 10.39  2.05 15.49  8.12  9.52  7.77 10.70  6.37  1.91
 8.60 22.22  1.74  5.84 12.90 13.06  5.08  2.09  6.41  1.40
15.60  2.36  3.97  6.17  0.62  8.56  9.36 10.19  7.16  2.37
12.91  0.95  0.89  3.82  7.86  5.33 12.92  2.64  7.92 14.06
;
run;
```

It is decided to fit a *folded normal distribution* to the offset measurements. A variable X has a folded normal distribution if $X = |Y|$, where Y is distributed as $N(\mu, \sigma)$. The fitted density is

$$h(x) = \frac{1}{\sqrt{2\pi}\sigma} \left[\exp\left(-\frac{(x-\mu)^2}{2\sigma^2}\right) + \exp\left(-\frac{(x+\mu)^2}{2\sigma^2}\right) \right]$$

where $x \geq 0$.

You can use SAS/IML to compute preliminary estimates of μ and σ based on a method of moments given by Elandt (1961). These estimates are computed by solving equation (19) Elandt (1961), which is given by

$$f(\theta) = \frac{\left(\frac{2}{\sqrt{2\pi}} e^{-\theta^2/2} - \theta[1 - 2\Phi(\theta)]\right)^2}{1 + \theta^2} = A$$

where $\Phi(\cdot)$ is the standard normal distribution function and

$$A = \frac{\bar{x}^2}{\frac{1}{n}\sum_{i=1}^{n} x_i^2}$$

Then the estimates of σ and μ are given by

$$\hat{\sigma}_0 = \sqrt{\frac{\frac{1}{n}\sum_{i=1}^{n} x_i^2}{1 + \hat{\theta}^2}}$$
$$\hat{\mu}_0 = \hat{\theta} \cdot \hat{\sigma}_0$$

Begin by using PROC MEANS to compute the first and second moments and by using the following DATA step to compute the constant A:

```
proc means data = Assembly noprint;
   var Offset;
   output out=stat mean=m1 var=var n=n min = min;
run;

* Compute constant A from equation (19) of \citet{elan_r:61};
data stat;
   keep m2 a min;
   set stat;
   a  = (m1*m1);
   m2 = ((n-1)/n)*var + a;
   a  = a/m2;
run;
```

Next, use the SAS/IML subroutine NLPDD to solve equation (19) by minimizing $(f(\theta) - A)^2$, and compute $\hat{mu}_0$ and $\hat{\sigma}_0$:

```
proc iml;
   use stat;
   read all var {m2}  into m2;
   read all var {a}   into a;
   read all var {min} into min;

   * f(t) is the function in equation (19) of \citet{elan_r:61};
   start f(t) global(a);
      y = .39894*exp(-0.5*t*t);
      y = (2*y-(t*(1-2*probnorm(t))))**2/(1+t*t);
      y = (y-a)**2;
      return(y);
   finish;

   * Minimize (f(t)-A)**2 and estimate mu and sigma;
   if ( min < 0 ) then do;
      print "Warning: Observations are not all nonnegative.";
      print "    The folded normal is inappropriate.";
      stop;
   end;
   if ( a < 0.637 ) then do;
      print "Warning: the folded normal may be inappropriate";
      end;
   opt = { 0 0 };
   con = { 1e-6 };
   x0  = { 2.0 };
   tc  = { . . . . . 1e-8 . . . . . . . };
   call nlpdd(rc,etheta0,"f",x0,opt,con,tc);
   esig0 = sqrt(m2/(1+etheta0*etheta0));
   emu0  = etheta0*esig0;

   create prelim var {emu0 esig0 etheta0};
   append;
   close prelim;

   * Define the log likelihood of the folded normal;
   start g(p) global(x);
      y = 0.0;
      do i = 1 to nrow(x);
         z = exp( (-0.5/p[2])*(x[i]-p[1])*(x[i]-p[1]) );
         z = z + exp( (-0.5/p[2])*(x[i]+p[1])*(x[i]+p[1]) );
         y = y + log(z);
      end;
      y = y - nrow(x)*log( sqrt( p[2] ) );
      return(y);
   finish;
   * Maximize the log likelihood with subroutine NLPDD;
   use assembly;
   read all var {offset} into x;
   esig0sq = esig0*esig0;
```

```
   x0      = emu0||esig0sq;
   opt     = { 1 0 };
   con     = { . 0.0, . . };
   call nlpdd(rc,xr,"g",x0,opt,con);
   emu     = xr[1];
   esig    = sqrt(xr[2]);
   etheta  = emu/esig;
   create parmest var{emu esig etheta};
   append;
   close parmest;
quit;
```

The preliminary estimates are saved in the data set Prelim, as shown in Output 4.25.1.

Output 4.25.1 Preliminary Estimates of μ, σ, and θ

	Three-Parameter Lognormal Fit		
Obs	EMU0	ESIG0	ETHETA0
1	6.51735	6.54953	0.99509

Now, using $\hat{mu}_0$ and $\hat{\sigma}_0$ as initial estimates, call the NLPDD subroutine to maximize the log likelihood, $l(\mu,\sigma)$, of the folded normal distribution, where, up to a constant,

$$l(\mu,\sigma) = -n\log\sigma + \sum_{i=1}^{n}\log\left[\exp\left(-\frac{(x_i-\mu)^2}{2\sigma^2}\right) + \exp\left(-\frac{(x_i+\mu)^2}{2\sigma^2}\right)\right]$$

```
* Define the log likelihood of the folded normal;
start g(p) global(x);
   y = 0.0;
   do i = 1 to nrow(x);
      z = exp( (-0.5/p[2])*(x[i]-p[1])*(x[i]-p[1]) );
      z = z + exp( (-0.5/p[2])*(x[i]+p[1])*(x[i]+p[1]) );
      y = y + log(z);
   end;
   y = y - nrow(x)*log( sqrt( p[2] ) );
   return(y);
finish;
* Maximize the log likelihood with subroutine NLPDD;
use assembly;
read all var {offset} into x;
esig0sq = esig0*esig0;
x0      = emu0||esig0sq;
opt     = { 1 0 };
con     = { . 0.0, . . };
call nlpdd(rc,xr,"g",x0,opt,con);
emu     = xr[1];
esig    = sqrt(xr[2]);
etheta  = emu/esig;
create parmest var{emu esig etheta};
append;
close parmest;
quit;
```

The data set ParmEst contains the maximum likelihood estimates $\hat{mu}$ and $\hat{\sigma}$ (as well as $\hat{mu}/\hat{\sigma}$), as shown in Output 4.25.2.

Output 4.25.2 Final Estimates of μ, σ, and θ

	Three-Parameter Lognormal Fit		
Obs	EMU	ESIG	ETHETA
1	6.66761	6.39650	1.04239

To annotate the curve on a histogram, begin by computing the width and endpoints of the histogram intervals. The following statements save these values in a data set called OutCalc. Note that a plot is not produced at this point.

```
proc univariate data = Assembly noprint;
   histogram Offset / outhistogram = out normal(noprint) noplot;
run;

data OutCalc (drop = _MIDPT_);
   set out (keep = _MIDPT_) end = eof;
   retain _MIDPT1_ _WIDTH_;
   if _N_ = 1 then _MIDPT1_ = _MIDPT_;
   if eof then do;
      _MIDPTN_ = _MIDPT_;
      _WIDTH_ = (_MIDPTN_ - _MIDPT1_) / (_N_ - 1);
      output;
   end;
run;
```

Output 4.25.3 provides a listing of the data set OutCalc. The width of the histogram bars is saved as the value of the variable _WIDTH_; the midpoints of the first and last histogram bars are saved as the values of the variables _MIDPT1_ and _MIDPTN_.

Output 4.25.3 The Data Set OutCalc

	Three-Parameter Lognormal Fit		
Obs	_MIDPT1_	_WIDTH_	_MIDPTN_
1	1.5	3	22.5

The following statements create an annotate data set named Anno, which contains the coordinates of the fitted curve:

```
data Anno;
   merge ParmEst OutCalc;
   length function color $ 8;
   function = 'point';
   color    = 'black';
   size     =  2;
   xsys     = '2';
   ysys     = '2';
   when     = 'a';
   constant = 39.894*_width_;;
   left     =  _midpt1_ - .5*_width_;
   right    =  _midptn_ + .5*_width_;
   inc      = (right-left)/100;
   do x = left to right by inc;
      z1 = (x-emu)/esig;
      z2 = (x+emu)/esig;
      y  = (constant/esig)*(exp(-0.5*z1*z1)+exp(-0.5*z2*z2));
      output;
      function = 'draw';
   end;
run;
```

The following statements read the ANNOTATE= data set and display the histogram and fitted curve:

```
title 'Folded Normal Distribution';
proc univariate data=assembly noprint;
   histogram Offset / annotate = anno;
run;
```

Output 4.25.4 displays the histogram and fitted curve.

Output 4.25.4 Histogram with Annotated Folded Normal Curve

Folded Normal Distribution

[Histogram with x-axis "Offset (in mm)" ranging from 1.5 to 22.5, y-axis "Percent" ranging from 0 to 30, with annotated folded normal curve overlay]

A sample program for this example, *uniex15.sas*, is available in the SAS Sample Library for Base SAS software.

Example 4.26: Creating Lognormal Probability Plots

This example is a continuation of the example explored in the section "Modeling a Data Distribution" on page 229.

In the normal probability plot shown in Output 4.6, the nonlinearity of the point pattern indicates a departure from normality in the distribution of Deviation. Because the point pattern is curved with slope increasing from left to right, a theoretical distribution that is skewed to the right, such as a lognormal distribution, should provide a better fit than the normal distribution. See the section "Interpretation of Quantile-Quantile and Probability Plots" on page 363.

You can explore the possibility of a lognormal fit with a lognormal probability plot. When you request such a plot, you must specify the shape parameter σ for the lognormal distribution. This value must be positive, and typical values of σ range from 0.1 to 1.0. You can specify values for σ with the SIGMA= secondary option in the LOGNORMAL primary option, or you can specify that σ is to be estimated from the data.

The following statements illustrate the first approach by creating a series of three lognormal probability plots for the variable Deviation introduced in the section "Modeling a Data Distribution" on page 229:

```
symbol v=plus height=3.5pct;
title 'Lognormal Probability Plot for Position Deviations';
proc univariate data=Aircraft noprint;
   probplot Deviation /
      lognormal(theta=est zeta=est sigma=0.7 0.9 1.1)
      href  = 95
      lhref = 1
      square;
run;
```

The LOGNORMAL primary option requests plots based on the lognormal family of distributions, and the SIGMA= secondary option requests plots for σ equal to 0.7, 0.9, and 1.1. These plots are displayed in Output 4.26.1, Output 4.26.2, and Output 4.26.3, respectively. Alternatively, you can specify σ to be estimated using the sample standard deviation by using the option SIGMA=EST.

The SQUARE option displays the probability plot in a square format, the HREF= option requests a reference line at the 95th percentile, and the LHREF= option specifies the line type for the reference line.

Output 4.26.1 Probability Plot Based on Lognormal Distribution with $\sigma = 0.7$

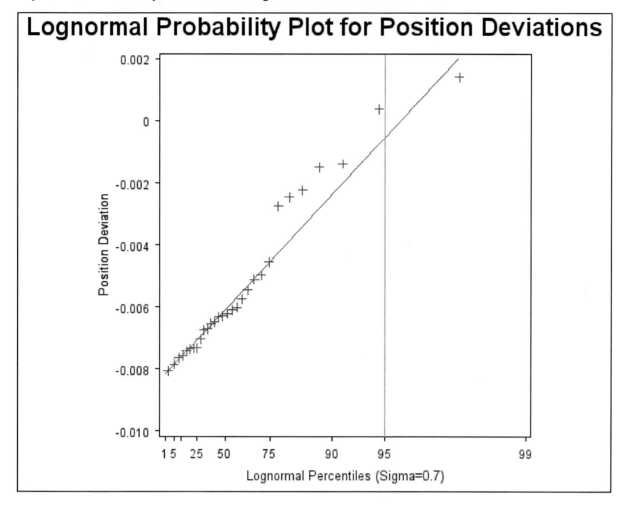

Output 4.26.2 Probability Plot Based on Lognormal Distribution with $\sigma = 0.9$

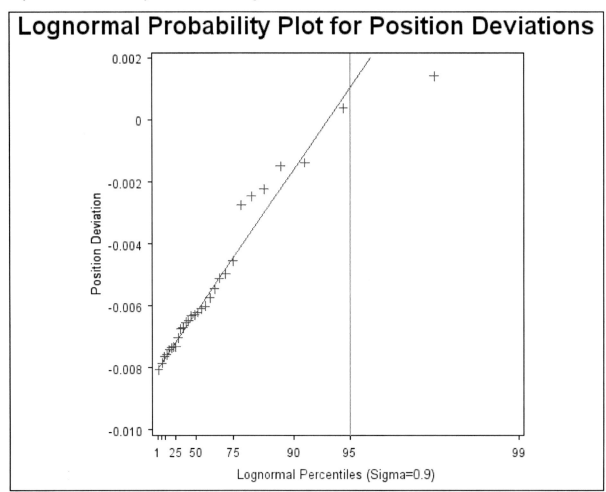

Output 4.26.3 Probability Plot Based on Lognormal Distribution with $\sigma = 1.1$

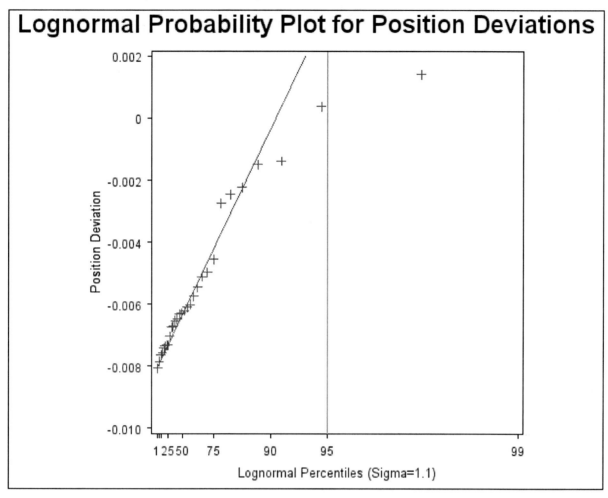

The value $\sigma = 0.9$ in Output 4.26.2 most nearly linearizes the point pattern. The 95th percentile of the position deviation distribution seen in Output 4.26.2 is approximately 0.001, because this is the value corresponding to the intersection of the point pattern with the reference line.

NOTE: After the σ that produces the most linear fit is found, you can then estimate the threshold parameter θ and the scale parameter ζ. See Example 4.31.

The following statements illustrate how you can create a lognormal probability plot for Deviation by using a local maximum likelihood estimate for σ.

```
symbol v=plus height=3.5pct;
title 'Lognormal Probability Plot for Position Deviations';
proc univariate data=Aircraft noprint;
   probplot Deviation / lognormal(theta=est zeta=est sigma=est)
                       href    = 95
                       square;
run;
```

The plot is displayed in Output 4.26.4. Note that the maximum likelihood estimate of σ (in this case, 0.882) does not necessarily produce the most linear point pattern.

Output 4.26.4 Probability Plot Based on Lognormal Distribution with Estimated σ

A sample program for this example, *uniex16.sas*, is available in the SAS Sample Library for Base SAS software.

Example 4.27: Creating a Histogram to Display Lognormal Fit

This example uses the data set Aircraft from Example 4.26 to illustrate how to display a lognormal fit with a histogram. To determine whether the lognormal distribution is an appropriate model for a distribution, you should consider the graphical fit as well as conduct goodness-of-fit tests. The following statements fit a lognormal distribution and display the density curve on a histogram:

```
title 'Distribution of Position Deviations';
ods select Lognormal.ParameterEstimates Lognormal.GoodnessOfFit MyPlot;
proc univariate data=Aircraft;
   var Deviation;
   histogram / lognormal(w=3 theta=est)
               vaxis = axis1
               name  = 'MyPlot';
   inset n mean (5.3) std='Std Dev' (5.3) skewness (5.3) /
         pos = ne  header = 'Summary Statistics';
   axis1 label=(a=90 r=0);
run;
```

The ODS SELECT statement restricts the output to the "ParameterEstimates" and "GoodnessOfFit" tables; see the section "ODS Table Names" on page 376. The LOGNORMAL primary option superimposes a fitted curve on the histogram in Output 4.27.1. The W= option specifies the line width for the curve. The INSET statement specifies that the mean, standard deviation, and skewness be displayed in an inset in the northeast corner of the plot. Note that the default value of the threshold parameter θ is zero. In applications where the threshold is not zero, you can specify θ with the THETA= option. The variable Deviation includes values that are less than the default threshold; therefore, the option THETA= EST is used.

Output 4.27.1 Normal Probability Plot Created with Graphics Device

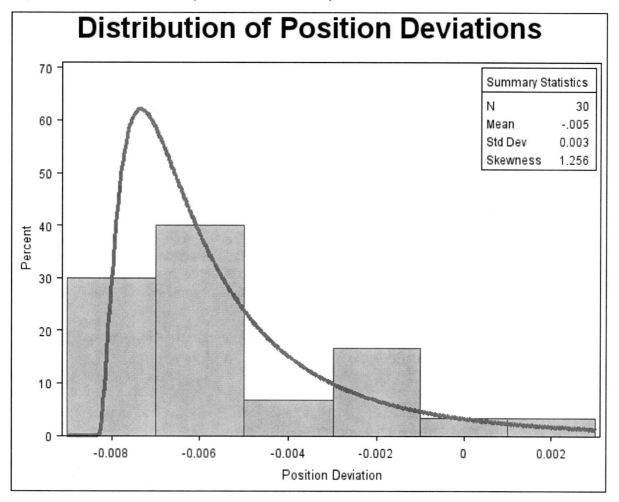

Output 4.27.2 provides three EDF goodness-of-fit tests for the lognormal distribution: the Anderson-Darling, the Cramér-von Mises, and the Kolmogorov-Smirnov tests. The null hypothesis for the three tests is that a lognormal distribution holds for the sample data.

Output 4.27.2 Summary of Fitted Lognormal Distribution

```
                    Distribution of Position Deviations

                           The UNIVARIATE Procedure
                  Fitted Lognormal Distribution for Deviation

                     Parameters for Lognormal Distribution

                      Parameter    Symbol    Estimate

                      Threshold    Theta     -0.00834
                      Scale        Zeta      -6.14382
                      Shape        Sigma      0.882225
                      Mean                   -0.00517
                      Std Dev                 0.003438


                  Goodness-of-Fit Tests for Lognormal Distribution

           Test                 ----Statistic-----     ------p Value------

           Kolmogorov-Smirnov    D      0.09419634     Pr > D       >0.500
           Cramer-von Mises      W-Sq   0.02919815     Pr > W-Sq    >0.500
           Anderson-Darling      A-Sq   0.21606642     Pr > A-Sq    >0.500
```

The *p*-values for all three tests are greater than 0.5, so the null hypothesis is not rejected. The tests support the conclusion that the two-parameter lognormal distribution with scale parameter $\hat{\zeta} = -6.14$ and shape parameter $\hat{\sigma} = 0.88$ provides a good model for the distribution of position deviations. For further discussion of goodness-of-fit interpretation, see the section "Goodness-of-Fit Tests" on page 356.

A sample program for this example, *uniex16.sas*, is available in the SAS Sample Library for Base SAS software.

Example 4.28: Creating a Normal Quantile Plot

This example illustrates how to create a normal quantile plot. An engineer is analyzing the distribution of distances between holes cut in steel sheets. The following statements save measurements of the distance between two holes cut into 50 steel sheets as values of the variable Distance in the data set Sheets:

```
data Sheets;
   input Distance @@;
   label Distance = 'Hole Distance (cm)';
   datalines;
 9.80 10.20 10.27  9.70  9.76
10.11 10.24 10.20 10.24  9.63
 9.99  9.78 10.10 10.21 10.00
 9.96  9.79 10.08  9.79 10.06
```

```
10.10  9.95  9.84 10.11  9.93
10.56 10.47  9.42 10.44 10.16
10.11 10.36  9.94  9.77  9.36
 9.89  9.62 10.05  9.72  9.82
 9.99 10.16 10.58 10.70  9.54
10.31 10.07 10.33  9.98 10.15
;
run;
```

The engineer decides to check whether the distribution of distances is normal. The following statements create a Q-Q plot for Distance, shown in Output 4.28.1:

```
symbol v=plus;
title 'Normal Quantile-Quantile Plot for Hole Distance';
proc univariate data=Sheets noprint;
   qqplot Distance;
run;
```

The plot compares the ordered values of Distance with quantiles of the normal distribution. The linearity of the point pattern indicates that the measurements are normally distributed. Note that a normal Q-Q plot is created by default.

Output 4.28.1 Normal Quantile-Quantile Plot for Distance

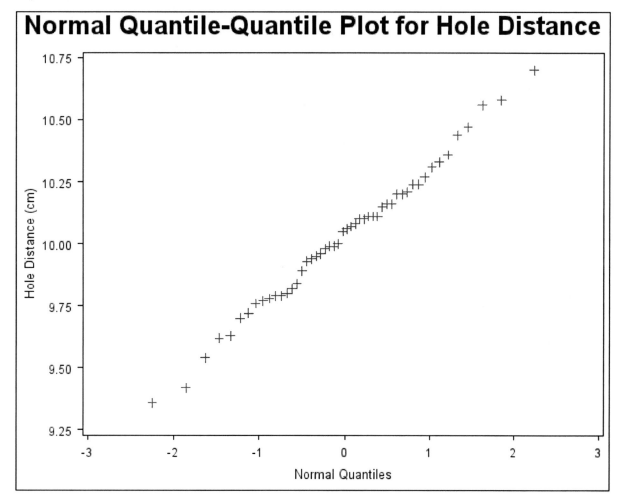

A sample program for this example, *uniex17.sas*, is available in the SAS Sample Library for Base SAS software.

Example 4.29: Adding a Distribution Reference Line

This example, which is a continuation of Example 4.28, illustrates how to add a reference line to a normal Q-Q plot, which represents the normal distribution with mean μ_0 and standard deviation σ_0. The following statements reproduce the Q-Q plot in Output 4.28.1 and add the reference line:

```
symbol v=plus;
title 'Normal Quantile-Quantile Plot for Hole Distance';
proc univariate data=Sheets noprint;
   qqplot Distance / normal(mu=est sigma=est color=red l=2)
                    square;
run;
```

The plot is displayed in Output 4.29.1.

Specifying MU=EST and SIGMA=EST with the NORMAL primary option requests the reference line for which μ_0 and σ_0 are estimated by the sample mean and standard deviation. Alternatively, you can specify numeric values for μ_0 and σ_0 with the MU= and SIGMA= secondary options. The COLOR= and L= options specify the color and type of the line, and the SQUARE option displays the plot in a square format. The NOPRINT options in the PROC UNIVARIATE statement and after the NORMAL option suppress all the tables of statistical output produced by default.

Output 4.29.1 Adding a Distribution Reference Line to a Q-Q Plot

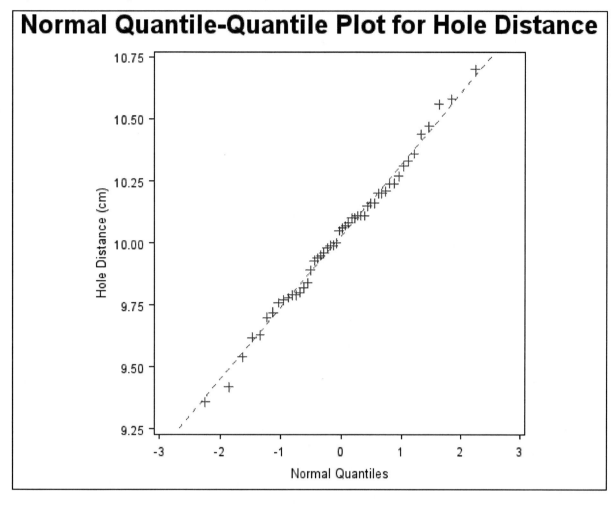

The data clearly follow the line, which indicates that the distribution of the distances is normal.

A sample program for this example, *uniex17.sas*, is available in the SAS Sample Library for Base SAS software.

Example 4.30: Interpreting a Normal Quantile Plot

This example illustrates how to interpret a normal quantile plot when the data are from a non-normal distribution. The following statements create the data set Measures, which contains the measurements of the diameters of 50 steel rods in the variable Diameter:

```
data Measures;
   input Diameter @@;
   label Diameter = 'Diameter (mm)';
   datalines;
5.501  5.251  5.404  5.366  5.445  5.576  5.607
5.200  5.977  5.177  5.332  5.399  5.661  5.512
5.252  5.404  5.739  5.525  5.160  5.410  5.823
5.376  5.202  5.470  5.410  5.394  5.146  5.244
5.309  5.480  5.388  5.399  5.360  5.368  5.394
5.248  5.409  5.304  6.239  5.781  5.247  5.907
5.208  5.143  5.304  5.603  5.164  5.209  5.475
5.223
;
run;
```

The following statements request the normal Q-Q plot in Output 4.30.1:

```
symbol v=plus;
title 'Normal Q-Q Plot for Diameters';
proc univariate data=Measures noprint;
   qqplot Diameter / normal
                     square
                     vaxis=axis1;
   axis1 label=(a=90 r=0);
run;
```

The nonlinearity of the points in Output 4.30.1 indicates a departure from normality. Because the point pattern is curved with slope increasing from left to right, a theoretical distribution that is skewed to the right, such as a lognormal distribution, should provide a better fit than the normal distribution. The mild curvature suggests that you should examine the data with a series of lognormal Q-Q plots for small values of the shape parameter σ, as illustrated in Example 4.31. For details on interpreting a Q-Q plot, see the section "Interpretation of Quantile-Quantile and Probability Plots" on page 363.

Output 4.30.1 Normal Quantile-Quantile Plot of Nonnormal Data

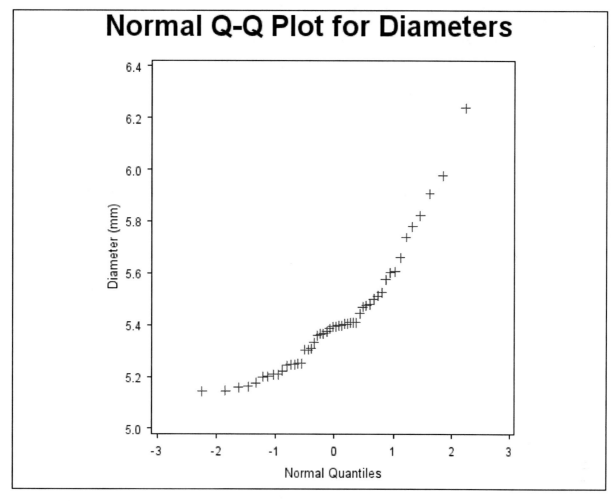

A sample program for this example, *uniex18.sas*, is available in the SAS Sample Library for Base SAS software.

Example 4.31: Estimating Three Parameters from Lognormal Quantile Plots

This example, which is a continuation of Example 4.30, demonstrates techniques for estimating the shape, location, and scale parameters, and the theoretical percentiles for a three-parameter lognormal distribution.

The three-parameter lognormal distribution depends on a threshold parameter θ, a scale parameter ζ, and a shape parameter σ. You can estimate σ from a series of lognormal Q-Q plots which use the SIGMA= secondary option to specify different values of σ; the estimate of σ is the value that linearizes the point pattern. You can then estimate the threshold and scale parameters from the intercept and slope of the point pattern. The following statements create the series of plots in Output 4.31.1, Output 4.31.2, and Output 4.31.3 for σ values of 0.2, 0.5, and 0.8, respectively:

```
symbol v=plus;
title 'Lognormal Q-Q Plot for Diameters';
proc univariate data=Measures noprint;
   qqplot Diameter / lognormal(sigma=0.2 0.5 0.8)
                     square;
run;
```

NOTE: You must specify a value for the shape parameter σ for a lognormal Q-Q plot with the SIGMA= option or its alias, the SHAPE= option.

Output 4.31.1 Lognormal Quantile-Quantile Plot (σ =0.2)

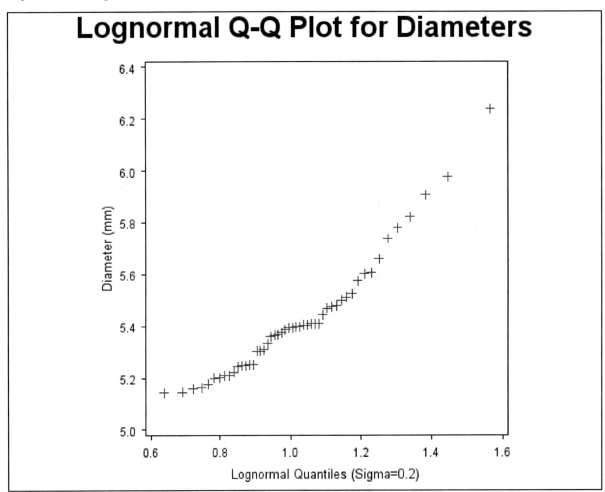

Output 4.31.2 Lognormal Quantile-Quantile Plot ($\sigma = 0.5$)

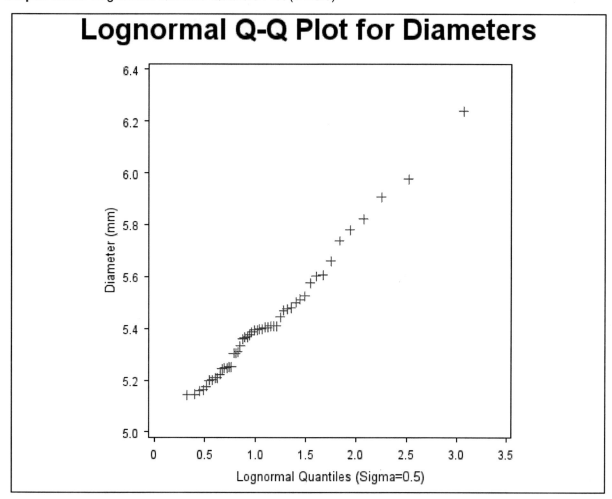

Output 4.31.3 Lognormal Quantile-Quantile Plot ($\sigma = 0.8$)

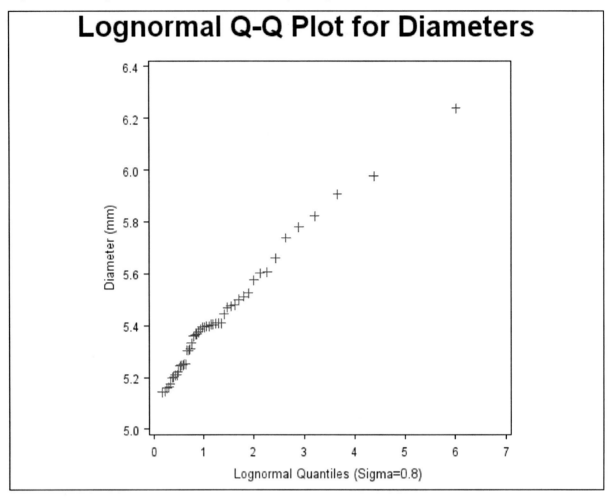

The plot in Output 4.31.2 displays the most linear point pattern, indicating that the lognormal distribution with $\sigma = 0.5$ provides a reasonable fit for the data distribution.

Data with this particular lognormal distribution have the following density function:

$$p(x) = \begin{cases} \frac{\sqrt{2}}{\sqrt{\pi}(x-\theta)} \exp\left(-2(\log(x-\theta) - \zeta)^2\right) & \text{for } x > \theta \\ 0 & \text{for } x \leq \theta \end{cases}$$

The points in the plot fall on or near the line with intercept θ and slope $\exp(\zeta)$. Based on Output 4.31.2, $\theta \approx 5$ and $\exp(\zeta) \approx \frac{1.2}{3} = 0.4$, giving $\zeta \approx \log(0.4) \approx -0.92$.

You can also request a reference line by using the SIGMA=, THETA=, and ZETA= options together. The following statements produce the lognormal Q-Q plot in Output 4.31.4:

```
symbol v=plus;
title 'Lognormal Q-Q Plot for Diameters';
proc univariate data=Measures noprint;
   qqplot Diameter / lognormal(theta=5 zeta=est sigma=est
                               color=black l=2)
                     square;
run;
```

Output 4.31.1 through Output 4.31.3 show that the threshold parameter θ is not equal to zero. Specifying THETA=5 overrides the default value of zero. The SIGMA=EST and ZETA=EST secondary options request estimates for σ and $\exp \zeta$ that use the sample mean and standard deviation.

Output 4.31.4 Lognormal Quantile-Quantile Plot (σ =est, ζ =est, θ =5)

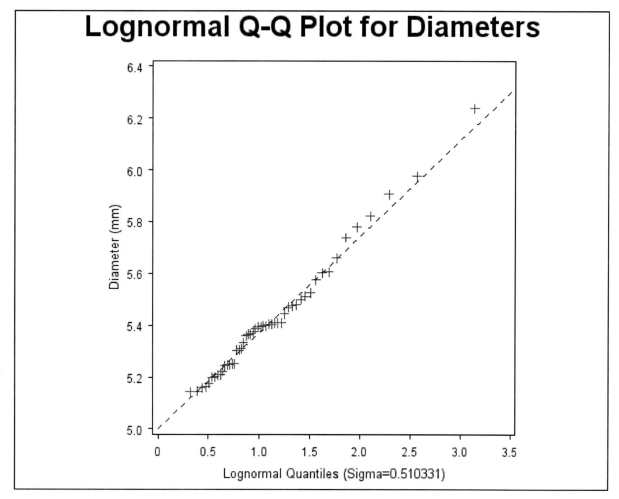

From the plot in Output 4.31.2, σ can be estimated as 0.51, which is consistent with the estimate of 0.5 derived from the plot in Output 4.31.2. Example 4.32 illustrates how to estimate percentiles by using lognormal Q-Q plots.

A sample program for this example, *uniex18.sas*, is available in the SAS Sample Library for Base SAS software.

Example 4.32: Estimating Percentiles from Lognormal Quantile Plots

This example, which is a continuation of Example 4.31, shows how to use a Q-Q plot to estimate percentiles such as the 95th percentile of the lognormal distribution. A probability plot can also be used for this purpose, as illustrated in Example 4.26.

The point pattern in Output 4.31.4 has a slope of approximately 0.39 and an intercept of 5. The following statements reproduce this plot, adding a lognormal reference line with this slope and intercept:

```
symbol v=plus;
title 'Lognormal Q-Q Plot for Diameters';
proc univariate data=Measures noprint;
   qqplot Diameter / lognormal(sigma=0.5 theta=5 slope=0.39)
                    pctlaxis(grid)
                    vref  = 5.8 5.9 6.0
                    square;
run;
```

The result is shown in Output 4.32.1:

Output 4.32.1 Lognormal Q-Q Plot Identifying Percentiles

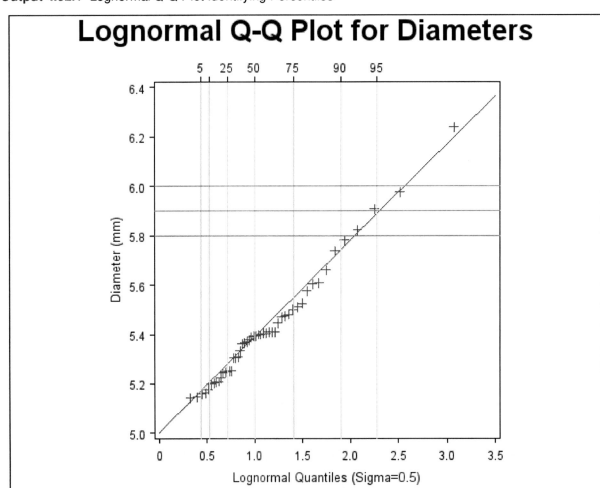

The PCTLAXIS option labels the major percentiles, and the GRID option draws percentile axis reference lines. The 95th percentile is 5.9, because the intersection of the distribution reference line and the 95th reference line occurs at this value on the vertical axis.

Alternatively, you can compute this percentile from the estimated lognormal parameters. The αth percentile of the lognormal distribution is

$$P_\alpha = \exp(\sigma \Phi^{-1}(\alpha) + \zeta) + \theta$$

where $\Phi^{-1}(\cdot)$ is the inverse cumulative standard normal distribution. Consequently,

$$\hat{P}_{0.95} = \exp\left(\tfrac{1}{2}\Phi^{-1}(0.95) + \log(0.39)\right) + 5 = 5.89$$

A sample program for this example, *uniex18.sas*, is available in the SAS Sample Library for Base SAS software.

Example 4.33: Estimating Parameters from Lognormal Quantile Plots

This example, which is a continuation of Example 4.31, demonstrates techniques for estimating the shape, location, and scale parameters, and the theoretical percentiles for a two-parameter lognormal distribution.

If the threshold parameter is known, you can construct a two-parameter lognormal Q-Q plot by subtracting the threshold from the data values and making a normal Q-Q plot of the log-transformed differences, as illustrated in the following statements:

```
data ModifiedMeasures;
   set Measures;
   LogDiameter = log(Diameter-5);
   label LogDiameter = 'log(Diameter-5)';
run;
symbol v=plus;
title 'Two-Parameter Lognormal Q-Q Plot for Diameters';
proc univariate data=ModifiedMeasures noprint;
   qqplot LogDiameter / normal(mu=est sigma=est)
                        square
                        vaxis=axis1;
   inset n mean (5.3) std (5.3)
            / pos = nw header = 'Summary Statistics';
   axis1 label=(a=90 r=0);
run;
```

Output 4.33.1 Two-Parameter Lognormal Q-Q Plot for Diameters

Two-Parameter Lognormal Q-Q Plot for Diameters

Summary Statistics	
N	50
Mean	-.990
Std Deviation	0.510

Because the point pattern in Output 4.33.1 is linear, you can estimate the lognormal parameters ζ and σ as the normal plot estimates of μ and σ, which are -0.99 and 0.51. These values correspond to the previous estimates of -0.92 for ζ and 0.5 for σ from Example 4.31. A sample program for this example, *uniex18.sas*, is available in the SAS Sample Library for Base SAS software.

Example 4.34: Comparing Weibull Quantile Plots

This example compares the use of three-parameter and two-parameter Weibull Q-Q plots for the failure times in months for 48 integrated circuits. The times are assumed to follow a Weibull distribution. The following statements save the failure times as the values of the variable Time in the data set Failures:

```
data Failures;
   input Time @@;
   label Time = 'Time in Months';
   datalines;
29.42 32.14 30.58 27.50 26.08 29.06 25.10 31.34
29.14 33.96 30.64 27.32 29.86 26.28 29.68 33.76
29.32 30.82 27.26 27.92 30.92 24.64 32.90 35.46
30.28 28.36 25.86 31.36 25.26 36.32 28.58 28.88
26.72 27.42 29.02 27.54 31.60 33.46 26.78 27.82
29.18 27.94 27.66 26.42 31.00 26.64 31.44 32.52
;
run;
```

If no assumption is made about the parameters of this distribution, you can use the WEIBULL option to request a three-parameter Weibull plot. As in the previous example, you can visually estimate the shape parameter c by requesting plots for different values of c and choosing the value of c that linearizes the point pattern. Alternatively, you can request a maximum likelihood estimate for c, as illustrated in the following statements:

```
symbol v=plus;
title 'Three-Parameter Weibull Q-Q Plot for Failure Times';
proc univariate data=Failures noprint;
   qqplot Time / weibull(c=est theta=est sigma=est)
                 square
                 href=0.5 1 1.5 2
                 vref=25 27.5 30 32.5 35
                 lhref=4 lvref=4;
run;
```

NOTE: When using the WEIBULL option, you must either specify a list of values for the Weibull shape parameter c with the C= option or specify C=EST.

Output 4.34.1 displays the plot for the estimated value $\hat{c} = 1.99$. The reference line corresponds to the estimated values for the threshold and scale parameters of $\hat{\theta} = 24.19$ and $\hat{\sigma}_0 = 5.83$, respectively.

Output 4.34.1 Three-Parameter Weibull Q-Q Plot

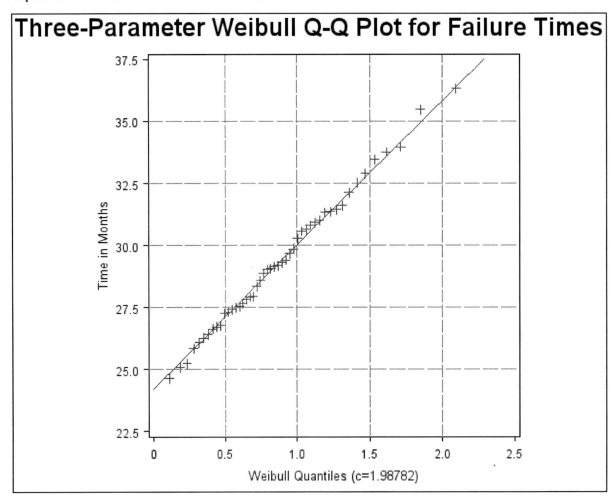

Now, suppose it is known that the circuit lifetime is at least 24 months. The following statements use the known threshold value $\theta_0 = 24$ to produce the two-parameter Weibull Q-Q plot shown in Output 4.31.4:

```
symbol v=plus;
title 'Two-Parameter Weibull Q-Q Plot for Failure Times';
proc univariate data=Failures noprint;
   qqplot Time / weibull(theta=24 c=est sigma=est)
                 square
                 vref= 25 to 35 by 2.5
                 href= 0.5 to 2.0 by 0.5
                 lhref=4 lvref=4;
run;
```

The reference line is based on maximum likelihood estimates $\hat{c} = 2.08$ and $\hat{\sigma} = 6.05$.

Output 4.34.2 Two-Parameter Weibull Q-Q Plot for $\theta_0 = 24$

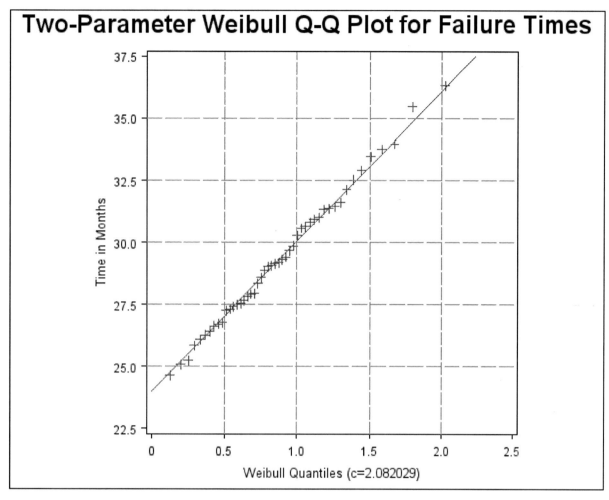

A sample program for this example, *uniex19.sas*, is available in the SAS Sample Library for Base SAS software.

Example 4.35: Creating a Cumulative Distribution Plot

A company that produces fiber-optic cord is interested in the breaking strength of the cord. The following statements create a data set named Cord, which contains 50 breaking strengths measured in pounds per square inch (psi):

```
data Cord;
   label Strength="Breaking Strength (psi)";
   input Strength @@;
datalines;
6.94 6.97 7.11 6.95 7.12 6.70 7.13 7.34 6.90 6.83
7.06 6.89 7.28 6.93 7.05 7.00 7.04 7.21 7.08 7.01
7.05 7.11 7.03 6.98 7.04 7.08 6.87 6.81 7.11 6.74
6.95 7.05 6.98 6.94 7.06 7.12 7.19 7.12 7.01 6.84
6.91 6.89 7.23 6.98 6.93 6.83 6.99 7.00 6.97 7.01
;
run;
```

You can use the CDFPLOT statement to fit any of six theoretical distributions (beta, exponential, gamma, lognormal, normal, and Weibull) and superimpose them on the cdf plot. The following statements use the NORMAL option to display a fitted normal distribution function on a cdf plot of breaking strengths:

```
title 'Cumulative Distribution Function of Breaking Strength';
ods graphics on;
proc univariate data=Cord noprint;
   cdf Strength / normal;
   inset normal(mu sigma);
run;
```

The NORMAL option requests the fitted curve. The INSET statement requests an inset containing the parameters of the fitted curve, which are the sample mean and standard deviation. For more information about the INSET statement, see "INSET Statement" on page 271. The resulting plot is shown in Output 4.35.1.

Output 4.35.1 Cumulative Distribution Function

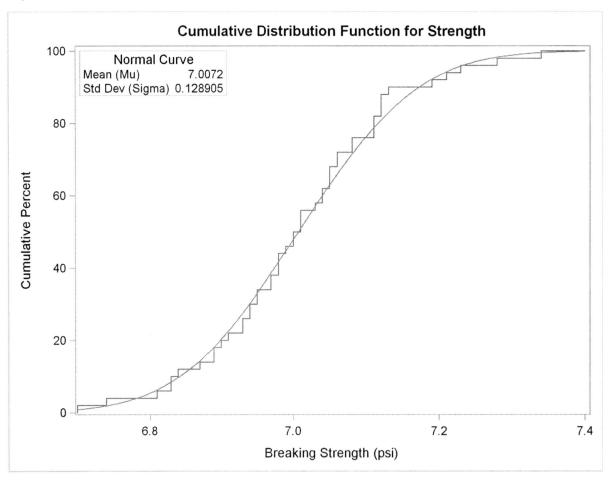

The plot shows a symmetric distribution with observations concentrated 6.9 and 7.1. The agreement between the empirical and the normal distribution functions in Output 4.35.1 is evidence that the normal distribution is an appropriate model for the distribution of breaking strengths.

Example 4.36: Creating a P-P Plot

The distances between two holes cut into 50 steel sheets are measured and saved as values of the variable Distance in the following data set:

```
data Sheets;
   input Distance @@;
   label Distance='Hole Distance in cm';
   datalines;
 9.80 10.20 10.27  9.70  9.76
10.11 10.24 10.20 10.24  9.63
 9.99  9.78 10.10 10.21 10.00
 9.96  9.79 10.08  9.79 10.06
10.10  9.95  9.84 10.11  9.93
10.56 10.47  9.42 10.44 10.16
10.11 10.36  9.94  9.77  9.36
 9.89  9.62 10.05  9.72  9.82
 9.99 10.16 10.58 10.70  9.54
10.31 10.07 10.33  9.98 10.15
;
run;
```

It is decided to check whether the distances are normally distributed. The following statements create a P-P plot, shown in Output 4.36.1, which is based on the normal distribution with mean $\mu = 10$ and standard deviation $\sigma = 0.3$:

```
title 'Normal Probability-Probability Plot for Hole Distance';
ods graphics on;
proc univariate data=Sheets noprint;
   ppplot Distance / normal(mu=10 sigma=0.3)
                    square;
run;
```

The NORMAL option in the PPPLOT statement requests a P-P plot based on the normal cumulative distribution function, and the MU= and SIGMA= *normal-options* specify μ and σ. Note that a P-P plot is always based on a *completely specified* distribution—in other words, a distribution with specific parameters. In this example, if you did not specify the MU= and SIGMA= *normal-options*, the sample mean and sample standard deviation would be used for μ and σ.

Output 4.36.1 Normal P-P Plot with Diagonal Reference Line

P-P Plot for Distance

[Plot: Cumulative Distribution of Distance vs Normal(Mu=10 Sigma=0.3)]

The linearity of the pattern in Output 4.36.1 is evidence that the measurements are normally distributed with mean 10 and standard deviation 0.3. The SQUARE option displays the plot in a square format.

References

Blom, G. (1958), *Statistical Estimates and Transformed Beta Variables*, New York: John Wiley & Sons.

Bowman, K. O. and Shenton, L. R. (1983), "Johnson's System of Distributions," in S. Kotz, N. L. Johnson, and C. B. Read, eds., *Encyclopedia of Statistical Sciences*, volume 4, 303–314, New York: John Wiley & Sons.

Chambers, J. M., Cleveland, W. S., Kleiner, B., and Tukey, P. A. (1983), *Graphical Methods for Data Analysis*, Belmont, CA: Wadsworth International Group.

Cohen, A. C. (1951), "Estimating Parameters of Logarithmic-Normal Distributions by Maximum Likelihood," *Journal of the American Statistical Association*, 46, 206–212.

Conover, W. J. (1999), *Practical Nonparametric Statistics*, Third Edition, New York: John Wiley & Sons.

Croux, C. and Rousseeuw, P. J. (1992), "Time-Efficient Algorithms for Two Highly Robust Estimators of Scale," *Computational Statistics*, 1, 411–428.

D'Agostino, R. and Stephens, M. (1986), *Goodness-of-Fit Techniques*, New York: Marcel Dekker.

Dixon, W. J. and Tukey, J. W. (1968), "Approximate Behavior of the Distribution of Winsorized t (Trimming/Winsorization 2)," *Technometrics*, 10, 83–98.

Elandt, R. C. (1961), "The Folded Normal Distribution: Two Methods of Estimating Parameters from Moments," *Technometrics*, 3, 551–562.

Fisher, R. A. (1973), *Statistical Methods for Research Workers*, Fourteenth Edition, New York: Hafner Publishing.

Fowlkes, E. B. (1987), *A Folio of Distributions: A Collection of Theoretical Quantile-Quantile Plots*, New York: Marcel Dekker.

Hahn, G. J. and Meeker, W. Q. (1991), *Statistical Intervals: A Guide for Practitioners*, New York: John Wiley & Sons.

Hampel, F. R. (1974), "The Influence Curve and Its Role in Robust Estimation," *Journal of the American Statistical Association*, 69, 383–393.

Iman, R. L. (1974), "Use of a t-Statistic as an Approximation to the Exact Distribution of the Wilcoxon Signed Rank Statistic," *Communications in Statistics*, 3, 795–806.

Johnson, N. L., Kotz, S., and Balakrishnan, N. (1994), *Continuous Univariate Distributions-1*, Second Edition, New York: John Wiley & Sons.

Johnson, N. L., Kotz, S., and Balakrishnan, N. (1995), *Continuous Univariate Distributions-2*, Second Edition, New York: John Wiley & Sons.

Jones, M. C., Marron, J. S., and Sheather, S. J. (1996), "A Brief Survey of Bandwidth Selection for Density Estimation," *Journal of the American Statistical Association*, 91, 401–407.

Lehmann, E. L. (1998), *Nonparametrics: Statistical Methods Based on Ranks*, San Francisco: Holden-Day.

Odeh, R. E. and Owen, D. B. (1980), *Tables for Normal Tolerance Limits, Sampling Plans, and Screening*, New York: Marcel Dekker.

Owen, D. B. and Hua, T. A. (1977), "Tables of Confidence Limits on the Tail Area of the Normal Distribution," *Communication and Statistics, Part B—Simulation and Computation*, 6, 285–311.

Rousseeuw, P. J. and Croux, C. (1993), "Alternatives to the Median Absolute Deviation," *Journal of the American Statistical Association*, 88, 1273–1283.

Royston, J. P. (1992), "Approximating the Shapiro-Wilk's W Test for Nonnormality," *Statistics and Computing*, 2, 117–119.

Shapiro, S. S. and Wilk, M. B. (1965), "An Analysis of Variance Test for Normality (Complete Samples)," *Biometrika*, 52, 591–611.

Silverman, B. W. (1986), *Density Estimation*, New York: Chapman & Hall.

Slifker, J. F. and Shapiro, S. S. (1980), "The Johnson System: Selection and Parameter Estimation," *Technometrics*, 22, 239–246.

Terrell, G. R. and Scott, D. W. (1985), "Oversmoothed Nonparametric Density Estimates," *Journal of the American Statistical Association*, 80, 209–214.

Tukey, J. W. (1977), *Exploratory Data Analysis*, Reading, MA: Addison-Wesley.

Tukey, J. W. and McLaughlin, D. H. (1963), "Less Vulnerable Confidence and Significance Procedures for Location Based on a Single Sample: Trimming/Winsorization 1," *Sankhya A*, 25, 331–352.

Velleman, P. F. and Hoaglin, D. C. (1981), *Applications, Basics, and Computing of Exploratory Data Analysis*, Boston, MA: Duxbury Press.

Wainer, H. (1974), "The Suspended Rootogram and Other Visual Displays: An Empirical Validation," *The American Statistician*, 28, 143–145.

Subject Index

adjusted odds ratio, 162
agreement, measures of, 153
alpha level
    FREQ procedure, 81, 89
Anderson-Darling statistic, 359
Anderson-Darling test, 237
annotating
    histograms, 317
ANOVA (row mean scores) statistic, 161
association, measures of
    FREQ procedure, 124
asymmetric lambda, 124, 131

beta distribution, 348, 365
    cdf plots, 246
    deviation from theoretical distribution, 357
    EDF goodness-of-fit test, 357
    estimation of parameters, 261
    fitting, 261, 348
    formulas for, 348
    P-P plots, 289
    probability plots, 300, 365
    quantile plots, 310, 365
binomial proportion test, 133
    examples, 197
Bowker's test of symmetry, 153, 154
box plots, line printer, 238, 337
    side-by-side, 237, 338
Breslow-Day test, 164

case-control studies
    odds ratio, 148, 162
cdf plots, 241
    axes, specifying, 250
    beta distribution, 247
    creating, 459
    example, 459
    exponential distribution, 247
    gamma distribution, 248
    lognormal distribution, 248
    normal distribution, 249
    normal distribution, example, 459
    options summarized by function, 243
    parameters for distributions, 243
    suppressing empirical cdf, 249
    Weibull distribution, 250
cell count data, 112
    example (FREQ), 187
chi-square tests

    examples (FREQ), 193, 200, 203
    FREQ procedure, 119, 120
Cicchetti-Allison weights, 156
Cochran's Q test, 153, 158
Cochran's Q test, 212
Cochran-Armitage test for trend, 150, 206
Cochran-Mantel-Haenszel statistics (FREQ), 83, 158, see chi-square tests
    ANOVA (row mean scores) statistic, 161
    correlation statistic, 160
    examples, 205
    general association statistic, 161
cohort studies, 205
    relative risk, 149, 163
comparative plots, 251, 252, 343
    histograms, 266, 321, 322, 407, 409, 419
concordant observations, 124
confidence ellipse, 28
confidence limits
    asymptotic (FREQ), 125
    exact (FREQ), 79
    for percentiles, 329
    means, for, 333
    parameters of normal distribution, for, 333
    standard deviations, for, 333
    variances, for, 333
confidence limits for the correlation
    Fisher's z transformation, 25
contingency coefficient, 119, 124
contingency tables, 64, 85
continuity-adjusted chi-square, 119, 121
CORR procedure
    concepts, 17
    details, 17
    examples, 33
    missing values, 29
    ODS graph names, 33
    ODS table names, 32
    output, 30
    output data sets, 31
    overview, 4
    results, 29
    syntax, 8
    task tables, 8
corrected sums of squares and crossproducts, 9
correlation coefficients, 4
    limited combinations of, 16
    printing, for each variable, 9

suppressing probabilities, 9
correlation statistic, 160
covariances, 9
Cramer's V statistic, 119, 124
Cramér-von Mises statistic, 359
Cramér-von Mises test, 237
Cronbach's coefficient alpha, 27
    calculating and printing, 9
    example, 48
    for estimating reliability, 4
crosstabulation tables, 64, 85

data summarization tools, 224
density estimation, *see* kernel density estimation
descriptive statistics
    computing, 325
discordant observations, 124
distribution of variables, 224

EDF, *see* empirical distribution function
EDF goodness-of-fit tests, 357
    probability values of, 359
empirical distribution function
    definition of, 358
    EDF test statistics, 357, 358
exact tests
    computational algorithms (FREQ), 168
    computational resources (FREQ), 169
    confidence limits, 79
    FREQ procedure, 167, 207
    network algorithm (FREQ), 168
    p-value, definitions, 169
exponential distribution, 350, 365
    cdf plots, 247
    deviation from theoretical distribution, 357
    EDF goodness-of-fit test, 357
    estimation of parameters, 263
    fitting, 350
    formulas for, 350
    P-P plots, 290
    probability plots, 301, 365
    quantile plots, 311, 365
extreme observations, 271, 384
extreme values, 384

Fisher's exact test
    FREQ procedure, 119, 122, 123
Fisher's *z* transformation
    applications, 26
Fisher's *z* transformation, 9, 24
    confidence limits for the correlation, 25
fitted parametric distributions, 348
    beta distribution, 348
    exponential distribution, 350
    folded normal distribution, 431

    gamma distribution, 350
    Johnson S_B distribution, 353
    Johnson S_U distribution, 354
    lognormal distribution, 351
    normal distribution, 352
    Weibull distribution, 355
Fleiss-Cohen weights, 156
folded normal distribution, 431
Freeman-Halton test, 123
FREQ procedure
    alpha level, 81, 89
    binomial proportion, 133, 197
    Bowker's test of symmetry, 153
    Breslow-Day test, 164
    cell count data, 112, 187
    chi-square tests, 119, 120, 193, 200
    Cochran's Q test, 153
    Cochran-Mantel-Haenszel statistics, 205
    computation time, limiting, 81
    computational methods, 168
    computational resources, 169, 171
    contingency coefficient, 119
    contingency table, 200
    continuity-adjusted chi-square, 119, 121
    correlation statistic, 160
    Cramer's V statistic, 119
    default tables, 85, 86
    displayed output, 175
    exact *p*-values, 169
    EXACT statement, used with TABLES, 80
    exact tests, 79, 167, 207
    Fisher's exact test, 119
    Friedman's chi-square statistic, 210
    gamma statistic, 124
    general association statistic, 161
    grouping variables, 113
    input data sets, 76, 112
    kappa coefficient, 156, 157
    Kendall's tau-*b* statistic, 124
    lambda asymmetric, 124
    lambda symmetric, 124
    likelihood-ratio chi-square test, 119
    Mantel-Haenszel chi-square test, 119
    McNemar's test, 153
    measures of association, 124
    missing values, 114
    Monte Carlo estimation, 79, 81, 170
    multiway tables, 177, 179, 180
    network algorithm, 168
    odds ratio, 148, 162
    ODS graph names, 186
    ODS table names, 183
    one-way frequency tables, 119, 120, 175, 176, 193

order of variables, 78
output data sets, 82, 172–174, 187, 203
output variable prefixes, 174
overall kappa coefficient, 153
Pearson chi-square test, 119, 120
Pearson correlation coefficient, 124
phi coefficient, 119
polychoric correlation coefficient, 124
relative risk, 149, 162, 163
row mean scores statistic, 161
scores, 118
simple kappa coefficient, 153
Somers' D statistics, 124
Spearman rank correlation coefficient, 124
statistical computations, 118
stratified table, 205
Stuart's tau-c statistic, 124
two-way frequency tables, 119, 120, 200
uncertainty coefficients, 124
weighted kappa coefficient, 153
frequency tables, 64, 85
    creating (UNIVARIATE), 386
    one-way (FREQ), 119, 120, 175, 176, 193
    two-way (FREQ), 119, 120, 200
Friedman's chi-square statistic, 210

gamma distribution, 350, 365
    cdf plots, 248
    deviation from theoretical distribution, 357
    EDF goodness-of-fit test, 357
    estimation of parameters, 264
    fitting, 264, 350
    formulas for, 350
    P-P plots, 290, 291
    probability plots, 301, 365
    quantile plots, 311, 365
gamma statistic, 124, 126
general association statistic, 161
Gini's mean difference, 335
goodness-of-fit tests, 237, 356, *see* empirical distribution function, 423
    Anderson-Darling, 359
    Cramér-von Mises, 359
    EDF, 357, 359
    Kolmogorov D, 358
    Shapiro-Wilk, 357
graphics, 224
    annotating, 235
    descriptions, 319
    high-resolution, 224, 339
    insets, 271, 344–346
    line printer, 337
    naming, 321
    probability plots, 294

    quantile plots, 305
    saving, 236
high-resolution graphics, 224, 339
histograms, 253, 377
    adding a grid, 264
    annotating, 317
    appearance, 261, 262, 265, 268–270, 317–323
    axis color, 317
    axis scaling, 270
    bar labels, 261
    bar width, 261, 267
    bars, suppressing, 268
    beta curve, superimposed, 261
    binning, 414
    color, options, 262, 317–319
    comparative, 266, 321, 322, 407, 409, 419
    creating, 405
    endpoints of intervals, 269
    exponential curve, superimposed, 263
    extreme observations, 384
    filling area under density curve, 263
    gamma curve, superimposed, 264
    hanging, 264
    insets, 411
    intervals, 268, 378
    Johnson S_B curve, superimposed, 269
    Johnson S_U curve, superimposed, 269
    kernel density estimation, options, 262, 265, 266, 270
    kernel density estimation, superimposed, 360, 428
    line type, 321
    lognormal curve, superimposed, 266
    lognormal distribution, 442
    midpoints, 267
    multiple distributions, example, 423
    normal curve, superimposed, 268
    normal distribution, 267
    options summarized by function, 256
    output data sets, 372, 373
    parameters for fitted density curves, 261, 267, 269, 270
    plots, suppressing, 268
    quantiles, 378
    reference lines, options, 264, 318–323
    saving histogram intervals, 268
    tables of statistical output, 377
    tables of statistical output, suppressing, 268
    three-parameter lognormal distribution, superimposed, 429
    three-parameter Weibull distribution, superimposed, 430

tick marks on horizontal axis, 320
tiles for comparative plots, 320
Weibull curve, superimposed, 270
Hoeffding's measure of dependence, 4, 21
    calculating and printing, 9
    example, 33
    output data set with, 9
    probability values, 21
hypothesis tests
    exact (FREQ), 79

insets, 271, 411
    appearance, 279
    appearance, color, 278
    positioning, 278, 279, 344
    positioning in margins, 345
    positioning with compass point, 344
    positioning with coordinates, 346
    statistics associated with distributions, 274–276
insets for descriptive statistics, *see* insets
interquartile range, 335

Johnson S_B distribution, 353
    estimation of parameters, 269
    fitting, 269, 353
    formulas for, 353
Johnson S_U distribution, 354
    estimation of parameters, 269
    fitting, 269, 354
    formulas for, 354
Jonckheere-Terpstra test, 151

kappa coefficient, 154, 155
    tests, 157
    weights, 156
Kendall correlation statistics, 9
    Kendall's partial tau-b, 4, 15
    Kendall's tau-b, 4, 20
    probability values, 20
Kendall's tau-*b* statistic, 124, 126
kernel density estimation, 360, 428
    adding density curve to histogram, 266
    bandwidth parameter, specifying, 262
    kernel function, specifying type of, 265
    line type for density curve, 321
    lower bound, specifying, 266
    upper bound, specifying, 270
kernel function, *see* kernel density estimation
key cell for comparative plots, 252
Kolmogorov D statistic, 358
Kolmogorov-Smirnov test, 237

lambda asymmetric, 124, 131
lambda symmetric, 124, 132

likelihood-ratio chi-square test, 119
likelihood-ratio test
    chi-square (FREQ), 121
line printer plots, 337
    box plots, 337, 338
    normal probability plots, 338
    stem-and-leaf plots, 337
listwise deletion, 29
location estimates
    robust, 239, 240
location parameters, 368
    probability plots, estimation with, 368
    quantile plots, estimation with, 368
location, tests for
    UNIVARIATE procedure, 403
lognormal distribution, 351, 366
    cdf plots, 248
    deviation from theoretical distribution, 357
    EDF goodness-of-fit test, 357
    estimation of parameters, 266
    fitting, 266, 351
    formulas for, 351
    histograms, 429, 442
    P-P plots, 291, 292
    probability plots, 301, 366, 437
    quantile plots, 311, 366, 455

Mantel-Haenszel chi-square test, 119, 122
McNemar's test, 153
measures of agreement, 153
measures of association, 33
    nonparametric, 4
measures of location
    means, 325
    modes, 328, 382
    trimmed means, 335
    Winsorized means, 334
median absolute deviation (MAD), 335
Mehta and Patel, network algorithm, 168
missing values
    UNIVARIATE procedure, 323
mode calculation, 328
modified ridit scores, 119
Monte Carlo estimation
    FREQ procedure, 79, 81, 170

network algorithm, 168
Newton-Raphson approximation
    gamma shape parameter, 317
    Weibull shape parameter, 317
nonparametric density estimation, *see* kernel density estimation
nonparametric measures of association, 4
normal distribution, 352, 366

cdf plots, 249
cdf plots, example, 459
deviation from theoretical distribution, 357
EDF goodness-of-fit test, 357
estimation of parameters, 268
fitting, 268, 352
formulas for, 352
histograms, 267
P-P plots, 292
probability plots, 295, 302, 366
quantile plots, 312, 366, 444
normal probability plots, *see* probability plots
line printer, 238, 338

odds ratio
adjusted, 162
Breslow-Day test, 164
case-control studies, 148, 162
logit estimate, 162
Mantel-Haenszel estimate, 162
ODS (Output Delivery System)
CORR procedure and, 32
UNIVARIATE procedure table names, 376
ODS graph names
CORR procedure, 33
FREQ procedure, 186
output data sets
saving correlations in, 51
overall kappa coefficient, 153, 157

P-P plots, 284
beta distribution, 289
distribution options, 285, 287
distribution reference line, 286, 462
exponential distribution, 290
gamma distribution, 290, 291
lognormal distribution, 291, 292
normal distribution, 292
options summarized by function, 287, 288
plot layout, 287
Weibull distribution, 293
paired data, 331, 404
pairwise deletion, 29
parameters for fitted density curves, 261, 267, 269, 270
partial correlations, 22
probability values, 24
Pearson chi-square test, 119, 120
Pearson correlation coefficient, 124, 128
Pearson correlation statistics, 4
example, 33
in output data set, 9
Pearson partial correlation, 4, 15

Pearson product-moment correlation, 4, 9, 17, 33
Pearson weighted product-moment correlation, 4, 16
probability values, 19
suppressing, 9
percent plots, *see* See P-P plots
percentiles
axes, quantile plots, 312, 369
calculating, 328
confidence limits for, 329, 399
defining, 238, 328
empirical distribution function, 328
options, 282, 283
probability plots and, 294
quantile plots and, 305
saving to an output data set, 396
visual estimates, probability plots, 369
visual estimates, quantile plots, 369
weighted, 329
weighted average, 328
phi coefficient, 119, 123
plot statements, UNIVARIATE procedure, 223
plots
box plots, 237, 238, 337, 338
comparative, 251, 252, 343
comparative histograms, 266, 321, 322, 407, 409, 419
line printer plots, 337
normal probability plots, 238, 338
probability plots, 294, 364
quantile plots, 305, 364
size of, 238
stem-and-leaf, 238, 337
suppressing, 268
polychoric correlation coefficient, 99, 124, 130
prediction ellipse, 28
probability plots, 294
appearance, 300–302
beta distribution, 300, 365
distribution reference lines, 303
distributions for, 364
exponential distribution, 301, 365
gamma distribution, 301, 365
location parameters, estimation of, 368
lognormal distribution, 301, 366, 437, 442
normal distribution, 295, 302, 366
options summarized by function, 296
overview, 294
parameters for distributions, 296, 300–304
percentile axis, 302
percentiles, estimates of, 369
reference lines, 301
reference lines, options, 301

scale parameters, estimation of, 368
shape parameters, 368
three-parameter Weibull distribution, 367
threshold parameter, 303
threshold parameters, estimation of, 368
two-parameter Weibull distribution, 367
Weibull distribution, 304
prospective study, 205

Q-Q plots, *see* quantile plots
Q_n, 335
quantile plots, 305
appearance, 311, 314
axes, percentile scale, 312, 369
beta distribution, 310, 365
creating, 362
diagnostics, 363
distribution reference lines, 313, 446
distributions for, 364
exponential distribution, 311, 365
gamma distribution, 311, 365
interpreting, 363
legends, suppressing (UNIVARIATE), 446
location parameters, estimation of, 368
lognormal distribution, 311, 366, 449, 455
lognormal distribution, percentiles, 454
nonnormal data, 447
normal distribution, 312, 366, 444
options summarized by function, 307
overview, 305
parameters for distributions, 307, 310, 311, 313–315
percentiles, estimates of, 369
reference lines, 311, 312
reference lines, options, 311
scale parameters, estimation of, 368
shape parameters, 368
three-parameter Weibull distribution, 367
threshold parameter, 314
threshold parameters, estimation of, 368
two-parameter Weibull distribution, 367
Weibull distribution, 314, 456
quantile-quantile plots, *see* quantile plots
quantiles
defining, 328
empirical distribution function, 328
histograms and, 378
weighted average, 328

rank scores, 119
relative risk, 162
cohort studies, 149
logit estimate, 163
Mantel-Haenszel estimate, 163

reliability estimation, 4
ridit scores, 119
risks and risk differences, 140
robust estimates, 238–240
robust estimators, 333, 401
Gini's mean difference, 335
interquartile range, 335
median absolute deviation (MAD), 335
Q_n, 335
S_n, 335
trimmed means, 335
Winsorized means, 334
robust measures of scale, 335
rounding, 238, 324
UNIVARIATE procedure, 324
row mean scores statistic, 161

saving correlations
example, 51
scale estimates
robust, 238
scale parameters, 368
probability plots, 368
quantile plots, 368
shape parameters, 368
Shapiro-Wilk statistic, 357
Shapiro-Wilk test, 237
sign test, 331, 332
paired data and, 404
signed rank statistic, computing, 332
simple kappa coefficient, 153, 154
singularity of variables, 9
smoothing data distribution, *see* kernel density estimation
S_n, 335
Somers' *D* statistics, 124, 127
Spearman correlation statistics, 9
probability values, 19
Spearman partial correlation, 4, 15
Spearman rank-order correlation, 4, 19, 33
Spearman rank correlation coefficient, 124, 129
standard deviation, 9
specifying, 249
stem-and-leaf plots, 238, 337
stratified analysis
FREQ procedure, 64, 85
stratified table
example, 205
Stuart's tau-*c* statistic, 124, 127
Student's *t* test, 331
summary statistics
insets of, 271
saving, 238, 374
sums of squares and crossproducts, 9

t test
    Student's, 331
table scores, 119
tables
    frequency and crosstabulation (FREQ), 64, 85
    multiway, 177, 179, 180
    one-way frequency, 175, 176
    one-way, tests, 119, 120
    two-way, tests, 119, 120
Tarone's adjustment
    Breslow-Day test, 164
tests for location, 331, 403
    paired data, 331, 404
    sign test, 332
    Student's *t* test, 331
    Wilcoxon signed rank test, 332
tetrachoric correlation coefficient, 99, 130
theoretical distributions, 364
three-parameter Weibull distribution, 367
    probability plots, 367
    quantile plots, 367
threshold parameter
    probability plots, 303
    quantile plots, 314
threshold parameters
    probability plots, 368
    quantile plots, 368
tiles for comparative plots
    histograms, 320
trend test, 150, 206
trimmed means, 239, 335
two-parameter Weibull distribution, 367
    probability plots, 367
    quantile plots, 367

uncertainty coefficients, 124, 132, 133
univariate analysis
    for multiple variables, 380
UNIVARIATE procedure
    calculating modes, 382
    classification levels, 251
    comparative plots, 251, 252, 343
    computational resources, 379
    concepts, 323
    confidence limits, 235, 333, 397
    descriptive statistics, 325, 380
    examples, 380
    extreme observations, 271, 384
    extreme values, 384
    fitted continuous distributions, 348
    frequency variables, 393
    goodness-of-fit tests, 356
    high-resolution graphics, 339
    histograms, 377, 384
    insets for descriptive statistics, 271
    keywords for insets, 271
    keywords for output data sets, 280
    line printer plots, 337, 387
    missing values, 251, 323
    mode calculation, 328
    normal probability plots, 338
    ODS graph names, 378
    ODS table names, 376
    output data sets, 280, 370, 394
    overview, 224
    percentiles, 294, 305, 328
    percentiles, confidence limits, 235, 236, 399
    plot statements, 223
    probability plots, 294, 364
    quantile plots, 305, 364
    quantiles, confidence limits, 235, 236
    results, 376
    robust estimates, 401
    robust estimators, 333
    robust location estimates, 239, 240
    robust scale estimates, 238
    rounding, 324
    sign test, 332, 404
    specifying analysis variables, 315
    task tables, 305
    testing for location, 403
    tests for location, 331
    weight variables, 315
UNIVARIATE procedure, OUTPUT statement
    output data set, 370

variances, 9

Weibull distribution, 355
    cdf plots, 250
    deviation from theoretical distribution, 357
    EDF goodness-of-fit test, 357
    estimation of parameters, 270
    fitting, 270, 355
    formulas for, 355
    histograms, 430
    P-P plots, 293
    probability plots, 304
    quantile plots, 314, 456
    three-parameter, 367
    two-parameter, 367
weight values, 236
weighted kappa coefficient, 153, 155
weighted percentiles, 329
Wilcoxon signed rank test, 331, 332
Winsorized means, 240, 334

Yule's *Q* statistic, 126

zeros, structural and random
  FREQ procedure, 158

Syntax Index

AGREE option
    EXACT statement (FREQ), 80
    OUTPUT statement (FREQ), 83
    TABLES statement (FREQ), 88
    TEST statement (FREQ), 111
AJCHI option
    OUTPUT statement (FREQ), 83
ALL option
    OUTPUT statement (FREQ), 83
    PLOTS option (CORR), 12
    PROC UNIVARIATE statement, 234
    TABLES statement (FREQ), 89
ALPHA option
    PROC CORR statement, 9
ALPHA= option
    EXACT statement (FREQ), 81
    HISTOGRAM statement (UNIVARIATE), 261, 349
    PLOTS=SCATTER option (CORR), 13
    PROBPLOT statement (UNIVARIATE), 300
    PROC UNIVARIATE statement, 235
    QQPLOT statement (UNIVARIATE), 310
    TABLES statement (FREQ), 89
ALPHADELTA= option
    plot statements (UNIVARIATE), 317
ALPHAINITIAL= option
    plot statements (UNIVARIATE), 317
ANNOKEY option
    plot statements (UNIVARIATE), 317
ANNOTATE= option
    HISTOGRAM statement (UNIVARIATE), 431
    plot statements (UNIVARIATE), 317
    PROC UNIVARIATE statement, 235, 370

BARLABEL= option
    HISTOGRAM statement (UNIVARIATE), 261
BARWIDTH= option
    HISTOGRAM statement (UNIVARIATE), 261
BDCHI option
    OUTPUT statement (FREQ), 83
BDT option
    TABLES statement (FREQ), 89
BEST= option
    PROC CORR statement, 10
BETA option
    HISTOGRAM statement (UNIVARIATE), 261, 348, 421
    PROBPLOT statement (UNIVARIATE), 300, 365
    QQPLOT statement (UNIVARIATE), 310, 365
BETA= option
    HISTOGRAM statement (UNIVARIATE), 261, 349
    PROBPLOT statement (UNIVARIATE), 300
    QQPLOT statement (UNIVARIATE), 310
BINOMIAL option
    EXACT statement (FREQ), 80
    OUTPUT statement (FREQ), 83
    TABLES statement (FREQ), 89
BINOMIALC option
    TABLES statement (FREQ), 93
BY statement
    CORR procedure, 14
    FREQ procedure, 78
    UNIVARIATE procedure, 240

C= option
    HISTOGRAM statement (UNIVARIATE), 261, 262, 360, 361, 428
    PROBPLOT statement (UNIVARIATE), 300
    QQPLOT statement (UNIVARIATE), 310, 456
CAXIS= option
    plot statements (UNIVARIATE), 317
CBARLINE= option
    HISTOGRAM statement (UNIVARIATE), 262
CDFPLOT statement
    examples, 459
    options summarized by function, 243
    UNIVARIATE procedure, 241
CELLCHI2 option
    TABLES statement (FREQ), 93
CFILL= option
    HISTOGRAM statement (UNIVARIATE), 262
    INSET statement (UNIVARIATE), 278
CFILLH= option
    INSET statement (UNIVARIATE), 278
CFRAME= option
    INSET statement (UNIVARIATE), 278
    plot statements (UNIVARIATE), 317

CFRAMESIDE= option
    plot statements (UNIVARIATE), 318
CFRAMETOP= option
    plot statements (UNIVARIATE), 318
CGRID= option
    HISTOGRAM statement (UNIVARIATE), 262
    PROBPLOT statement (UNIVARIATE), 300
    QQPLOT statement (UNIVARIATE), 310
CHEADER= option
    INSET statement (UNIVARIATE), 278
CHISQ option
    EXACT statement (FREQ), 80, 200
    OUTPUT statement (FREQ), 83
    TABLES statement (FREQ), 93, 120, 200
CHREF= option
    plot statements (UNIVARIATE), 318
CIBASIC option
    PROC UNIVARIATE statement, 235, 397
CIPCTLDF option
    PROC UNIVARIATE statement, 235
CIPCTLNORMAL option
    PROC UNIVARIATE statement, 236
CIQUANTDF option
    PROC UNIVARIATE statement, 399
CIQUANTNORMAL option
    PROC UNIVARIATE statement, 236, 399
CL option
    TABLES statement (FREQ), 93
CLASS statement
    UNIVARIATE procedure, 251
CLIPREF option
    HISTOGRAM statement (UNIVARIATE), 262
CMH option
    OUTPUT statement (FREQ), 83
    TABLES statement (FREQ), 94
CMH1 option
    OUTPUT statement (FREQ), 83
    TABLES statement (FREQ), 94
CMH2 option
    OUTPUT statement (FREQ), 83
    TABLES statement (FREQ), 94
CMHCOR option
    OUTPUT statement (FREQ), 83
CMHGA option
    OUTPUT statement (FREQ), 83
CMHRMS option
    OUTPUT statement (FREQ), 83
COCHQ option
    OUTPUT statement (FREQ), 83
COMOR option
    EXACT statement (FREQ), 80
COMPRESS option
    PROC FREQ statement, 76
CONTENTS= option
    HISTOGRAM statement (UNIVARIATE), 262
    plot statements (UNIVARIATE), 318
    TABLES statement (FREQ), 94
CONTGY option
    OUTPUT statement (FREQ), 83
CONVERGE= option
    TABLES statement (FREQ), 95
CORR procedure
    syntax, 8
CORR procedure, BY statement, 14
CORR procedure, FREQ statement, 15
CORR procedure, ID statement, 15
CORR procedure, PARTIAL statement, 15
CORR procedure, PLOTS option
    ALL option, 12
    HISTOGRAM option, 13
    MATRIX option, 12
    NONE option, 12
    NVAR= option, 13
    NWITH= option, 13
    ONLY option, 12
    SCATTER option, 13
CORR procedure, PLOTS=SCATTER option
    ALPHA=, 13
    ELLIPSE=, 13
    NOINSET, 13
CORR procedure, PROC CORR statement, 8
    ALPHA option, 9
    BEST= option, 10
    COV option, 10
    CSSCP option, 10
    DATA= option, 10
    EXCLNPWGT option, 10
    FISHER option, 10
    HOEFFDING option, 11
    KENDALL option, 11
    NOCORR option, 11
    NOMISS option, 11
    NOPRINT option, 11
    NOPROB option, 11
    NOSIMPLE option, 11
    OUT= option, 12
    OUTH= option, 11
    OUTK= option, 11
    OUTP= option, 12
    OUTS= option, 12
    PEARSON option, 12
    RANK option, 14
    SINGULAR= option, 14
    SPEARMAN option, 14
    SSCP option, 14

VARDEF= option, 14
CORR procedure, VAR statement, 16
CORR procedure, WEIGHT statement, 16
CORR procedure, WITH statement, 16
COV option
    PROC CORR statement, 10
CPROP= option
    HISTOGRAM statement (UNIVARIATE), 419
    plot statements (UNIVARIATE), 318
CRAMV option
    OUTPUT statement (FREQ), 83
CROSSLIST option
    TABLES statement (FREQ), 95
CSHADOW= option
    INSET statement (UNIVARIATE), 278
CSSCP option
    PROC CORR statement, 10
CTEXT= option
    INSET statement (UNIVARIATE), 278
    plot statements (UNIVARIATE), 319
CTEXTSIDE= option
    plot statements (UNIVARIATE), 319
CTEXTTOP= option
    plot statements (UNIVARIATE), 319
CUMCOL option
    TABLES statement (FREQ), 95
CVREF= option
    plot statements (UNIVARIATE), 319

DATA option
    INSET statement (UNIVARIATE), 278
DATA= option
    INSET statement (UNIVARIATE), 273
    PROC CORR statement, 10
    PROC FREQ statement, 76
    PROC UNIVARIATE statement, 236, 370
DESCENDING option
    BY statement (UNIVARIATE), 241
DESCRIPTION= option
    plot statements (UNIVARIATE), 319
DEVIATION option
    TABLES statement (FREQ), 96

ELLIPSE= option
    PLOTS=SCATTER option (CORR), 13
ENDPOINTS= option
    HISTOGRAM statement (UNIVARIATE), 262, 414
EQKAP option
    OUTPUT statement (FREQ), 83
EQOR option
    EXACT statement (FREQ), 80
    OUTPUT statement (FREQ), 83

EQWKP option
    OUTPUT statement (FREQ), 83
EXACT option
    OUTPUT statement (FREQ), 83
EXACT statement
    FREQ procedure, 79
EXCLNPWGT option
    PROC CORR statement, 10
    PROC UNIVARIATE statement, 236
EXPECTED option
    TABLES statement (FREQ), 96
EXPONENTIAL option
    HISTOGRAM statement (UNIVARIATE), 263, 350
    PROBPLOT statement (UNIVARIATE), 301, 365
    QQPLOT statement (UNIVARIATE), 311, 365

FILL option
    HISTOGRAM statement (UNIVARIATE), 263
FISHER option
    EXACT statement (FREQ), 80
    OUTPUT statement (FREQ), 83
    PROC CORR statement, 10
    TABLES statement (FREQ), 96
FITINTERVAL= option
    plot statements (UNIVARIATE), 319
FITMETHOD= option
    plot statements (UNIVARIATE), 319
FITTOLERANCE= option
    plot statements (UNIVARIATE), 319
FONT= option
    INSET statement (UNIVARIATE), 279
    plot statements (UNIVARIATE), 319
FORCEHIST option
    HISTOGRAM statement (UNIVARIATE), 264
FORMAT= option
    INSET statement (UNIVARIATE), 279
    TABLES statement (FREQ), 96
FORMCHAR= option
    PROC FREQ statement, 77
FREQ option
    PROC UNIVARIATE statement, 236, 386
FREQ procedure
    syntax, 75
FREQ procedure, BY statement, 78
FREQ procedure, EXACT statement, 79
    AGREE option, 80
    ALPHA= option, 81
    BINOMIAL option, 80
    CHISQ option, 80, 200

COMOR option, 80
EQOR option, 80
FISHER option, 80
JT option, 80
KAPPA option, 80
LRCHI option, 80
MAXTIME= option, 81
MC option, 81
MCNEM option, 80
MEASURES option, 80
MHCHI option, 80
N= option, 81
OR option, 80, 200
PCHI option, 80
PCORR option, 80
POINT option, 81
RISKDIFF option, 80
RISKDIFF1 option, 80
RISKDIFF2 option, 80
SCORR option, 80
SEED= option, 82
TREND option, 80, 207
WTKAP option, 80
FREQ procedure, OUTPUT statement, 82
AGREE option, 83
AJCHI option, 83
ALL option, 83
BDCHI option, 83
BINOMIAL option, 83
CHISQ option, 83
CMH option, 83
CMH1 option, 83
CMH2 option, 83
CMHCOR option, 83
CMHGA option, 83
CMHRMS option, 83
COCHQ option, 83
CONTGY option, 83
CRAMV option, 83
EQKAP option, 83
EQOR option, 83
EQWKP option, 83
EXACT option, 83
FISHER option, 83
GAMMA option, 83
JT option, 84
KAPPA option, 84
KENTB option, 84
LAMCR option, 84
LAMDAS option, 84
LAMRC option, 84
LGOR option, 84
LGRRC1 option, 84
LGRRC2 option, 84

LRCHI option, 84, 203
MCNEM option, 84
MEASURES option, 84
MHCHI option, 84
MHOR option, 84
MHRRC1 option, 84
MHRRC2 option, 84
N option, 84
NMISS option, 84
OR option, 84
OUT= option, 82
PCHI option, 84, 203
PCORR option, 84
PHI option, 84
PLCORR option, 84
RDIF1 option, 84
RDIF2 option, 84
RELRISK option, 84
RISKDIFF option, 84
RISKDIFF1 option, 84
RISKDIFF2 option, 84
RRC1 option, 84
RRC2 option, 84
RSK1 option, 84
RSK11 option, 84
RSK12 option, 84
RSK2 option, 84
RSK21 option, 84
RSK22 option, 85
SCORR option, 85
SMDCR option, 85
SMDRC option, 85
STUTC option, 85
TREND option, 85
TSYMM option, 85
U option, 85
UCR option, 85
URC option, 85
WTKAP option, 85
FREQ procedure, PROC FREQ statement, 76
COMPRESS option, 76
DATA= option, 76
FORMCHAR= option, 77
NLEVELS option, 77
NOPRINT option, 77
ORDER= option, 78
PAGE option, 78
FREQ procedure, TABLES statement, 85
AGREE option, 88
ALL option, 89
ALPHA= option, 89
BDT option, 89
BINOMIAL option, 89
BINOMIALC option, 93

CELLCHI2 option, 93
CHISQ option, 93, 120, 200
CL option, 93
CMH option, 94
CMH1 option, 94
CMH2 option, 94
CONTENTS= option, 94
CONVERGE= option, 95
CROSSLIST option, 95
CUMCOL option, 95
DEVIATION option, 96
EXPECTED option, 96
FISHER option, 96
FORMAT= option, 96
JT option, 96
LIST option, 97
MAXITER= option, 97
MEASURES option, 97
MISSING option, 97
MISSPRINT option, 98
NOCOL option, 98
NOCUM option, 98
NOFREQ option, 98
NOPERCENT option, 98
NOPRINT option, 98
NOROW option, 98
NOSPARSE option, 98
NOWARN option, 99
OUT= option, 99
OUTCUM option, 99
OUTEXPECT option, 99, 188
OUTPCT option, 99
PLCORR option, 99
PLOTS= option, 100
PRINTKWT option, 105
RELRISK option, 105, 200
RISKDIFF option, 105
RISKDIFFC option, 108
SCORES= option, 108, 210
SCOROUT option, 108
SPARSE option, 109, 188
TESTF= option, 109, 120
TESTP= option, 109, 120, 194
TOTPCT option, 109
TREND option, 110, 207
FREQ procedure, TEST statement, 110
AGREE option, 111
GAMMA option, 111
KAPPA option, 111
KENTB option, 111
MEASURES option, 111
PCORR option, 111
SCORR option, 111
SMDCR option, 111, 207

SMDRC option, 111
STUTC option, 111
WTKAP option, 111
FREQ procedure, WEIGHT statement, 111
ZEROS option, 112
FREQ statement
CORR procedure, 15
UNIVARIATE procedure, 253
FRONTREF option
HISTOGRAM statement (UNIVARIATE), 264

GAMMA option
HISTOGRAM statement (UNIVARIATE), 264, 350, 353, 354, 423
OUTPUT statement (FREQ), 83
PROBPLOT statement (UNIVARIATE), 301, 365
QQPLOT statement (UNIVARIATE), 311, 365
TEST statement (FREQ), 111
GOUT= option
PROC UNIVARIATE statement, 236
GRID option
HISTOGRAM statement (UNIVARIATE), 264
PROBPLOT statement (UNIVARIATE), 301
QQPLOT statement (UNIVARIATE), 311, 312, 454

HAXIS= option
plot statements (UNIVARIATE), 319
HEADER= option
INSET statement (UNIVARIATE), 279
HEIGHT= option
INSET statement (UNIVARIATE), 279
plot statements (UNIVARIATE), 319
HISTOGRAM
PLOTS option (CORR), 13
HISTOGRAM statement
UNIVARIATE procedure, 253
HMINOR= option
plot statements (UNIVARIATE), 320
HOEFFDING option
PROC CORR statement, 11
HOFFSET= option
HISTOGRAM statement (UNIVARIATE), 265
HREF= option
plot statements (UNIVARIATE), 320
HREFLABELS= option
plot statements (UNIVARIATE), 320
HREFLABPOS= option
plot statements (UNIVARIATE), 320

ID statement
    CORR procedure, 15
    UNIVARIATE procedure, 271
IDOUT option
    PROC UNIVARIATE statement, 236
INFONT= option
    plot statements (UNIVARIATE), 320
INHEIGHT= option
    plot statements (UNIVARIATE), 320
INSET statement
    UNIVARIATE procedure, 271
INTERBAR= option
    HISTOGRAM statement (UNIVARIATE), 265
INTERTILE= option
    HISTOGRAM statement (UNIVARIATE), 419
    plot statements (UNIVARIATE), 320

JT option
    EXACT statement (FREQ), 80
    OUTPUT statement (FREQ), 84
    TABLES statement (FREQ), 96

K= option
    HISTOGRAM statement (UNIVARIATE), 265, 360, 361
KAPPA option
    EXACT statement (FREQ), 80
    OUTPUT statement (FREQ), 84
    TEST statement (FREQ), 111
KENDALL option
    PROC CORR statement, 11
KENTB option
    OUTPUT statement (FREQ), 84
    TEST statement (FREQ), 111
KERNEL option
    HISTOGRAM statement (UNIVARIATE), 266, 360, 361, 428
KEYLEVEL= option
    CLASS statement (UNIVARIATE), 252
    PROC UNIVARIATE statement, 409

L= option
    plot statements (UNIVARIATE), 321
LABEL= option
    QQPLOT statement (UNIVARIATE), 312
LAMCR option
    OUTPUT statement (FREQ), 84
LAMDAS option
    OUTPUT statement (FREQ), 84
LAMRC option
    OUTPUT statement (FREQ), 84
LGOR option
    OUTPUT statement (FREQ), 84

LGRID= option
    HISTOGRAM statement (UNIVARIATE), 266
    PROBPLOT statement (UNIVARIATE), 301
    QQPLOT statement (UNIVARIATE), 311, 312
LGRRC1 option
    OUTPUT statement (FREQ), 84
LGRRC2 option
    OUTPUT statement (FREQ), 84
LHREF= option
    plot statements (UNIVARIATE), 321
LIST option
    TABLES statement (FREQ), 97
LOCCOUNT option
    PROC UNIVARIATE statement, 236, 403
LOGNORMAL option
    HISTOGRAM statement (UNIVARIATE), 266, 351, 423, 429, 442
    PROBPLOT statement (UNIVARIATE), 301, 366, 437
    QQPLOT statement (UNIVARIATE), 311, 366
LOWER= option
    HISTOGRAM statement (UNIVARIATE), 266
LRCHI option
    EXACT statement (FREQ), 80
    OUTPUT statement (FREQ), 84, 203
LVREF= option
    plot statements (UNIVARIATE), 321

MATRIX option
    PLOTS option (CORR), 12
MAXITER= option
    plot statements (UNIVARIATE), 321
    TABLES statement (FREQ), 97
MAXNBIN= option
    HISTOGRAM statement (UNIVARIATE), 266
MAXSIGMAS= option
    HISTOGRAM statement (UNIVARIATE), 266
MAXTIME= option
    EXACT statement (FREQ), 81
MC option
    EXACT statement (FREQ), 81
MCNEM option
    EXACT statement (FREQ), 80
    OUTPUT statement (FREQ), 84
MEASURES option
    EXACT statement (FREQ), 80
    OUTPUT statement (FREQ), 84
    TABLES statement (FREQ), 97

TEST statement (FREQ), 111
MHCHI option
    EXACT statement (FREQ), 80
    OUTPUT statement (FREQ), 84
MHOR option
    OUTPUT statement (FREQ), 84
MHRRC1 option
    OUTPUT statement (FREQ), 84
MHRRC2 option
    OUTPUT statement (FREQ), 84
MIDPERCENTS option
    HISTOGRAM statement (UNIVARIATE), 267, 417
MIDPOINTS= option
    HISTOGRAM statement (UNIVARIATE), 267, 411, 414
MISSING option
    CLASS statement (UNIVARIATE), 251
    TABLES statement (FREQ), 97
MISSPRINT option
    TABLES statement (FREQ), 98
MODES option
    PROC UNIVARIATE statement, 236, 382
MU0= option
    PROC UNIVARIATE statement, 237
MU= option
    HISTOGRAM statement (UNIVARIATE), 267, 417
    PROBPLOT statement (UNIVARIATE), 301
    QQPLOT statement (UNIVARIATE), 311, 446

N option
    OUTPUT statement (FREQ), 84
N= option
    EXACT statement (FREQ), 81
NADJ= option
    PROBPLOT statement (UNIVARIATE), 302
    QQPLOT statement (UNIVARIATE), 312, 362
NAME= option
    plot statements (UNIVARIATE), 321
NCOLS= option
    plot statements (UNIVARIATE), 321
NENDPOINTS= option
    HISTOGRAM statement (UNIVARIATE), 267
NEXTROBS= option
    PROC UNIVARIATE statement, 237, 384
NEXTRVAL= option
    PROC UNIVARIATE statement, 237, 384
NLEVELS option
    PROC FREQ statement, 77
NMIDPOINTS= option

HISTOGRAM statement (UNIVARIATE), 268
NMISS option
    OUTPUT statement (FREQ), 84
NOBARS option
    HISTOGRAM statement (UNIVARIATE), 268
NOBYPLOT option
    PROC UNIVARIATE statement, 237
NOCOL option
    TABLES statement (FREQ), 98
NOCORR option
    PROC CORR statement, 11
NOCUM option
    TABLES statement (FREQ), 98
NOFRAME option
    INSET statement (UNIVARIATE), 279
    plot statements (UNIVARIATE), 321
NOFREQ option
    TABLES statement (FREQ), 98
NOHLABEL option
    plot statements (UNIVARIATE), 321
NOINSET option
    PLOTS=SCATTER option (CORR), 13
NOKEYMOVE option
    CLASS statement (UNIVARIATE), 253
NOMISS option
    PROC CORR statement, 11
NONE option
    PLOTS option (CORR), 12
NOPERCENT option
    TABLES statement (FREQ), 98
NOPLOT option
    HISTOGRAM statement (UNIVARIATE), 268
NOPRINT option
    HISTOGRAM statement (UNIVARIATE), 268
    PROC CORR statement, 11
    PROC FREQ statement, 77
    PROC UNIVARIATE statement, 237
    TABLES statement (FREQ), 98
NOPROB option
    PROC CORR statement, 11
NORMAL option
    HISTOGRAM statement (UNIVARIATE), 268, 352, 417
    PROBPLOT statement (UNIVARIATE), 302, 366
    PROC UNIVARIATE statement, 237
    QQPLOT statement (UNIVARIATE), 312, 366
NORMALTEST option
    PROC UNIVARIATE statement, 237

NOROW option
    TABLES statement (FREQ), 98
NOSIMPLE option
    PROC CORR statement, 11
NOSPARSE option
    TABLES statement (FREQ), 98
NOTABCONTENTS option
    HISTOGRAM statement (UNIVARIATE), 268
    PROC UNIVARIATE statement, 237
NOTSORTED option
    BY statement (UNIVARIATE), 241
NOVARCONTENTS option
    PROC UNIVARIATE statement, 238
NOVLABEL option
    plot statements (UNIVARIATE), 321
NOVTICK option
    plot statements (UNIVARIATE), 321
NOWARN option
    TABLES statement (FREQ), 99
NROWS= option
    HISTOGRAM statement (UNIVARIATE), 407
    plot statements (UNIVARIATE), 322
NVAR= option
    PLOTS option (CORR), 13
NWITH= option
    PLOTS option (CORR), 13

ONLY option
    PLOTS option (CORR), 12
OR option
    EXACT statement (FREQ), 80, 200
    OUTPUT statement (FREQ), 84
ORDER= option
    CLASS statement (UNIVARIATE), 251
    PROC FREQ statement, 78
OUT= option
    OUTPUT statement (FREQ), 82
    OUTPUT statement (UNIVARIATE), 280
    PROC CORR statement, 12
    TABLES statement (FREQ), 99
OUTCUM option
    TABLES statement (FREQ), 99
OUTEXPECT option
    TABLES statement (FREQ), 99, 188
OUTH= option
    PROC CORR statement, 11
OUTHISTOGRAM= option
    HISTOGRAM statement (UNIVARIATE), 268, 372, 414
OUTK= option
    PROC CORR statement, 11
OUTKERNEL= option
    HISTOGRAM statement (UNIVARIATE), 373
OUTP= option
    PROC CORR statement, 12
OUTPCT option
    TABLES statement (FREQ), 99
OUTPUT statement
    FREQ procedure, 82
    UNIVARIATE procedure, 280, 315
OUTS= option
    PROC CORR statement, 12
OUTTABLE= option
    PROC UNIVARIATE statement, 238, 374
OVERLAY option
    plot statements (UNIVARIATE), 322

PAGE option
    PROC FREQ statement, 78
PARTIAL statement
    CORR procedure, 15
PCHI option
    EXACT statement (FREQ), 80
    OUTPUT statement (FREQ), 84, 203
PCORR option
    EXACT statement (FREQ), 80
    OUTPUT statement (FREQ), 84
    TEST statement (FREQ), 111
PCTLAXIS option
    QQPLOT statement (UNIVARIATE), 312, 369, 454
PCTLDEF= option
    PROC UNIVARIATE statement, 238, 328
PCTLMINOR option
    PROBPLOT statement (UNIVARIATE), 302
    QQPLOT statement (UNIVARIATE), 312
PCTLNAME= option
    OUTPUT statement (UNIVARIATE), 283
PCTLORDER= option
    PROBPLOT statement (UNIVARIATE), 302
PCTLPRE= option
    OUTPUT statement (UNIVARIATE), 283
PCTLPTS= option
    OUTPUT statement (UNIVARIATE), 282
PCTLSCALE option
    QQPLOT statement (UNIVARIATE), 312, 369
PEARSON option
    PROC CORR statement, 12
PERCENTS= option
    HISTOGRAM statement (UNIVARIATE), 268
PFILL= option
    HISTOGRAM statement (UNIVARIATE), 268

PHI option
    OUTPUT statement (FREQ), 84
PLCORR option
    OUTPUT statement (FREQ), 84
    TABLES statement (FREQ), 99
PLOT option
    PROC UNIVARIATE statement, 387
plot statements
    UNIVARIATE procedure, 317
PLOTS option
    PROC UNIVARIATE statement, 238
PLOTS= option
    TABLES statement (FREQ), 100
PLOTSIZE= option
    PROC UNIVARIATE statement, 238
POINT option
    EXACT statement (FREQ), 81
POSITION= option
    INSET statement (UNIVARIATE), 279
PPPLOT statement
    options dictionary, 289
    options summarized by function, 287, 288
    UNIVARIATE procedure, 284
PRINTKWT option
    TABLES statement (FREQ), 105
PROBPLOT statement
    UNIVARIATE procedure, 294
PROC CORR statement, 8, *see* CORR procedure
    CORR procedure, 8
PROC FREQ statement, *see* FREQ procedure
PROC UNIVARIATE statement, 233, *see*
    UNIVARIATE procedure

QQPLOT statement
    UNIVARIATE procedure, 305

RANK option
    PROC CORR statement, 14
RANKADJ= option
    PROBPLOT statement (UNIVARIATE), 302
    QQPLOT statement (UNIVARIATE), 312, 362
RDIF1 option
    OUTPUT statement (FREQ), 84
RDIF2 option
    OUTPUT statement (FREQ), 84
REFPOINT= option
    INSET statement (UNIVARIATE), 279
RELRISK option
    OUTPUT statement (FREQ), 84
    TABLES statement (FREQ), 105, 200
RISKDIFF option
    EXACT statement (FREQ), 80
    OUTPUT statement (FREQ), 84

TABLES statement (FREQ), 105
RISKDIFF1 option
    EXACT statement (FREQ), 80
    OUTPUT statement (FREQ), 84
RISKDIFF2 option
    EXACT statement (FREQ), 80
    OUTPUT statement (FREQ), 84
RISKDIFFC option
    TABLES statement (FREQ), 108
ROBUSTSCALE option
    PROC UNIVARIATE statement, 238, 401
ROTATE option
    PROBPLOT statement (UNIVARIATE), 302
    QQPLOT statement (UNIVARIATE), 313
ROUND= option
    PROC UNIVARIATE statement, 238
RRC1 option
    OUTPUT statement (FREQ), 84
RRC2 option
    OUTPUT statement (FREQ), 84
RSK1 option
    OUTPUT statement (FREQ), 84
RSK11 option
    OUTPUT statement (FREQ), 84
RSK12 option
    OUTPUT statement (FREQ), 84
RSK2 option
    OUTPUT statement (FREQ), 84
RSK21 option
    OUTPUT statement (FREQ), 84
RSK22 option
    OUTPUT statement (FREQ), 85
RTINCLUDE option
    HISTOGRAM statement (UNIVARIATE), 268, 414

SB option
    HISTOGRAM statement (UNIVARIATE), 269
SCALE= option
    HISTOGRAM statement (UNIVARIATE), 350, 351, 421
SCATTER option
    PLOTS option (CORR), 13
SCORES= option
    TABLES statement (FREQ), 108, 210
SCOROUT option
    TABLES statement (FREQ), 108
SCORR option
    EXACT statement (FREQ), 80
    OUTPUT statement (FREQ), 85
    TEST statement (FREQ), 111
SEED= option
    EXACT statement (FREQ), 82

SIGMA= option
  HISTOGRAM statement (UNIVARIATE), 269, 349, 417
  PROBPLOT statement (UNIVARIATE), 302, 437
  QQPLOT statement (UNIVARIATE), 313, 446, 449
SINGULAR= option
  PROC CORR statement, 14
SLOPE= option
  PROBPLOT statement (UNIVARIATE), 303
  QQPLOT statement (UNIVARIATE), 313
SMDCR option
  OUTPUT statement (FREQ), 85
  TEST statement (FREQ), 111, 207
SMDRC option
  OUTPUT statement (FREQ), 85
  TEST statement (FREQ), 111
SPARSE option
  TABLES statement (FREQ), 109, 188
SPEARMAN option
  PROC CORR statement, 14
SQUARE option
  PROBPLOT statement (UNIVARIATE), 303, 437
  QQPLOT statement, 446
  QQPLOT statement (UNIVARIATE), 314
SSCP option
  PROC CORR statement, 14
STUTC option
  OUTPUT statement (FREQ), 85
  TEST statement (FREQ), 111
SU option
  HISTOGRAM statement (UNIVARIATE), 269
SUMMARYCONTENTS= option
  PROC UNIVARIATE statement, 239

TABLES statement
  FREQ procedure, 85
TEST statement
  FREQ procedure, 110
TESTF= option
  TABLES statement (FREQ), 109, 120
TESTP= option
  TABLES statement (FREQ), 109, 120, 194
THETA= option
  HISTOGRAM statement (UNIVARIATE), 269, 349, 421, 429, 442
  PROBPLOT statement (UNIVARIATE), 303
  QQPLOT statement (UNIVARIATE), 314
THRESHOLD= option
  HISTOGRAM statement (UNIVARIATE), 269, 351

PROBPLOT statement (UNIVARIATE), 303
QQPLOT statement (UNIVARIATE), 314
TOTPCT option
  TABLES statement (FREQ), 109
TREND option
  EXACT statement (FREQ), 80, 207
  OUTPUT statement (FREQ), 85
  TABLES statement (FREQ), 110, 207
TRIMMED= option
  PROC UNIVARIATE statement, 239, 401
TSYMM option
  OUTPUT statement (FREQ), 85
TURNVLABELS option
  plot statements (UNIVARIATE), 322

U option
  OUTPUT statement (FREQ), 85
UCR option
  OUTPUT statement (FREQ), 85
UNIVARIATE procedure
  syntax, 232
UNIVARIATE procedure, BY statement, 240
  DESCENDING option, 241
  NOTSORTED option, 241
UNIVARIATE procedure, CDFPLOT statement, 241
  ALPHA= beta-option, 246
  ALPHA= gamma-option, 246
  BETA beta-option, 246
  BETA= option, 247
  C= option, 247
  DELTA= option, 262
  EXPONENTIAL option, 247
  GAMMA option, 248
  GAMMA= option, 264
  LOGNORMAL option, 248
  MU= option, 249
  NOECDF option, 249
  NORMAL option, 249
  SIGMA= option, 249
  THETA= option, 250
  THRESHOLD= option, 250
  VSCALE= option, 250
  WEIBULL Weibull-option, 250
  ZETA= option, 251
UNIVARIATE procedure, CLASS statement, 251
  KEYLEVEL= option, 252
  MISSING option, 251
  NOKEYMOVE option, 253
  ORDER= option, 251
UNIVARIATE procedure, FREQ statement, 253
UNIVARIATE procedure, HISTOGRAM statement, 253
  ALPHA= option, 261, 349

ANNOTATE= option, 431
BARLABEL= option, 261
BARWIDTH= option, 261
BETA option, 261, 348, 421
BETA= option, 261, 349
C= option, 261, 262, 360, 361, 428
CBARLINE= option, 262
CFILL= option, 262
CGRID= option, 262
CLIPREF option, 262
CONTENTS= option, 262
CPROP= option, 419
ENDPOINTS= option, 262, 414
EXPONENTIAL option, 263, 350
FILL option, 263
FORCEHIST option, 264
FRONTREF option, 264
GAMMA option, 264, 350, 423
GRID option, 264
HANGING option, 264
HOFFSET= option, 265
INTERBAR= option, 265
INTERTILE= option, 419
K= option, 265, 360, 361
KERNEL option, 266, 360, 361, 428
LGRID= option, 266
LOGNORMAL option, 266, 351, 423, 429, 442
LOWER= option, 266
MAXNBIN= option, 266
MAXSIGMAS= option, 266
MIDPERCENTS option, 267, 417
MIDPOINTS= option, 267, 411, 414
MU= option, 267, 417
NENDPOINTS= option, 267
NMIDPOINTS= option, 268
NOBARS option, 268
NOPLOT option, 268
NOPRINT option, 268
NORMAL option, 268, 352, 417
NOTABCONTENTS option, 268
NROWS= option, 407
OUTHISTOGRAM= option, 268, 372, 414
OUTKERNEL= option, 373
PERCENTS= option, 268
PFILL= option, 268
RTINCLUDE option, 268, 414
SB option, 269, 353
SCALE= option, 350, 351, 421
SIGMA= option, 269, 349, 417
SU option, 269, 354
THETA= option, 269, 349, 421, 429, 442
THRESHOLD= option, 269, 351
UPPER= option, 269

VOFFSET= option, 270
VSCALE= option, 270
WBARLINE= option, 270
WEIBULL option, 270, 355, 423
WGRID= option, 270
ZETA= option, 270
UNIVARIATE procedure, ID statement, 271
UNIVARIATE procedure, INSET statement, 271
CFILL= option, 278
CFILLH= option, 278
CFRAME= option, 278
CHEADER= option, 278
CSHADOW= option, 278
CTEXT= option, 278
DATA option, 278
DATA= option, 273
FONT= option, 279
FORMAT= option, 279
HEADER= option, 279
HEIGHT= option, 279
NOFRAME option, 279
POSITION= option, 279
REFPOINT= option, 279
UNIVARIATE procedure, OUTPUT statement, 280, 315
OUT= option, 280
PCTLNAME= option, 283
PCTLPRE= option, 283
PCTLPTS= option, 282
UNIVARIATE procedure, plot statements, 317
ALPHADELTA= gamma-option, 317
ALPHAINITIAL= gamma-option, 317
ANNOKEY option, 317
ANNOTATE= option, 317
CAXIS= option, 317
CDELTA= option, 317
CFRAME= option, 317
CFRAMESIDE= option, 318
CFRAMETOP= option, 318
CHREF= option, 318
CINITIAL= option, 318
COLOR= option, 318
CONTENTS= option, 318
CPROP= option, 318
CTEXT= option, 319
CTEXTSIDE= option, 319
CTEXTTOP= option, 319
CVREF= option, 319
DESCRIPTION= option, 319
FITINTERVAL= option, 319
FITMETHOD= option, 319
FITTOLERANCE= option, 319
FONT= option, 319
HAXIS= option, 319

HEIGHT= option, 319
HMINOR= option, 320
HREF= option, 320
HREFLABELS= option, 320
HREFLABPOS= option, 320
INFONT= option, 320
INHEIGHT= option, 320
INTERTILE= option, 320
L= option, 321
LHREF= option, 321
LVREF= option, 321
MAXITER= option, 321
NAME= option, 321
NCOLS= option, 321
NOFRAME option, 321
NOHLABEL option, 321
NOVLABEL option, 321
NOVTICK option, 321
NROWS= option, 322
OVERLAY option, 322
SCALE= option, 322
SHAPE= option, 322
TURNVLABELS option, 322
VAXIS= option, 322
VAXISLABEL= option, 322
VMINOR= option, 322
VREF= option, 322
VREFLABELS= option, 323
VREFLABPOS= option, 323
W= option, 323
WAXIS= option, 323
UNIVARIATE procedure, PPPLOT statement, 284
    ALPHA= option, 289, 291
    BETA option, 286, 289
    BETA= option, 290
    C= option, 290, 294
    EXPONENTIAL option, 286, 290
    GAMMA option, 287, 290
    LOGNORMAL option, 287, 291
    MU= option, 286, 292, 293
    NOLINE option, 292
    NORMAL option, 287, 292
    SCALE= option, 291, 292
    SHAPE= option, 291, 292
    SIGMA= option, 286, 291–294
    SQUARE option, 293, 462
    THETA= option, 291–294
    THRESHOLD= option, 291–293
    WEIBULL option, 287, 293
    ZETA= option, 292, 294
UNIVARIATE procedure, PROBPLOT statement, 294
    ALPHA= option, 300

BETA option, 300, 365
BETA= option, 300
C= option, 300
CGRID= option, 300
EXPONENTIAL option, 301, 365
GAMMA option, 301, 365
GRID option, 301
LGRID= option, 301
LOGNORMAL option, 301, 366, 437
MU= option, 301
NADJ= option, 302
NORMAL option, 302, 366
PCTLMINOR option, 302
PCTORDER= option, 302
RANKADJ= option, 302
ROTATE option, 302
SIGMA= option, 302, 437
SLOPE= option, 303
SQUARE option, 303, 437
THETA= option, 303
THRESHOLD= option, 303
WEIBULL option, 304, 367
WEIBULL2 option, 367
WEIBULL2 statement, 304
WGRID= option, 304
ZETA= option, 304
UNIVARIATE procedure, PROC UNIVARIATE statement, 233
    ALL option, 234
    ALPHA= option, 235
    ANNOTATE= option, 235, 370
    CIBASIC option, 235, 397
    CIPCTLDF option, 235
    CIPCTLNORMAL option, 236
    CIQUANTDF option, 399
    CIQUANTNORMAL option, 236, 399
    DATA= option, 236, 370
    EXCLNPWGT option, 236
    FREQ option, 236, 386
    GOUT= option, 236
    IDOUT option, 236
    KEYLEVEL= option, 409
    LOCCOUNT option, 236, 403
    MODES option, 236, 382
    MU0= option, 237
    NEXTROBS= option, 237, 384
    NEXTRVAL= option, 237, 384
    NOBYPLOT option, 237
    NOPRINT option, 237
    NORMAL option, 237
    NORMALTEST option, 237
    NOTABCONTENTS option, 237
    NOVARCONTENTS option, 238
    OUTTABLE= option, 238, 374

PCTLDEF= option, 238, 328
PLOT option, 387
PLOTS option, 238
PLOTSIZE= option, 238
ROBUSTSCALE option, 238, 401
ROUND= option, 238
SUMMARYCONTENTS= option, 239
TRIMMED= option, 239, 401
VARDEF= option, 239
WINSORIZED= option, 240, 401
UNIVARIATE procedure, QQPLOT statement, 305
   ALPHA= option, 310
   BETA option, 310, 365
   BETA= option, 310
   C= option, 310, 456
   CGRID= option, 310
   EXPONENTIAL option, 311, 365
   GAMMA option, 311, 365
   GRID option, 311, 312, 454
   LABEL= option, 312
   LGRID= option, 311, 312
   LOGNORMAL option, 311, 366
   MU= option, 311, 446
   NADJ= option, 312, 362
   NORMAL option, 312, 366
   PCTLAXIS option, 312, 369, 454
   PCTLMINOR option, 312
   PCTLSCALE option, 312, 369
   RANKADJ= option, 312, 362
   ROTATE option, 313
   SIGMA= option, 313, 446, 449
   SLOPE= option, 313
   SQUARE option, 314, 446
   THETA= option, 314
   THRESHOLD= option, 314
   WEIBULL option, 314, 367, 456
   WEIBULL2 option, 367
   WEIBULL2 statement, 314
   WGRID= option, 315
   ZETA= option, 315, 449
UNIVARIATE procedure, VAR statement, 315
UNIVARIATE procedure, WEIGHT statement, 315
UPPER= option
   HISTOGRAM statement (UNIVARIATE), 269
URC option
   OUTPUT statement (FREQ), 85

VAR statement
   CORR procedure, 16
   UNIVARIATE procedure, 315
VARDEF= option
   PROC CORR statement, 14
   PROC UNIVARIATE statement, 239
VAXISLABEL= option
   plot statements (UNIVARIATE), 322
VMINOR= option
   plot statements (UNIVARIATE), 322
VOFFSET= option
   HISTOGRAM statement (UNIVARIATE), 270
VREF= option
   plot statements (UNIVARIATE), 322
VREFLABELS= option
   plot statements (UNIVARIATE), 323
VREFLABPOS= option
   plot statements (UNIVARIATE), 323
VSCALE= option
   HISTOGRAM statement (UNIVARIATE), 270

W= option
   plot statements (UNIVARIATE), 323
WAXIS= option
   plot statements (UNIVARIATE), 323
WBARLINE= option
   HISTOGRAM statement (UNIVARIATE), 270
WEIBULL option
   HISTOGRAM statement (UNIVARIATE), 270, 355, 423
   PROBPLOT statement (UNIVARIATE), 304, 367
   QQPLOT statement (UNIVARIATE), 314, 367, 456
WEIBULL2 option
   PROBPLOT statement (UNIVARIATE), 304, 367
   QQPLOT statement (UNIVARIATE), 314, 367
WEIGHT statement
   CORR procedure, 16
   FREQ procedure, 111
   UNIVARIATE procedure, 315
WGRID= option
   HISTOGRAM statement (UNIVARIATE), 270
   PROBPLOT statement (UNIVARIATE), 304
   QQPLOT statement (UNIVARIATE), 315
WINSORIZED= option
   PROC UNIVARIATE statement, 240, 401
WITH statement
   CORR procedure, 16
WTKAP option
   EXACT statement (FREQ), 80
   OUTPUT statement (FREQ), 85

TEST statement (FREQ), 111

ZEROS option
    WEIGHT statement (FREQ), 112
ZETA= option
    HISTOGRAM statement (UNIVARIATE), 270
    PROBPLOT statement (UNIVARIATE), 304
    QQPLOT statement (UNIVARIATE), 315, 449

Your Turn

We welcome your feedback.

- If you have comments about this book, please send them to **yourturn@sas.com**. Include the full title and page numbers (if applicable).

- If you have comments about the software, please send them to **suggest@sas.com**.